Advances in Kinanthropometry: Techniques and Applications in Sports and Health

Advances in Kinanthropometry: Techniques and Applications in Sports and Health

Guest Editors

Stefania Toselli
Luciana Zaccagni
Natascia Rinaldo

MDPI

Basel • Beijing • Wuhan • Barcelona • Belgrade • Novi Sad • Cluj • Manchester

Guest Editors
Stefania Toselli
Department of Biomedical and Neuromotor Science
University of Bologna
Bologna
Italy

Luciana Zaccagni
Department of Neuroscience and Rehabilitation
University of Ferrara
Ferrara
Italy

Natascia Rinaldo
Department of Neuroscience and Rehabilitation
University of Ferrara
Ferrara
Italy

Editorial Office
MDPI AG
Grosspeteranlage 5
4052 Basel, Switzerland

This is a reprint of the Special Issue, published open access by the journal *Journal of Functional Morphology and Kinesiology* (ISSN 2411-5142), freely accessible at: https://www.mdpi.com/journal/jfmk/special_issues/T0C559Z6Y2.

For citation purposes, cite each article independently as indicated on the article page online and as indicated below:

Lastname, A.A.; Lastname, B.B. Article Title. *Journal Name* **Year**, *Volume Number*, Page Range.

ISBN 978-3-7258-7621-1 (Hbk)
ISBN 978-3-7258-7622-8 (PDF)
https://doi.org/10.3390/books978-3-7258-7622-8

Contents

About the Editors . . . . **vii**

Stefania Toselli, Natascia Rinaldo and Luciana Zaccagni
Special Issue "Advances in Kinanthropometry: Techniques and Applications in Sports and Health"
Reprinted from: *J. Funct. Morphol. Kinesiol.* **2026**, *11*, 56,
https://doi.org/10.3390/jfmk11010056 . . . . **1**

Sílvio A. Carvalho, Pedro Bezerra, José E. Teixeira, Pedro Forte, Rui M. Silva and José Mª Cancela-Carral
Associations Between Neurofeedback, Anthropometrics, Technical, Physical, and Tactical Performance in Young Women's Football Players
Reprinted from: *J. Funct. Morphol. Kinesiol.* **2025**, *10*, 423,
https://doi.org/10.3390/jfmk10040423 . . . . **6**

Alex Ojeda-Aravena, María Alejandra Camacho-Villa, Fernando Millan-Domingo, Ronald Quintero-Bernal, Jeimy Andrea Merchán, Fabio Villafrades and Adrián De la Rosa
Sex-Specific Morphological and Neuromuscular Profiles of U-15 Colombian Basketball Players
Reprinted from: *J. Funct. Morphol. Kinesiol.* **2025**, *10*, 422,
https://doi.org/10.3390/jfmk10040422 . . . . **22**

Pedro Forte, Samuel G. Encarnação, José E. Teixeira, Luís Branquinho, Tiago M. Barbosa, António M. Monteiro and Daniel Pecos-Martín
Predicting Sleep Quality Based on Metabolic, Body Composition, and Physical Fitness Variables in Aged People: Exploratory Analysis with a Conventional Machine Learning Model
Reprinted from: *J. Funct. Morphol. Kinesiol.* **2025**, *10*, 337,
https://doi.org/10.3390/jfmk10030337 . . . . **39**

Ximena Martínez-Mireles, José Omar Lagunes-Carrasco, Vianney Curiel-Cervantes, Ximena Ortega-Salinas, Mauro E. Valencia, Ricardo López-García, et al.
A National Study of Somatotypes in Mexican Athletes Across 43 Sports
Reprinted from: *J. Funct. Morphol. Kinesiol.* **2025**, *10*, 329,
https://doi.org/10.3390/jfmk10030329 . . . . **52**

Barry Gerber and Anita Elizabeth Pienaar
Age-Related Changes in Predictors of BMI in 6, 9 and 12-Year-Old Boys and Girls: The NW-CHILD Study
Reprinted from: *J. Funct. Morphol. Kinesiol.* **2025**, *10*, 320,
https://doi.org/10.3390/jfmk10030320 . . . . **80**

Francesca Campoli, Elvira Padua, Lucio Caprioli, Saeid Edriss, Giuseppe Annino, Vincenzo Bonaiuto and Mauro Lombardo
From Overweight to Severe Obesity: Physical Activity and Behavioural Profiles in a Large Clinical Cohort
Reprinted from: *J. Funct. Morphol. Kinesiol.* **2025**, *10*, 283,
https://doi.org/10.3390/jfmk10030283 . . . . **102**

Ana M. González-Martin, Edgar Samid Limón-Villegas, Zyanya Reyes-Castillo, Francisco Esparza-Ros, Luis Alexis Hernández-Palma, Minerva Saraí Santillán-Rivera, et al.
Anthropometric Measurements for Predicting Low Appendicular Lean Mass Index for the Diagnosis of Sarcopenia: A Machine Learning Model
Reprinted from: *J. Funct. Morphol. Kinesiol.* **2025**, *10*, 276,
https://doi.org/10.3390/jfmk10030276 . . . . **114**

Luciana Zaccagni, Mattia Reggiani, Stefania Toselli and Emanuela Gualdi-Russo
Changes in Body Composition and Body Image Perception in Adolescent Soccer Players Examined with Repeated Measurements During Pre-Season and In-Season Training
Reprinted from: *J. Funct. Morphol. Kinesiol.* **2025**, *10*, 219,
https://doi.org/10.3390/jfmk10020219 . . . . **136**

Adrián De la Rosa, María Alejandra Camacho-Villa, Fernando Millan-Domingo, Juan Carlos Saavedra, Marina Politi Okoshi and Luana Urbano Pagan
Positional Profiling of Anthropometric, Baropodometric, and Grip Strength Traits in Male Volleyball Players: Insights from a National Colombian Study
Reprinted from: *J. Funct. Morphol. Kinesiol.* **2025**, *10*, 197,
https://doi.org/10.3390/jfmk10020197 . . . . **149**

Wiliam Carvajal-Veitía, Carlos Abraham Herrera-Amante, Rodrigo Yáñez-Sepúlveda, Vladimir Gainza-Pérez, Yanell Deturnell-Campos, Carlos Cristi-Montero, et al.
Anthropometric Characteristics and Body Composition Changes in a Five-Time Olympic Champion in Greco-Roman Wrestling: A Longitudinal Case Study Towards the Paris 2024 Olympic Games
Reprinted from: *J. Funct. Morphol. Kinesiol.* **2025**, *10*, 176,
https://doi.org/10.3390/jfmk10020176 . . . . **166**

Drazen Cular, Tea Beslija, Marijana Cavala, Matej Babic and Ana Kezic
Comparative Analysis of Maturation Prediction Methods (Moore, Mirwald, BAUSport^{TM}): Croatian Female Volleyball Youth Team Example
Reprinted from: *J. Funct. Morphol. Kinesiol.* **2025**, *10*, 171,
https://doi.org/10.3390/jfmk10020171 . . . . **183**

Anida Kapo-Gurda, Amin Efendić, Indira Mahmutović, Siniša Kovač, Husnija Kajmović, Safet Kapo and Jožef Šimenko
Posture Status Differences Between Preschool Boys and Girls
Reprinted from: *J. Funct. Morphol. Kinesiol.* **2025**, *10*, 101,
https://doi.org/10.3390/jfmk10020101 . . . . **198**

Carlos Abraham Herrera-Amante, William Carvajal-Veitía, Rodrigo Yáñez-Sepúlveda, Fernando Alacid, Juan Gavala-González, José Francisco López-Gil, et al.
Body Asymmetry and Sports Specialization: An Exploratory Anthropometric Comparison of Adolescent Canoeists and Kayakers
Reprinted from: *J. Funct. Morphol. Kinesiol.* **2025**, *10*, 70,
https://doi.org/10.3390/jfmk10010070 . . . . **214**

Eduardo Guzmán-Muñoz, Yeny Concha-Cisternas, Guillermo Méndez-Rebolledo, Manuel Vásquez-Muñoz, Pablo Valdés-Badilla, Tomás Herrera-Valenzuela, et al.
Relationship Between Body Mass Index and Fat Mass Percentage with Proprioception in Children
Reprinted from: *J. Funct. Morphol. Kinesiol.* **2025**, *10*, 59,
https://doi.org/10.3390/jfmk10010059 . . . . **229**

Boryi A. Becerra-Patiño, Armando Monterrosa-Quintero, Jorge Olivares-Arancibia, José Francisco López-Gil and José Pino-Ortega
Differences in Anthropometric and Body Composition Factors of Blind 5-a-Side Soccer Players in Response to Playing Position: A Systematic Review
Reprinted from: *J. Funct. Morphol. Kinesiol.* **2025**, *10*, 238,
https://doi.org/10.3390/jfmk10030238 . . . . **241**

About the Editors

Stefania Toselli

Stefania Toselli is Full Professor of Physical Anthropology at the Department of Quality of Life Sciences, University of Bologna (Italy). Her expertise lies in Human Biology and Anthropometry, with emphasis on body composition, fat distribution, hydration status, and body image in relation to age, sex, and physical activity. She has authored 162 peer-reviewed publications. Her current h-index is 36, with over 12,600 citations.

Luciana Zaccagni

Luciana Zaccagni is currently an Associate Professor of Methods and Didactics of Sports Activities in the Department of Neuroscience and Rehabilitation at the University of Ferrara (Italy). Her expertise lies in the field of sport sciences, with an emphasis on sports performance, kinanthropometry, and public health. She has authored over 60 peer-reviewed publications. Her current h-index is 24 (Scopus), with over 10,400 citations.

Natascia Rinaldo

Natascia Rinaldo is currently an Associate Professor of Biological Anthropology at the Department of Quality of Life Sciences, University of Bologna (Italy). Her area of expertise is in anthropometry and osteology, with a particular focus on body image perception, anthropometric assessment across growth and aging, kinanthropometry, and methods for estimating the biological profile and health status in forensic anthropology. She has authored more than 60 peer-reviewed publications. Her current h-index is 16, with over 3500 citations.

Journal of
Functional Morphology and Kinesiology

MDPI

Editorial

Special Issue "Advances in Kinanthropometry: Techniques and Applications in Sports and Health"

Stefania Toselli [1], Natascia Rinaldo [2] and Luciana Zaccagni [2,*]

1 Department for Life Quality Studies, University of Bologna, 47921 Rimini, Italy; stefania.toselli@unibo.it
2 Department of Neuroscience and Rehabilitation, Faculty of Medicine, Pharmacy and Prevention, University of Ferrara, 44121 Ferrara, Italy; natascia.rinaldo@unife.it
* Correspondence: luciana.zaccagni@unife.it

1. Background and Current Challenges in Kinanthropometry

The field of kinanthropometry has developed substantially over recent years, reflecting the growing interest in understanding human structure, function, and performance through methods that capture the multidimensional nature of growth, training, health, and aging [1]. While traditional anthropometric measurements remain essential, they are now supported by technological innovations, computational approaches, and integrative perspectives that align morphology with neuromuscular function, cognition, metabolism, and behavior. This evolution is driven by an increasingly complex scientific landscape in which sport performance, health assessment and monitoring, and biological development can no longer be examined in isolation.

Although significant progress has been made in the study of kinanthropometry, notable gaps remain across multiple areas of the field. Updated population-specific morphological profiling tools are necessary, particularly in sports, in order to address differences in anthropometric characteristics. Accurately evaluating growth and maturation during childhood and adolescence remains challenging, as biological variability increases and the existing prediction tools do not always perform consistently across all populations. In sports science, understanding how structural traits interact with neuromuscular and cognitive variables to influence performance requires the development of more integrative frameworks. In public health, accessible and reliable methods for monitoring body composition and identifying sarcopenia or obesity remain a priority, particularly in low-resource settings. Moreover, additional research is necessary for special populations, including para-athletes, preschool children, and aging individuals, who often remain underrepresented in anthropometric investigations.

This Special Issue, "Advances in Kinanthropometry: Techniques and Applications in Sports and Health", addresses these challenges and emphasizes the importance of multidisciplinary work. The papers included offer an updated and comprehensive overview of contemporary research, covering methodological innovations, sport-specific profiling, developmental trajectories, health-related analyses, and systematic syntheses of evidence.

2. Advances in Methods and Sport-Specific Applications

A major direction emerging from this Special Issue is the incorporation of neuromotor and cognitive indicators into morphological profiling. Carvalho et al. provide an innovative perspective by examining the associations between neurofeedback outcomes and anthropometric, physical, technical, and tactical performance in young women football players. Their approach demonstrates how neural activity and cognitive processing may be combined with structural assessments to obtain a more complete representation of performance determinants. This expanded perspective illustrates the potential for future applications of kinanthropometry in models that incorporate central nervous system functioning alongside somatic traits, especially in team sports where decision-making and perceptual–motor coordination play a crucial role.

The importance of sport-specific morphological profiling is reinforced in several studies. Ojeda-Aravena et al. describe sex-based differences in young basketball players, revealing variations in neuromuscular and structural characteristics that are already substantial by mid-adolescence. These findings are essential for designing tailored training programs that consider maturation timing and sex-specific developmental pathways.

Similarly, De la Rosa et al. provide detailed anthropometric and neuromuscular profiles for male volleyball players, focusing on positional differences that significantly influence performance demands. Their results demonstrate the practical value of integrating anthropometric, baropodometric, and handgrip strength assessments into sport selection and training design.

The longitudinal case study by Carvajal-Veitía et al. documents morphological adaptations in a five-time Olympic Greco-Roman wrestling champion preparing for the Paris 2024 Games. Longitudinal data of this nature are rare in research on elite sports, particularly for athletes with extended careers at the highest competitive level. The study not only illustrates the dynamic nature of elite performance but also highlights the critical role of anthropometric monitoring in weight category sports, where body composition management is a cornerstone of competitive success.

The Special Issue further underscores the relevance of studying body asymmetry in sports characterized by unilateral loading. Herrera-Amante et al. investigate adolescent canoeists and kayakers, demonstrating structural and functional directional biases that reflect sport-specific demands. Their findings support the importance of monitoring asymmetry to prevent injuries and optimize performance, aligning with current trends in injury prevention research.

Large-scale studies remain essential for defining normative values across different populations. The extensive contribution by Martínez-Mireles et al. provides a comprehensive somatotype dataset covering 43 sports and a large national sample of Mexican athletes. This research is fundamental for understanding population-level variations and for contextualizing athlete selection and training design within specific cultural and environmental settings.

3. Growth, Development, and Health-Oriented Perspectives

In addition to performance in sports, this Special Issue includes an important series of papers on growth, maturation, and development during childhood and adolescence, periods in which anthropometric variability is particularly pronounced. Gerber and Pienaar show that the Body Mass Index (BMI) is often insufficient for capturing meaningful changes in children's body composition, reinforcing the need for additional indicators, especially during periods of rapid growth.

Zaccagni et al. study adolescent soccer players and examine how the physical changes induced by training interact with body image perception—an aspect that is too often overlooked but that may influence self-esteem, motivation, adherence to training, and overall well-being.

The methodological study by Cular et al. comparing different maturity prediction methods illustrates the persistence of inconsistencies between models. The increasing adoption of new technologies such as the BAUSport™ SonicBone system requires careful validation, especially when used for talent identification, training prescription, or injury risk estimation in youth athletes.

https://doi.org/10.3390/jfmk11010056

The Special Issue also includes studies exploring early childhood. Kapo-Gurda et al. reveal posture differences between preschool boys and girls, emphasizing that musculoskeletal development begins early and requires monitoring to prevent future postural deviations. Meanwhile, Guzmán-Muñoz et al. provide insight into the relationship between body composition and proprioception in children, illustrating how excess adiposity may impair sensory–motor functioning and potentially influence motor skill development.

The studies included in this Special Issue additionally address kinanthropometry's relevance to public health. Campoli et al. examine behavioral and lifestyle profiles across obesity classes in a large Italian clinical cohort, showing a clear heterogeneity in physical activity patterns, dietary preferences, and behavioral traits. These findings support the need for individualized approaches to obesity interventions and contribute to increasing evidence that anthropometric classifications alone cannot adequately describe health risks.

The application of machine learning marks another noteworthy innovation. Forte et al. used metabolic, anthropometric, and physical fitness variables to model sleep quality in older adults. Their findings illustrate how multidimensional models can help identify predictors of health-related outcomes that are influenced by complex physiological interactions.

Likewise, González-Martin et al. propose a machine learning model that accurately predicts low appendicular lean mass, offering a cost-effective approach to sarcopenia screening. This is particularly relevant in settings where advanced imaging tools such as DXA are unavailable.

The inclusion of research on underrepresented populations is another strength of this Special Issue. Becerra-Patiño et al. conducted a systematic review of morphological differences across playing positions in blind five-a-side soccer players, a sport that requires unique perceptual and motor adaptations. This work highlights gaps in parasport research and encourages more inclusive efforts to understand morphological and functional factors in athletes with disabilities.

4. Future Directions and Conclusions

Collectively, the papers included in this Special Issue address the main knowledge gaps identified in the field. They provide updated reference data from multiple countries, propose new approaches for linking structure and function, evaluate methodological tools for assessing growth and maturation, include new computational approaches for analyzing body composition and performance predictors, and investigate populations ranging from preschool children to elite adult athletes and older adults. This diversity of research questions and approaches reflects the vitality of kinanthropometry as a contemporary scientific discipline.

There are several significant future directions for the study of kinanthropometry. Longitudinal studies remain essential for understanding how morphological, neuromuscular, and metabolic traits evolve over time, particularly during critical developmental phases and in high-performance contexts. Future research must also prioritize validating and standardizing new technologies, especially portable imaging devices, 3D scanning systems, and automated measurement tools, to ensure that innovations are reliable, replicable, and suitable for use across different populations and settings.

In addition, the integration of anthropometric, neuromuscular, metabolic, psychological, and behavioral data will increasingly support multidimensional models that are able to capture the complexity of human performance and health. Cross-cultural research will remain crucial for developing more representative global reference values and reducing biases in diagnoses and performance assessment. The application of kinanthropometry in public health is also essential, particularly for the development of accessible tools for obesity and sarcopenia screening. Finally, future research should prioritize the inclusion of

https://doi.org/10.3390/jfmk11010056

underrepresented groups—including para-athletes, preschool children, and individuals from low- and middle-income countries—to provide a more comprehensive understanding of human morphology.

In summary, this first volume of the Special Issue "Advances in Kinanthropometry: Techniques and Applications in Sports and Health" offers an extensive and diverse representation of the current and emerging directions of study in the field of kinanthropometry. Through methodological innovations, sport profiling, developmental analyses, and health-oriented research, the contributors to the Special Issue help advance kinanthropometry as a fundamental tool in both applied and theoretical contexts. The findings presented here will help to stimulate further interdisciplinary collaborations and will contribute to the growing recognition of kinanthropometry as an essential discipline for understanding human development, performance, and health.

Author Contributions: Conceptualization: S.T., N.R. and L.Z.; writing—original draft, S.T.; writing—review and editing, S.T., N.R. and L.Z. All authors have read and agreed to the published version of the manuscript.

Funding: This research received no external funding.

Conflicts of Interest: The authors declare no conflicts of interest.

List of Contributions:

1. Carvalho, S.A.; Bezerra, P.; Teixeira, J.E.; Forte, P.; Silva, R.M.; Cancela-Carral, J.M. Associations between neurofeedback, anthropometrics, technical, physical, and tactical performance in young women's football players. *JFMK* **2025**, *10*, 423. https://doi.org/10.3390/jfmk10040423
2. Ojeda-Aravena, A.; Camacho-Villa, M.A.; Millan-Domingo, F.; Quintero-Bernal, R.; Merchán, J.A.; Villafrades, F.; De la Rosa, A. Sex-specific morphological and neuromuscular profiles of U-15 Colombian basketball players. *JFMK* **2025**, *10*, 422. https://doi.org/10.3390/jfmk10040422
3. De la Rosa, A.; Camacho-Villa, M.A.; Millan-Domingo, F.; Saavedra, J.C.; Okoshi, M.P.; Pagan, L.U. Positional profiling of anthropometric, baropodometric, and grip-strength traits in male volleyball players. *JFMK* **2025**, *10*, 197. https://doi.org/10.3390/jfmk10020197
4. Carvajal-Veitía, W.; Herrera-Amante, C.A.; Yáñez-Sepúlveda, R.; Gainza-Pérez, V.; Deturnell-Campos, Y.; Cristi-Montero, C.; Cortés-Roco, G.; Ramos-García, C.O. Anthropometric characteristics and body composition changes in a five-time Olympic champion in Greco-Roman wrestling. *JFMK* **2025**, *10*, 176. https://doi.org/10.3390/jfmk10020176
5. Herrera-Amante, C.A.; Carvajal-Veitía, W.; Yáñez-Sepúlveda, R.; Alacid, F.; Gavala-González, J.; López-Gil, J.F.; Olivares-Arancibia, J.; Ramos-García, C.O. Body asymmetry and sports specialization: An exploratory anthropometric comparison of adolescent canoeists and kayakers. *JFMK* **2025**, *10*, 70. https://doi.org/10.3390/jfmk10010070
6. Martínez-Mireles, X.; Lagunes-Carrasco, J.O.; Curiel-Cervantes, V.; Ortega-Salinas, X.; Valencia, M.E.; López-García, R.; García, S.; Bouzas, C.; Salas-García, R.; Ramírez, E.; Tur, J.A. A national study of somatotypes in Mexican athletes across 43 sports. *JFMK* **2025**, *10*, 329. https://doi.org/10.3390/jfmk10030329
7. Gerber, B.; Pienaar, A.E. Age-related changes in predictors of BMI in boys and girls: The NW-CHILD study. *JFMK* **2025**, *10*, 320. https://doi.org/10.3390/jfmk10030320
8. Zaccagni, L.; Reggiani, M.; Toselli, S.; Gualdi-Russo, E. Changes in body composition and body image perception in adolescent soccer players. *JFMK* **2025**, *10*, 219. https://doi.org/10.3390/jfmk10020219
9. Cular, D.; Beslija, T.; Cavala, M.; Babic, M.; Kezic, A. Comparative analysis of maturation prediction methods (Moore, Mirwald, BAUSportTM). *JFMK* **2025**, *10*, 171. https://doi.org/10.3390/jfmk10020171
10. Kapo-Gurda, A.; Efendić, A.; Mahmutović, I.; Kovač, S.; Kajmović, H.; Kapo, S.; Šimenko, J. Posture status differences between preschool boys and girls. *JFMK* **2025**, *10*, 101. https://doi.org/10.3390/jfmk10020101

11. Guzmán-Muñoz, E.; Concha-Cisternas, Y.; Méndez-Rebolledo, G.; Vásquez-Muñoz, M.; Valdés-Badilla, P.; Herrera-Valenzuela, T.; Núñez-Espinosa, C.; Hernández-Martínez, J. Relationship between body mass index and fat mass percentage with proprioception in children. *JFMK* **2025**, *10*, 59. https://doi.org/10.3390/jfmk10010059
12. Campoli, F.; Padua, E.; Caprioli, L.; Edriss, S.; Annino, G.; Bonaiuto, V.; Lombardo, M. From overweight to severe obesity: Physical activity and behavioural profiles in a large clinical cohort. *JFMK* **2025**, *10*, 283. https://doi.org/10.3390/jfmk10030283
13. Forte, P.; Encarnação, S.G.; Teixeira, J.E.; Branquinho, L.; Barbosa, T.M.; Monteiro, A.M.; Pecos-Martín, D. Predicting sleep quality based on metabolic, body composition, and physical fitness variables in aged people. *JFMK* **2025**, *10*, 337. https://doi.org/10.3390/jfmk10030337
14. González-Martín, A.M.; Limón-Villegas, E.S.; Reyes-Castillo, Z.; Esparza-Ros, F.; Hernández-Palma, L.A.; Santillán-Rivera, M.S.; Herrera-Amante, C.A.; Ramos-García, C.O.; Righini, N. Anthropometric measurements for predicting low appendicular lean mass index for the diagnosis of sarcopenia: A machine learning model. *JFMK* **2025**, *10*, 276. https://doi.org/10.3390/jfmk10030276
15. Becerra-Patiño, B.A.; Monterrosa-Quintero, A.; Olivares-Arancibia, J.; López-Gil, J.F.; Pino-Ortega, J. Differences in anthropometric and body composition factors of blind 5-a-side soccer players in response to playing position: A systematic review. *JFMK* **2025**, *10*, 238. https://doi.org/10.3390/jfmk10030238

Reference

1. Norton, K.; Olds, T. *Anthropometrica: A Textbook of Body Measurement for Sports and Health Courses*; UNSW Press: Sydney, Australia, 1996.

https://doi.org/10.3390/jfmk11010056

Journal of
Functional Morphology and Kinesiology

Article

Associations Between Neurofeedback, Anthropometrics, Technical, Physical, and Tactical Performance in Young Women's Football Players

Sílvio A. Carvalho [1,2,3], Pedro Bezerra [2,4], José E. Teixeira [2,5,6], Pedro Forte [3,7,8,9], Rui M. Silva [2,4] and José Mª Cancela-Carral [1,*]

1 Faculty of Educational Sciences and Sports Sciences, University of Vigo, 36005 Pontevedra, Spain; silvio.carvalho@live.com.pt
2 SPRINT—Sport Physical Activity and Health Research & Innovation Center, 4900-347 Viana do Castelo, Portugal; pbezerra@esdl.ipvc.pt (P.B.); jose.eduardo@ipg.pt (J.E.T.); ruimiguelfps@hotmail.com (R.M.S.)
3 Department of Sports, Higher Institute of Educational Sciences of the Douro, 4560-708 Penafiel, Portugal; pedromiguel.forte@iscedouro.pt
4 School of Sports and Leisure, Polytechnic Institute of Viana do Castelo, 4900-347 Viana do Castelo, Portugal
5 Sports Science Department, Polytechnic of Cávado and Ave., 4750-810 Guimarães, Portugal
6 Sports Science Department, Polytechnic Institute of Guarda, 6300-559 Guarda, Portugal
7 CI-ISCE, ISCE Douro, 4560-708 Penafiel, Portugal
8 Research Center for Active Living and Wellbeing (Livewell), Instituto Politécnico de Bragança, 5300-253 Bragança, Portugal
9 Sports Science Department, Instituto Politécnico de Bragança, 5300-253 Bragança, Portugal
* Correspondence: chemacc@uvigo.es

Abstract: Background: Neurofeedback training has emerged as a promising tool for enhancing performance by targeting specific brain activity patterns linked to motor skills, decision-making, and concentration. This study aimed to explore the associations between neurofeedback outcomes and football-specific performance metrics, including anthropometric, physical, technical, and tactical dimensions. **Methods**: A quasi-experimental design was used to examine the effects of a six-week neurofeedback training program on motor skills, tactical decision-making, and physical performance in young women's football players (n = 8, aged 14–18). Participants underwent 30-min sessions three times a week targeting sensorimotor rhythms (SMRs) in the 12–15 Hz range within virtual football scenarios. Pre- and post-intervention assessments included anthropometric measures, neurophysiological evaluations, Loughborough Soccer Shooting Test (LSST), and Yo-Yo Intermittent Recovery Test Level 1 (YYIR1). Tactical decision-making was evaluated with a FUT-SAT-based instrument, and biological maturity was estimated using the Mirwald equations. **Results**: Statistical analyses using Pearson's correlations revealed significant associations between neurofeedback outcomes, motor efficiency indices (MEIs), decision-making (DM), and football performance metrics. Correlation coefficients ranged from 0.504 to 0.998, with p-values from 0.010 to <0.001, indicating significant associations across physical, technical, and tactical dimensions. **Conclusions**: This study highlights the beneficial impact of neurofeedback on football performance in young female athletes.

Keywords: neurofeedback; soccer; performance; decision-making; motor skills

1. Introduction

Neurofeedback training has emerged as a promising tool to enhance cognitive and motor performance in sport contexts [1–3]. By providing real-time information on brain activity, neurofeedback allows athletes to modulate neural states associated with attention, arousal regulation, and executive control [4]. This modulation is particularly relevant in football, characterized by complex, rapidly changing environments that require efficient decision-making, tactical adaptability, and technical precision [3]. Previous research has demonstrated that neurofeedback can enhance cognitive processes such as sustained attention, working memory, and perceptual sensitivity, which are essential for optimal tactical performance during competition [5]. These skills must be executed under high-pressure conditions, particularly in youth players who are navigating the tumultuous period of adolescence [6,7].

Understanding the mechanisms that contribute to football performance is crucial for optimizing training protocols and fostering player development [8]. However, the application of neurofeedback in young women football players remains an area that warrants further exploration and research [9,10]. Based on this, new contributions are essential study intends to present new contributions for neurofeedback analysis in the performance analysis of youth women football. First, it focuses specifically on youth women football, a population largely underrepresented in the neurofeedback literature, thereby addressing an important gap in current scientific knowledge [11,12]. Second, it integrates neurofeedback outcomes with tactical performance indicators, offering an innovative perspective on the relationship between neurophysiological processes and tactical behavior in football [13,14]. Third, it employs a multidimensional correlational approach, combining neurofeedback measures with multiple football-specific performance dimensions to provide a more comprehensive understanding of the interactions between neural and performance-related variables. Neurofeedback using target sensorimotor rhythms (SMR) develops improvement strategies in different psychophysiological strategies such as focus, motor efficiency, and decision-making [3,15]. It enhances motor imagery, attention, and task execution, vital for physical, technical, and tactical skills [16,17]. For young women players, the neurofeedback could bridge cognitive adaptability with on-field performance, complementing traditional training to optimize skill development [18,19]. Despite its potential, neurofeedback remains underexplored in this demographic, highlighting the need for tailored interventions to address their specific developmental challenges [20,21].

Additionally, physical fitness and technical abilities are foundational to football performance. Aerobic endurance, for instance, is critical for sustaining high-intensity actions throughout a match and is often assessed using the Yo-Yo Intermittent Recovery Test Level 1 (YYIR1) [22]. This test provides valuable insights into a player's capacity to maintain performance levels during the physically demanding nature of football. Additionally, technical skills such as passing, dribbling, and shooting are essential for both individual and team success. The Loughborough Soccer Shooting Test (LSST) serves as a reliable measure for evaluating these technical abilities, highlighting the importance of integrating physical conditioning with technical proficiency in performance assessments [22–24]. The interplay between these aspects enables players to execute precise actions even when fatigued, emphasizing the necessity of a holistic approach in training and evaluation [18]. Also, anthropometric characteristics play a crucial role in football performance. These measures are not only linked to strength, speed, and endurance capacities, but also interact with technical efficiency and tactical execution during play [7,25]. In addition, anthropometric variables are essential to estimate biological maturation, which is a critical determinant in adolescence, affecting players' capacity to adapt physically and cognitively to training demands. Previous studies have shown that growth- and maturation-related differences influence aerobic performance, neuromuscular development, and decision-making processes in youth football players [26,27]. For this reason, anthropometric evaluation provides a fundamental baseline to contextualize physical, technical, and tactical performance in young women football players. Tactical proficiency in football encompasses decision-making, positioning, and the ability to balance offensive and defensive responsibilities. Metrics such as the Motor Efficiency Index (MEI) and Decision-Making (DM) provide quantitative insights into a player's capacity to optimize space, disrupt opponents, and maintain tactical equilibrium during gameplay. Tools like the FUT-SAT assess players' effectiveness in applying tactical principles in match-like scenarios, underscoring the cognitive demands inherent in football [28–30]. A comprehensive understanding of how players navigate their tactical roles while integrating physical and technical actions is vital for fostering overall performance development [30].

Integrating neurofeedback with conventional methods could provide holistic training programs, advancing both cognitive and physical performance while fostering a balanced approach to skill acquisition [31]. Young and adolescent players show notable differences in technical, tactical, physical, and cognitive domains due to growth, maturation, and training experience. Physically, aerobic and anaerobic capacities, strength, and speed increase but with large variability [25]. Technically, the refinement of core skills progresses with practice, though still influenced by coordination and motor control linked to maturation [18]. Tactically, the young women football players demonstrate greater ability to adapt positioning, anticipate actions, and balance offensive and defensive roles [28]. Cognitively, advances in executive functions and decision-making support more complex responses in game situations [32,33]. These developmental variations highlight the need to consider age and maturity when analyzing adolescent football performance. Considering maturation allows a more accurate interpretation of technical, physical, and tactical performance, since athletes of the same chronological age may be at different stages of biological development [34]. This variable is therefore essential to contextualize these insights in youth women football.

Indee, the previous studies on neurofeedback in women's football are practically non-existent, especially when considering the improvement of technical and tactical skills. The application of this methodology can serve as complementary training in the technical and tactical development of skills that can then improve individual and collective performance, but it can also enable eye-foot cognition training during prolonged periods of injury or during congested periods when recovery needs overlap with effective training. Thus, this study aims to investigate the associations between neurofeedback outcomes and football-specific performance metrics, including anthropometric, physical, technical, and tactical dimensions. The central hypothesis is that neurofeedback training targeting SMR will positively correlate with improvements in motor efficiency, decision-making, and overall football performance. By addressing this gap, the research seeks to advance the understanding of how neurofeedback can enhance athletic development in young women's football players.

2. Materials and Methods

2.1. Study Design

This observational cross sectional was integrated in a quasi-experimental design with a pre- and post-intervention comparison to investigate the associations with neurofeedback-monitored data from 6 weeks on young female football players' motor skills, tactical decision-making, and physical performance. Participants underwent neurofeedback sessions targeting SMR in the 12–15 Hz range. The study involved a baseline assessment phase, a six-week monitoring with three sessions per week. Each session lasted approximately 30 min and included visualization of football-specific scenarios and real-time neurofeedback tasks using virtual environments designed to simulate game conditions. Pre- and post-intervention evaluations included anthropometric measures, neurophysiological assessments, and performance tests such as the LSST and YYIR1. Tactical decision-making was evaluated using a FUT-SAT-based instrument, and maturity status was calculated using the Mirwald equations. The assessments were performed twice: at baseline (week 0) and post-intervention (week 6). Neurofeedback sessions occurred three times per week with at least 48–72 h between sessions, always under standardized conditions (same training facility, afternoon hours, controlled temperature, and after 24 h without strenuous exercise). After that, associations of neurofeedback mean scores and physical, technical, and tactical dimensions were possible to output.

The inclusion criteria were as follows: (i) women football players competing in youth federated teams (U15–U19); (ii) a minimum of three years of structured football training and competitive experience; (iii) voluntary participation with signed informed consent (and parental authorization when applicable); (iv) absence of musculoskeletal or neurological injuries within the previous six months that could affect physical, technical, or cognitive performance; and (v) attendance in at least 80% of the experimental sessions, including neurofeedback, anthropometric, technical, physical, and tactical assessments. The exclusion criteria were: (i) any diagnosed neurological, psychiatric, or epileptic disorders incompatible with neurofeedback procedures; (ii) regular use of medications, stimulants, or supplements that could influence cortical activity or performance outcomes; (iii) occurrence of musculoskeletal injury or concussion during the experimental period or within six weeks prior to testing; (iv) failure to complete the neurofeedback protocol or the scheduled physical, technical, or tactical evaluations; and (v) incomplete or inconsistent data, such as EEG signal loss or missing anthropometric or performance records.

2.2. Sample

The study included youth female football players aged 14 to 18 years, recruited from competitive teams affiliated with local football academies (Table 1). A total of 8 participants were selected based on the inclusion criteria, which required a minimum of three years of organized football experience, regular participation in competitive matches, and at least three weekly training sessions. Participants were segmented into six groups, with group assignments following the sequential order of assessments. Each group was evaluated only after the completion of the preceding group. Players with any recent injuries, neurological conditions, or contraindications for neurofeedback training were excluded. The sample had an average age of 16.2 years (±1.1), an average height of 162.5 cm (±5.8), and an average body mass of 58.3 kg (±6.9). Participants and their legal guardians provided informed consent, and ethical approval was obtained from the institutional review board prior to the study (CECSVS2023/11/viii). The study followed the recommendations of the Helsinki Declaration regarding human research.

Table 1. Descriptive Statistics of anthropometric variables and experience level.

Variable	Mean ± SD
Age (years)	13.75 ± 1.77
Body Mass (kg)	52.86 ± 10.39
Standing Height (cm)	159.95 ± 6.65
Seated Height (cm)	130.80 ± 4.01
Experience level (years)	3.65 ± 1.98

2.3. Procedures

2.3.1. Anthropometrics and Physical Fitness

Body measurements, including body mass, standing height, and seated height, were performed by qualified professionals to ensure precision. The procedures and materials outlined below align with contemporary standards and recommendations [35,36]. To measure body mass, a digital scale with an accuracy of ±0.1 kg is preferred (Tanita Corporation, Tokyo, Japan). A stadiometer mounted on a flat wall is recommended for standing height assessments, ensuring measurements are taken on a level, solid surface (Seca, Hamburg, Germany). For seated height and other dimensions, a flexible, non-elastic measuring tape is required. All measurements must adhere to standardized protocols to maintain reliability and validity.

Estimating biological maturity was conducted using validated predictive models tailored to youth populations. The Mirwald equations, commonly applied, estimate the maturity offset, defined as the number of years a child is from reaching their peak height velocity (PHV) [37]. Precise assessments of standing height, seated height, and body mass are critical for accurately applying these equations. Standing height is measured while the individual stands erect without shoes, ensuring the head is aligned in the Frankfort horizontal plane. Seated height is determined by measuring from the seating surface to the top of the head, with the individual sitting upright on a flat surface. Body mass is recorded with participants wearing light clothing and no footwear. Negative values indicate the years before PHV, while positive values reflect the years after PHV. Maturity timing was categorized into groups based on z-scores: values greater than 0.5 indicated early maturity, scores between −0.5 and 0.5 signified average maturity (indicating the individual was within the typical maturity range), and scores below −0.5 represented delayed maturity. These categorizations were based on validated references [25].

2.3.2. Neurofeedback

The six-week neurofeedback protocol for youth female football players focused on enhancing motor skills and tactical decision-making through targeted interventions using electroencephalography (EEG)-based neurofeedback systems [13]. Participants engaged in three sessions per week, each lasting approximately 30 min, where SMR in the 12–15 Hz range were targeted via full-cap EEG-based system with real-time feedback through visual interface. Full-cap EEG was used, but analysis focused on SMR at C3/C4. The EEG neurofeedback was chosen for ecological validity, faster setup in sports settings, and reduced discomfort for adolescents, despite lower signal-to-noise ratio compared to wet systems. Training included visualization of football-specific scenarios to improve motor imagery and decision-making. Sessions began with baseline assessments to establish individual brain activity patterns, followed by real-time neurofeedback tasks using virtual environments that simulated game-specific challenges, such as accurate passing, shooting, and defensive positioning. Feedback was provided visually through a computer screen to guide participants in modulating SMR activity. The protocol incorporated progressive difficulty, starting with static tasks and advancing to dynamic challenges mimicking competitive game scenarios. Researchers ensured individualized adjustments based on performance data to optimize neuroplasticity and motor skill development. This protocol was grounded in evidence from motor imagery and neurofeedback studies, showing improved cortical and corticospinal tract excitability [38]. Ethical considerations were addressed, and informed consent was obtained from all participants and guardians. Data were analyzed to assess changes in neurophysiological activity, skill performance, and decision-making quality over the intervention period.

Regarding the duration of each session, our recommendation is to allocate approximately 30 min in total, including 5 min for preparation, 5 min for calibration, and 20 min for effective training. After completing the first training session, we would like to further clarify how the difficulty levels should be adjusted. For each parameter—power and accuracy—the number of stars can be selected, with each star representing a difficulty level from 1 to 5 (1 being the lowest and 5 the highest). The objective is to ensure that each player trains at the appropriate difficulty level to achieve the greatest impact, similar to selecting the correct weights in a gym setting. Typically, football players start at level 2 for both parameters. After the demonstration training, the next difficulty level for each player should be set based on the following reasoning: (i) if a player achieves a score above 17 in any parameter (approximately an 80% success rate), the difficulty level should be increased by one; and (ii) if a player achieves a score below 6 or 7 (approximately a 30% success rate), the difficulty level should be decreased by one. It is important not to modify the difficulty levels after every session but to maintain them stable throughout the pilot phase. Ideally, each player should sustain a success rate between 30% and 80% for each parameter to maximize the effectiveness of the training [10,13,14]. The artifact handling (eye blinks, muscle noise) performed by iBrain software α i-Brain-Tech (https://ibrain.software/, accessed on 2 January 2024). The thresholds were individualized at 1 SD above baseline SMR, adjusted progressively. Prior studies showing effective SMR modulation within 20–30 min sessions [31].

The assessed variables were power, representing the participant's ability to make quick tactical decisions or demonstrate technical actions under time constraints. Accuracy indicates how precise the participant's decisions/actions were in alignment with the optimal tactical and technical responses. Goals, percentage of successful decisions/actions that directly or indirectly aligned with achieving a goal (e.g., scoring, advancing play effectively). Power DL (Difficulty Level), the complexity of tactical scenarios requiring quick decision-making or technical execution. Higher values suggest greater difficulty in decision-making under pressure. Accuracy DL (Difficulty Level), the complexity or ambiguity in scenarios requiring precise decisions or actions. Higher values imply more challenging situations where multiple options may seem viable. Power Sum, aggregated or weighted contribution of tactical decision-making speed and technical execution across all tasks, potentially normalized to account for difficulty. Accuracy Sum, aggregated or weighted contribution of tactical and technical precision across all tasks, also normalized for difficulty.

2.3.3. Technical and Physical Performance

Performance metrics were evaluated using standardized protocols, including the LSST and the YYIR1, both recognized for their reliability in assessing athletic performance. The LSST is designed to assess a player's shooting accuracy and proficiency under conditions that mimic match play [23,24]. During the test, players are required to execute a series of shots on goal from designated positions, aiming to hit specific targets within the goal area. The test is structured to evaluate both precision and decision-making speed. Performance is quantified based on the number of successful target hits and the time taken to complete the sequence. This test has been validated for use with female players, demonstrating reliability in distinguishing skill levels among participants [39,40]. The final variables evaluated were: LSST, these represent results from decision-making tasks where penalties or time delays reflect tactical and technical inefficiencies. LSST reflects performance in simulated scenarios, focusing on reaction time, average speed, and execution velocity in technical-tactical actions.

The YYIR1 evaluates an athlete's ability to repeatedly perform high-intensity aerobic activities with short recovery periods, mimicking the intermittent demands of football [23]. The test consists of consecutive 2 × 20-m shuttle runs at progressively faster speeds, separated by brief recovery intervals. Only YYIR1 distance was used as a variable, without direct VO_2max estimation. The total distance completed before the athlete reaches exhaustion is used as the performance metric. Although widely utilized in football to assess fitness levels, research suggests that the YYIR1 may not provide an accurate estimation of VO_2max in women football players, highlighting the need for gender-specific testing protocols [23,41].

2.3.4. Tactical Performance

Tactical performance was analyzed using a FUT-SAT-based instrument, which quantified correct and incorrect actions to assess motor efficiency and decision-making. The field test used in this system is called the "GK + 3 vs. 3 + GK" SSG format carried out for 4 min on a pitch measuring 36 m in length and 27 m in width. Each player performs the test twice, with a 2-min interval between repetitions. The matches are recorded by a video camera placed in a fixed elevated position, eliminating the need for equipment movement. The field dimensions were defined according to the official measurements established by the International Football Association Board (IFAB) and based on the ratio of space usage by outfield players. The duration was determined through a pilot study, which showed that 4 min was sufficient for players to execute all actions related to the tactical principles assessed by the observation instrument [14]. When compared to matches lasting up to 8 min. For the test, participants were randomly divided into two teams of three players, numbered 1 to 3 on one team and 4 to 6 on the other, to facilitate player identification in the video. During the games, players were instructed to follow the official laws of football, except for the offside rule, and recordings were made using a camera positioned diagonally in relation to the goal line and the sideline. In consequence this, the FUT-SAT-based instrument was adapted to small-sided conditions [26,28]. The final evaluated variables were: MEI encompasses various dimensions of tactical performance, including MEI Penetration, which measures the ability to make effective decisions for breaking through defensive lines; MEI Offensive Coverage, which evaluates support for the offensive unit through decision-making and positioning; MEI Depth Mobility, assessing off-ball movement to optimize depth in attacks; MEI Space, measuring the creation and utilization of space; MEI Offensive Unit, which reflects overall efficiency in offensive decision-making; MEI Delay, which indicates effectiveness in disrupting opponent transitions; MEI Defensive Coverage, evaluating decisions that bolster defensive structure; MEI Balance, assessing maintenance of tactical equilibrium in both attack and defense; MEI Concentration, reflecting focus in decision-making; and MEI Defensive Unit, capturing overall defensive efficiency. Similarly, DM metrics include DM Pen, evaluating decision-making related to penetration strategies; DM OC (Offensive Coverage), reflecting support for the offensive unit; DM Mobility, assessing movement-related positioning; DM Space, measuring spatial efficiency; DM Offensive Unit, a composite score for offensive decision-making; DM Delay, gauging efficiency in delaying transitions; DM DC (Defensive Coverage), evaluating defensive decisions; DM Balance, reflecting positional equilibrium; DM Concentration, highlighting critical focus in decision-making; and DM Defensive Unit, summarizing defensive decision-making performance. Additional variables include the DMI, which combines tactical decision-making scores, and Performance, representing overall tactical and technical efficiency.

2.4. Statistical Analysis

The statistical analysis was performed using Python (version 3.9) with the Pandas and Seaborn libraries for data manipulation and visualization [42,43]. Descriptive statistics included means, standard deviations, and minimum and maximum values. The normality of the data was verified using the Shapiro–Wilk test. A post hoc power analysis was conducted to evaluate the sensitivity of the correlational analyses performed. The statistical power ($1 - \beta$) for the Pearson's correlation was calculated based on the sample size and the observed correlation coefficients from the dataset. Additionally, an a priori estimation of the required sample size to achieve a statistical power of 0.80 (with $\alpha = 0.05$, two-tailed) was performed using the same analytical framework adopted in G*Power, Version 3.1.5.1 (Institut für Experimentelle Psychologie, Düsseldorf, Germany). Pairwise correlations between variables were computed using Pearson's correlation coefficient (r) associated with 95% confidence intervals (95% CI). The correlation magnitude was classified as: trivial if $r \leq 0.1$, small if $r = 0.1–0.3$, moderate if $r = 0.3–0.5$, large if $r = 0.5–0.7$, and very large if $r = 0.7–0.9$ and almost perfect if $r \geq 0.9$. The significance of correlations was determined at $p < 0.05$. Variables with significant correlations were visualized in a heatmap for clarity, emphasizing strong associations within the dataset. The graphical visualization of the correlations was performed using a correlation heatmap generated with seaborn (heatmap function) and matplotlib (pyplot) in Python. Pearson's correlation coefficients were computed to assess linear associations between continuous variables. The heatmap was configured with cmap="RdBu_r" to represent correlation strength on a blue-to-red scale, and annot=True was applied to display numerical r values within each cell. Visualization limits were set with vmin=0.5 and vmax=0.9 to enhance contrast, and the figure size was defined as plt.figure(figsize=(10, 8)) to ensure clarity. A colorbar was included to indicate the scale, and the figure was titled "Heatmap of Pairwise Correlations (Corrected)" for clear and standardized presentation.

3. Results

Table 2 presents the descriptive statistics for all variables, including mean, standard deviation (SD), minimum, and maximum values. This provides an overview of the distribution and variability within the dataset.

Table 2. Descriptive statistics of neurofeedback, anthropometrics, technical, physical, and tactical performance in sampled young women's football players.

Variable	Mean ± SD	Min	Max	Variable	Mean ± SD	Min	Max
Power	0.74 ± 0.15	0.50	0.95	DM Offensive Unit	0.68 ± 0.12	0.48	0.85
Accuracy	0.52 ± 0.10	0.35	0.72	DM Delay	0.60 ± 0.13	0.38	0.80
Goals	0.37 ± 0.08	0.25	0.55	DM DC	0.62 ± 0.14	0.40	0.85
Power Sum	0.40 ± 0.12	0.20	0.60	DM Balance	0.61 ± 0.13	0.42	0.82
Accuracy Sum	0.18 ± 0.05	0.12	0.30	DM Concentration	0.67 ± 0.14	0.45	0.88
MEI Penetration	0.68 ± 0.12	0.50	0.90	DM Defensive Unit	0.64 ± 0.13	0.45	0.85
MEI Offensive Coverage	0.62 ± 0.15	0.40	0.85	DMI	0.73 ± 0.11	0.50	0.90
MEI Depth Mobility	0.66 ± 0.14	0.45	0.85	Performance	0.71 ± 0.10	0.50	0.85
MEI Space	0.70 ± 0.13	0.48	0.88	Age	22.5 ± 2.1	18.0	26.0
Offensive Unit	0.65 ± 0.10	0.50	0.80	Body Mass (kg)	70.2 ± 7.5	55.0	85.0
MEI Delay	0.61 ± 0.14	0.40	0.82	Standing Height (cm)	175.4 ± 6.2	160.0	188.0
MEI Deffensive Coverage	0.59 ± 0.13	0.38	0.80	Sitting Height (cm)	90.5 ± 3.5	84.0	97.0
MEI Balance	0.63 ± 0.12	0.42	0.82	Leg Length (cm)	85.0 ± 5.0	75.0	92.0
MEI Concentration	0.67 ± 0.14	0.45	0.88	Maturity Offset	2.3 ± 0.5	1.5	3.5
MEI Deffensive Unit	0.60 ± 0.12	0.40	0.80	LSST_Total Time (s/+ with Penalties) T1	72.3 ± 5.2	62.0	80.5

Table 2. *Cont.*

Variable	Mean ± SD	Min	Max	Variable	Mean ± SD	Min	Max
MEI	0.64 ± 0.13	0.45	0.85	LSST_Total Time (s) Mean	10.2 ± 0.6	9.5	11.5
DM Pen	0.70 ± 0.14	0.45	0.90	LSST_Shot Speed (km/h) Mean	35.5 ± 6.3	25.0	45.0
DM OC	0.62 ± 0.15	0.40	0.85	LSST_Shot Points (mean of 10 shots) Mean	2.1 ± 0.4	1.5	2.8
DM Mobility	0.65 ± 0.14	0.42	0.85	YYIR1 (m)	560.0 ± 180.0	200.0	920.0
DM Space	0.63 ± 0.13	0.40	0.82				

Abbreviations: DM—Decision Making; OC—Offensive Coverage; Pen—Penetration; DC—Defensive Coverage; MEI—Match Evaluation Index; DMI—Decision Making Index; LSST—Loughborough Soccer Shooting Test; YYIR1—Yo-Yo Intermittent Recovery Test Level 1.

Table 3 presents the significant pairwise correlations between key performance variables, including neurofeedback measures, MEI, DM, and technical and physical performance metrics. The correlations (*r*) provide insights into the strength and direction of the associations, and the *p*-values reveal the statistical significance. Strong correlations highlight associations between neurofeedback power, tactical success, DM efficiency, and overall performance. Table 2 presents the pairwise correlations between key variables, including demographic data, physical metrics, decision-making, motor efficiency indices, and technical/tactical variables. Significant correlations ($p < 0.05$) are highlighted to emphasize key associations.

Table 3. Pairwise Correlations between neurofeedback, physical, decision-making, motor efficiency indices, and technical/tactical variables.

Variable 1	Variable 2	*r*	*p*
Power	Goals	0.75 **	<0.001
Power	Power Sum	0.89 **	<0.001
Accuracy	Accuracy Sum	0.99 **	<0.001
Goals	DM OC	0.51 *	0.020
DM OC	DM Mobility	0.72 **	<0.001
MEI Offensive Coverage	Performance	0.87 **	<0.001
MEI Delay	DM Pen	0.57 *	0.010
DM Offensive Unit	Performance	0.50 *	0.030
LSST_Finishing points	YYIR1 (m)	0.52 *	0.015
Power Sum	MEI Concentration	0.82 **	<0.001
DM Balance	MEI Balance	0.55 *	0.012
DM Delay	MEI Penetration	0.57 *	0.010
Performance	DM Mobility	0.72 **	<0.001

Abbreviations: DM—Decision Making; OC—Offensive Coverage; Pen—Penetration; MEI—Match Evaluation Index; LSST—Loughborough Soccer Shooting Test; YYIR1—Yo-Yo Intermittent Recovery Test Level 1. * $p < 0.05$; ** $p < 0.001$.

The heatmap visualizes significant pairwise correlations between neurofeedback measures, MEI, DM, technical performance, and physical performance variables in youth female football players. The intensity of the colour represents the strength and direction of the correlations, with stronger positive correlations indicated by darker shades and weaker correlations represented by lighter shades (Figure 1).

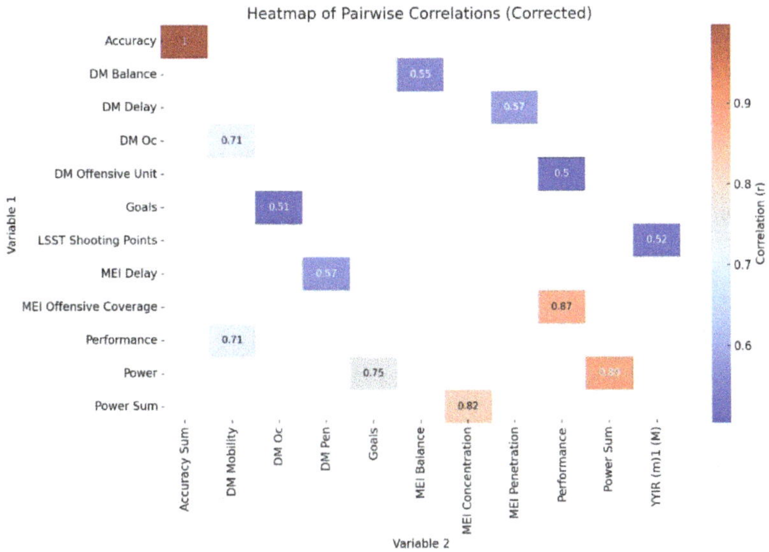

Figure 1. Heatmap of significant associations between Neurofeedback values, motor efficiency index, decision-making index, and technical performance.

4. Discussion

This study examined the associations between neurofeedback training outcomes and anthropometric, physical, technical, and tactical performance variables in young football players. The results demonstrated significant correlations between neurofeedback-derived he results demonstrated significant correlations between neurofeedback-derived indices and key performance measures, supporting the central hypothesis that enhanced neurophysiological regulation is linked to improved tactical decision-making, technical execution, and physical performance. These findings are consistent with previous evidence indicating that sensorimotor rhythm.

4.1. Neurofeedback, Power Efficiency, and Tactical Performance

The observed associations between neurofeedback power and tactical DM metrics suggest highlight the role of neural efficiency in supporting cognitive control and attentional stability during dynamic game scenarios. Players exhibiting higher power efficiency demonstrated improved tactical precision and reaction speed, aligning with previous studies showing that neurofeedback targeting SMR enhances cognitive adaptability and motor imagery in sport contexts [20,23,28]. Similar findings were reported in EEG-based and near-infrared neurofeedback research, where increased self-regulation of cortical rhythms correlated with improved tactical behavior and technical outcomes [11,12,40]. This aligns with evidence suggesting that SMR-based neurofeedback enhances attentional control, motor precision, and cognitive stability in athletes across sports domains [28,32,33]. The strong correlations between power sum and MEI concentration reinforce that sustained attentional focus facilitates tactical clarity and consistency across multiple contexts. Comparable results have been documented in studies linking executive functions and match performance in elite and youth football players [1,3,30,39,41].

Similarly to results reported in EEG-based and near-infrared neurofeedback studies [44,45], the players who demonstrated higher neurofeedback efficiency also showed improved indicators of tactical and technical execution. This neurophysiological-cognitive interaction is essential for supporting decision-making efficiency and tactical adaptability, particularly in high-pressure contexts typical of competitive football. Previous studies have reported that athletes who can effectively modulate cortical rhythms show enhanced sensorimotor integration and cognitive adaptability, allowing for more stable performance during complex decision-making scenarios [46,47]. Furthermore, the positive association between neurofeedback power and goals underscores the practical relevance of cognitive speed and precision for match effectiveness, aligning with evidence relating executive functioning to goal-oriented actions [29,40,45]. Current study associations reinforce the potential of neurofeedback as a complementary method for training cognitive readiness and self-regulatory capacity in youth football players [35,48].

The strong correlation between neurofeedback power sum and MEI concentration suggests that players with greater power efficiency maintain better focus, enabling more precise decisions in high-pressure scenarios. The positive association between neurofeedback power and goals emphasizes the importance of tactical decision speed in achieving successful outcomes, such as scoring. Lastly, the robust connection between neurofeedback power and power sum reflects the consistency required in applying tactical decisions across tasks, reinforcing the player's overall execution efficiency. These findings collectively underline that power, as a measure of tactical speed and efficiency, directly supports decision-making focus and successful outcomes in soccer. The correlation between neurofeedback power and executive functions suggests that players exhibiting greater power efficiency are better able to maintain focus, which is critical for making precise decisions under pressure [32,45]. Furthermore, while the association between neurofeedback power and goals scored is not directly established in the literature, the significance of tactical decision speed in achieving successful outcomes, such as scoring, is emphasized in studies on executive functioning in football players [7,32]. This is further reinforced by the connection between executive functions and performance, indicating the consistency required in applying tactical decisions across various tasks, which enhances overall execution efficiency [44,45]. Collectively, these findings highlight that power, as a measure of tactical speed and efficiency, is integral to supporting decision-making focus and achieving successful outcomes in football [46,47]. Also, current research reinforces that neurofeedback represents a neurophysiological correlation of tactical speed and precision, supporting the integration of cognitive training methods to complement tactical development.

4.2. Aerobic Endurance and Technical Execution

The moderate positive correlation between LSST Shooting Points and YYIR1 highlights the link between aerobic endurance and technical consistency under fatigue. Aerobic endurance supports sustained physical and cognitive performance by delaying fatigue, critical for maintaining precision in high-pressure, game-like scenarios. This link highlights the interplay between physical fitness and technical skill execution, where improved endurance enables players to maintain composure and accuracy even under physically demanding conditions. The moderate positive correlation between LSST shooting points and YYIR1 distance indicates that players with superior aerobic endurance, as assessed by the YYIR1, are more capable of maintaining shooting accuracy under pressure during the LSST. These results are in line with prior studies demonstrating that greater aerobic capacity supports the maintenance of technical execution quality, such as shooting accuracy, during intense or prolonged match scenarios [13,32,38,43]. This association emphasizes that aerobic conditioning not only delays fatigue but also preserves cognitive processing efficiency, a key factor in decision accuracy during match play [5,44]. Previous research has similarly demonstrated that endurance capacity plays a critical role in sustaining technical performance throughout competitive matches, especially in female athletes [36–38]. Aerobic endurance is crucial for sustaining both physical and cognitive performance, as it helps delay fatigue, which is essential for maintaining precision in high-pressure, game-like scenarios [35,48]. The aerobic fitness positively influences technical consistency and passing or shooting accuracy during match play. Similarly, Almeida et al. [35] highlighted that aerobic endurance delays cognitive fatigue, which is essential for maintaining decision-making and precision throughout the game. This association underscores the interplay between physical fitness and technical skill execution, where enhanced endurance allows players to remain composed and accurate even in demanding conditions [48]. Furthermore, the ability to execute technical skills, such as shooting, is significantly influenced by a player's physical fitness level. Studies have shown that improved aerobic capacity is linked to better performance in technical tasks, as players with higher fitness levels can execute skills more effectively during matches [8]. This connection highlights the importance of integrating aerobic training into football practice regimens to optimize both physical conditioning and technical skill performance [49]. Ultimately, the findings suggest that enhancing aerobic endurance can lead to improved technical execution, particularly in high-stakes situations where accuracy is paramount. Furthermore, the current findings thus emphasize the interplay between aerobic conditioning and technical skill, suggesting that endurance development contributes to sustained technical performance in match-relevant scenarios.

4.3. Decision-Making, Tactical Execution, and Cognitive-Motor Integration

The significant associations between DM Balance, MEI Balance, and performance indicators illustrate that equilibrium in decision-making aligns with tactical adaptability. This means that players who display balanced cognitive strategies are better at maintaining positional harmony between offensive and defensive actions, facilitating fluid transitions and team coordination. Players who maintain decision-making balance are better positioned to coordinate offensive and defensive actions, supporting fluid tactical transitions and collective team behavior. The associations emphasize the critical link between decision-making, tactical execution, and overall performance in football. The positive correlation between DM Balance and MEI Balance suggests that players who achieve equilibrium in their decisions are better at maintaining tactical balance, which is essential for effective transitions between offensive and defensive phases. These findings are consistent with evidence showing that tactical understanding, cognitive control, and spatial awareness are essential for optimizing performance [19,46,47]. The strong association between MEI Offensive Coverage and Performance highlights how precise decision-making and positioning in offensive plays directly enhance a player's overall contribution to team success. These results are aligned with studies emphasizing the integration of perceptual-cognitive skills and tactical decision-making in football, especially in youth development contexts [17,18,48]. Similarly, the correlation between DM Mobility and Performance underscores the importance of movement-related positioning, where dynamic adaptability supports tactical and technical execution. Collectively, these findings justify the integral role of cognitive and spatial skills in optimizing football performance. The intricate association between DM, tactical execution, and performance in football is underscored by various studies. The positive correlation between DM Balance and MEI Balance indicates that players who maintain equilibrium in their decision-making are more adept at tactical balance, which is crucial for seamless transitions between offense and defense [28]. Furthermore, the strong link between MEI Offensive Coverage and overall performance emphasizes that precise decision-making and positioning during offensive plays significantly enhance a player's contribution to team success [50,51]. Similarly, the association between DM Mobility and performance highlights the necessity of dynamic adaptability, where effective movement supports both tactical and technical execution [50,52].

Collectively, these findings affirm the essential role of cognitive and spatial skills in optimizing football performance, as they facilitate better anticipation, positioning, and execution of plays [28,53].

4.4. Practical Applications, Study Limitations, and Future Research

While the 6-week neurofeedback intervention yielded meaningful associations across multiple performance domains, it is important to acknowledge that these effects may reflect short-term adaptations. This study is among the first to examine the associations between neurofeedback metrics and multiple performance domains—anthropometric, physical, technical, and tactical—in youth female football players. A key strength is the multidimensional correlational design, allowing for a more integrated interpretation of cognitive, physical, and tactical interactions. However, the small sample size limits generalizability and statistical power. The exclusive focus on female youth athletes, while scientifically valuable, may not fully reflect broader populations, including male or mixed cohorts. Furthermore, the study's short duration precludes conclusions regarding long-term adaptations to neurofeedback. Also, the research study design does not allow for the assessment of the long-term effects of neurofeedback training on performance; therefore, future research should adopt longitudinal designs to evaluate sustained impacts. The reliance on specific neurofeedback parameters suggests exploring alternative neurofeedback protocols to optimize outcomes. Longer-term or longitudinal interventions could help determine the persistence and consolidation of neurofeedback-induced improvements. Future research should also expand on the sample size and include male and mixed cohorts to improve external validity. Additionally, exploring different neurofeedback protocols beyond SMR (e.g., theta/beta ratio, alpha coherence) may help identify more specific cognitive mechanisms linked to football performance. Integrating neurofeedback with perceptual-cognitive or tactical training could further elucidate its potential as a complementary tool in player development. In addition, there is a lack of a control group and the short-term nature of the intervention (this study is part of a pre- and post-intervention study, but this correlational analysis only reports on the first phase, carried out pre-intervention, and is intended to be a correlational study). Lastly, integrating neurofeedback with diverse training methodologies could provide a more comprehensive understanding of its role in enhancing football performance. These directions could help establish evidence-based practices for neurofeedback in sports training. Future research should include larger and more diverse samples, use randomized controlled designs, and explore longitudinal interventions to assess the persistence of neurofeedback-induced effects. Integrating other neurofeedback protocols (e.g., theta/beta ratio, alpha coherence) may help uncover specific neural mechanisms underlying tactical behavior. Additionally, combining neurofeedback with tactical or perceptual-cognitive training may offer synergistic effects on performance. Emerging approaches using artificial intelligence and machine learning [1,54] may enhance the precision of EEG data interpretation and provide individualized training feedback.

5. Conclusions

This study provides evidence that SMR-based neurofeedback training is associated with improvements in cognitive and performance-related domains in female youth football players. The main findings indicate that neurofeedback Power relates to attentional focus, decision-making efficiency, and tactical skills in dynamic match contexts. Additionally, aerobic endurance was linked to technical stability under fatigue, while decision-making balance and mobility correlated with overall performance consistency. Anthropometric and maturational variables, although not directly predictive, may modulate neurofeedback responsiveness and training outcomes, suggesting that individual developmental status should be considered in future applications. By targeting neural efficiency and cognitive control, neurofeedback may enhance players' readiness and adaptability—core determinants of high-level football performance.

Author Contributions: Conceptualization, S.A.C. and P.B.; methodology, S.A.C. and J.E.T.; software, R.M.S.; validation, P.F., J.E.T. and R.M.S.; formal analysis, S.A.C.; investigation, S.A.C. and J.E.T.; resources, J.M.C.-C. and P.B.; data curation, R.M.S. and P.F.; writing—original draft preparation, S.A.C.; writing—review and editing, P.F., J.E.T., R.M.S., P.B. and J.M.C.-C.; visualization, P.F.; supervision, P.B. and J.M.C.-C.; project administration, P.B. and J.M.C.-C.; funding acquisition, J.M.C.-C. All authors have read and agreed to the published version of the manuscript.

Funding: This project was supported by the National Funds through the FCT Portuguese Foundation for Science and Technology, SPRINT—Sport Physical Activity and Health Research and Innovation Center, Portugal.

Institutional Review Board Statement: The ethical approval was obtained from the institutional review board prior to the study (CECSVS2023/11/viii) of the Polytechnic Institute of Viana do Castelo on 2 September 2024.

Informed Consent Statement: Informed consent was obtained from all subjects and parents involved in the study, and to publish this paper.

Data Availability Statement: The raw data supporting the conclusions of this article will be made available by the authors on request.

Acknowledgments: The authors would like to thank the Football Association of Bragança and i-BrainTech Ltd.

Conflicts of Interest: The authors declare no conflicts of interest.

References

1. Cheng, M.Y.; Yu, C.L.; An, X.; Wang, L.; Tsai, C.L.; Qi, F.; Wang, K.P. Evaluating EEG Neurofeedback in Sport Psychology: A Systematic Review of RCT Studies for Insights into Mechanisms and Performance Improvement. *Front. Psychol.* **2024**, *15*, 1331997. [CrossRef] [PubMed]
2. Wang, K.-P.; Cheng, M.-Y.; Elbanna, H.; Schack, T. A New EEG Neurofeedback Training Approach in Sports: The Effects Function-Specific Instruction of Mu Rhythm and Visuomotor Skill Performance. *Front. Psychol.* **2023**, *14*, 1273186. [CrossRef] [PubMed]
3. Rydzik, Ł.; Wąsacz, W.; Ambroży, T.; Javdaneh, N.; Brydak, K.; Kopańska, M. The Use of Neurofeedback in Sports Training: Systematic Review. *Brain Sci.* **2023**, *13*, 660. [CrossRef] [PubMed]
4. Gong, A.; Gu, F.; Nan, W.; Qu, Y.; Jiang, C.; Fu, Y. A Review of Neurofeedback Training for Improving Sport Performance from the Perspective of User Experience. *Front. Neurosci.* **2021**, *15*, 638369. [CrossRef]
5. Vestberg, T.; Jafari, R.; Almeida, R.; Maurex, L.; Ingvar, M.; Petrovic, P. Level of Play and Coach-Rated Game Intelligence Are Related to Performance on Design Fluency in Elite Soccer Players. *Sci. Rep.* **2020**, *10*, 9852. [CrossRef]
6. Rabano-Munoz, A.; Suarez-Arrones, L.; Requena, B.; Asian-Clemente, J. Relative Match Load in Young Professional Soccer Players during Soccer-7 and Soccer-11. *Int. J. Sports Med.* **2023**, *45*, 134–140. [CrossRef]
7. Verburgh, L.; Scherder, E.J.A.; van Lange, P.A.M.; Oosterlaan, J. Executive Functioning in Highly Talented Soccer Players. *PLoS ONE* **2014**, *9*, e91254. [CrossRef]
8. Özcan, İ.; Eniseler, N.; Şahan, Ç. Effects of small-sided games and conventional aerobic interval training on various physiological characteristics and defensive and offensive skills used in soccer. *Kinesiology* **2018**, *50*, 104–111. [CrossRef]
9. Jeunet, C.; Glize, B.; McGonigal, A.; Batail, J.-M.; Micoulaud-Franchi, J.-A. Using EEG-Based Brain Computer Interface and Neurofeedback Targeting Sensorimotor Rhythms to Improve Motor Skills: Theoretical Background, Applications and Prospects. *Neurophysiol. Clin.* **2019**, *49*, 125–136. [CrossRef]
10. Ring, C.; Cooke, A.; Kavussanu, M.; McIntyre, D.; Masters, R. Investigating the Efficacy of Neurofeedback Training for Expediting Expertise and Excellence in Sport. *Psychol. Sport. Exerc.* **2015**, *16*, 118–127. [CrossRef]
11. Dornowski, M.; Wilczyńska, D.; Lachowicz, M.; Sokolowska, I.; Szot, T.; Urbański, R.; Maznychenko, A.; Szwarc, A.; Gotner, K.; Duda, D.; et al. The Effect of EEG Neurofeedback on Lowering the Stress. *Medicine* **2024**, *103*, e37042. [CrossRef]

12. Pourbehbahani, Z.; Saemi, E.; Cheng, M.-Y.; Dehghan, M.R. Both Sensorimotor Rhythm Neurofeedback and Self-Controlled Practice Enhance Motor Learning and Performance in Novice Golfers. *Behav. Sci.* **2023**, *13*, 65. [CrossRef] [PubMed]
13. Li, D.; Elbanna, H.; Lin, F.-Y.; Lu, C.-J.; Chen, L.-J.; Lu, G.; Yu, C.-L.; Wang, K.P. Neuromotor Mechanisms of Successful Football Penalty Kicking: An EEG Pilot Study. *Front. Psychol.* **2025**, *16*, 1452443. [CrossRef] [PubMed]
14. Mihara, M.; Miyai, I.; Hattori, N.; Hatakenaka, M.; Yagura, H.; Kawano, T.; Okibayashi, M.; Danjo, N.; Ishikawa, A.; Inoue, Y.; et al. Neurofeedback Using Real-Time Near-Infrared Spectroscopy Enhances Motor Imagery Related Cortical Activation. *PLoS ONE* **2012**, *7*, e32234. [CrossRef] [PubMed]
15. Köse, B.; Temizkan, E.; Kara, Ö.K.; Kara, K.; Şahin, S. How Does Visual Praxis Based Occupational Therapy Program Effect Motor Skills in Children with Hyperactivity and Attention Disorder?: Single Blind Randomised Study Design. *Deney. Ve Klin. Tıp Derg.* **2022**, *39*, 803–808. [CrossRef]
16. Lee, Y.-S.; Bae, S.-H.; Lee, S.-H.; Kim, K.-Y. Neurofeedback Training Improves the Dual-Task Performance Ability in Stroke Patients. *Tohoku J. Exp. Med.* **2015**, *236*, 81–88. [CrossRef]
17. Nijboer, F.; Birbaumer, N.; Kübler, A. The Influence of Psychological State and Motivation on Brain-Computer Interface Performance in Patients with Amyotrophic Lateral Sclerosis—A Longitudinal Study. *Front. Neuropharmacol.* **2010**, *4*, 55. [CrossRef]
18. Wang, K.-P.; Cheng, M.-Y.; Chen, T.-T.; Chang, Y.-K.; Huang, C.-J.; Feng, J.; Hung, T.-M.; Ren, J. Experts' Successful Psychomotor Performance Was Characterized by Effective Switch of Motor and Attentional Control. *Psychol. Sport. Exerc.* **2019**, *43*, 374–379. [CrossRef]
19. Moreau, D.; Kao, S.-C.; Wang, C.-H. Probing Brain Dynamics Correlates of Motor Expertise with Mobile Recordings. *Sport. Exerc. Perform. Psychol.* **2025**, *14*, 231–249. [CrossRef]
20. Yadav, H.; Maini, S. Electroencephalogram Based Brain-Computer Interface: Applications, Challenges, and Opportunities. *Multimed. Tools Appl.* **2023**, *82*, 47003–47047. [CrossRef]
21. Anton, N.E.; Ziliak, M.C.; Stefanidis, D. Augmenting Mental Imagery for Robotic Surgery Using Neurofeedback: Results of a Randomized Controlled Trial. *Glob. Surg. Educ. J. Assoc. Surg. Educ.* **2023**, *2*, 62. [CrossRef]
22. Benjamin, C.L.; Hosokawa, Y.; Curtis, R.M.; Schaefer, D.A.; Bergin, R.T.; Abegg, M.R.; Casa, D.J. Environmental Conditions, Preseason Fitness Levels, and Game Workload: Analysis of a Female NCAA DI National Championship Soccer Season. *J. Strength. Cond. Res.* **2020**, *34*, 988–994. [CrossRef] [PubMed]
23. Grgic, J.; Mikulic, I.; Mikulic, P. Negative Effects of Mental Fatigue on Performance in the Yo-Yo Test, Loughborough Soccer Passing and Shooting Tests: A Meta-Analysis. *J. Funct. Morphol. Kinesiol.* **2022**, *7*, 10. [CrossRef] [PubMed]
24. Serpiello, F.R.; Cox, A.; Oppici, L.; Hopkins, W.G.; Varley, M.C. The Loughborough Soccer Passing Test Has Impractical Criterion Validity in Elite Youth Football. *Sci. Med. Footb.* **2017**, *1*, 60–64. [CrossRef]
25. Teixeira, J.E.; Alves, A.R.; Ferraz, R.; Forte, P.; Leal, M.; Ribeiro, J.; Silva, A.J.; Barbosa, T.M.; Monteiro, A.M. Effects of Chronological Age, Relative Age, and Maturation Status on Accumulated Training Load and Perceived Exertion in Young Sub-Elite Football Players. *Front. Physiol.* **2022**, *13*, 832202. [CrossRef]
26. Vilar, L.; Duarte, R.; Silva, P.; Chow, J.Y.; Davids, K. The Influence of Pitch Dimensions on Performance during Small-Sided and Conditioned Soccer Games. *J. Sports Sci.* **2014**, *32*, 1751–1759. [CrossRef]
27. Vickers, J.N. Mind over Muscle: The Role of Gaze Control, Spatial Cognition, and the Quiet Eye in Motor Expertise. *Cogn. Process.* **2011**, *12*, 219–222. [CrossRef]
28. Matos, R.; Moreira, C.; Alves, E.; Teixeira, J.E.; Rodrigues, F.; Monteiro, D.; Antunes, R.; Forte, P. Tactical Knowledge by Decision Making and Motor Efficiency of Young Football Players in Different Playing Positions during a Three-a-Side Small-Sided Game. *Behav. Sci.* **2023**, *13*, 310. [CrossRef]
29. Vukelić, M.; Gharabaghi, A. Self-Regulation of Circumscribed Brain Activity Modulates Spatially Selective and Frequency Specific Connectivity of Distributed Resting State Networks. *Front. Behav. Neurosci.* **2015**, *9*, 181. [CrossRef]
30. Tinaz, S.; Para, K.; Vives-Rodriguez, A.; Martinez-Kaigi, V.; Nalamada, K.; Sezgin, M.; Scheinost, D.; Hampson, M.; Louis, E.D.; Constable, R.T. Insula as the Interface between Body Awareness and Movement: A Neurofeedback-Guided Kinesthetic Motor Imagery Study in Parkinson's Disease. *Front. Hum. Neurosci.* **2018**, *12*, 496. [CrossRef]
31. Corrado, S.; Tosti, B.; Mancone, S.; Di Libero, T.; Rodio, A.; Andrade, A.; Diotaiuti, P. Improving Mental Skills in Precision Sports by Using Neurofeedback Training: A Narrative Review. *Sports* **2024**, *12*, 70. [CrossRef]
32. Rodrigues, A.C.; Lima, M.D.M.; de Souza, L.C.; Furtado, C.; Marques, C.E.; Gonçalves, L.; Lima, M.V.; Lasmar, R.P.; Caramelli, P. No Evidence of Association Between Soccer Heading and Cognitive Performance in Professional Soccer Players: Cross-Sectional Results. *Front. Neurol.* **2019**, *10*, 209. [CrossRef] [PubMed]
33. Vestberg, T.; Reinebo, G.; Maurex, L.; Ingvar, M.; Petrovic, P. Core Executive Functions Are Associated with Success in Young Elite Soccer Players. *PLoS ONE* **2017**, *12*, e0170845. [CrossRef]
34. Teixeira, J.E.; Forte, P.; Ferraz, R.; Branquinho, L.; Silva, A.J.; Monteiro, A.M.; Barbosa, T.M. Integrating Physical and Tactical Factors in Football Using Positional Data: A Systematic Review. *PeerJ* **2022**, *10*, e14381. [CrossRef] [PubMed]

35. De Almeida, A.M.; Silva, P.R.S.; Pedrinelli, A.; Hernandez, A.J. Aerobic Fitness in Professional Soccer Players after Anterior Cruciate Ligament Reconstruction. *PLoS ONE* **2018**, *13*, e0194432. [CrossRef] [PubMed]
36. Aristizabal, J.; Restrepo, M.; Estrada, A. Body Composition Assessment by Anthropometry and Bioelectrical Impedance. *Biomedica* **2007**, *27*, 216–224.
37. Kozieł, S.M.; Malina, R.M. Modified Maturity Offset Prediction Equations: Validation in Independent Longitudinal Samples of Boys and Girls. *Sports Med.* **2017**, *48*, 221–236. [CrossRef]
38. Patel, H.H.; Berlinberg, E.J.; Nwachukwu, B.; Williams, R.J.; Mandelbaum, B.; Sonkin, K.; Forsythe, B. Quadriceps Weakness Is Associated with Neuroplastic Changes Within Specific Corticospinal Pathways and Brain Areas After Anterior Cruciate Ligament Reconstruction: Theoretical Utility of Motor Imagery-Based Brain-Computer Interface Technology for Rehabilitation. *Arthrosc. Sports Med. Rehabil.* **2023**, *5*, e207–e216. [CrossRef]
39. Oppici, L.; Panchuk, D.; Serpiello, F.R.; Farrow, D. Frontiers | Long-Term Practice with Domain-Specific Task Constraints Influences Perceptual Skills. *Front. Psychol.* **2017**, *8*, 1387. [CrossRef]
40. Ali, A.; Foskett, A.; Gant, N. Validation of a Soccer Skill Test for Use with Females. *Int. J. Sports Med.* **2008**, *29*, 917–921. [CrossRef]
41. Martínez-Lagunas, V.; Niessen, M.; Hartmann, U. Women's Football: Player Characteristics and Demands of the Game. *J. Sport. Health Sci.* **2014**, *3*, 258–272. [CrossRef]
42. Hao, J.; Ho, T. Machine Learning Made Easy: A Review of Scikit-Learn Package in Python Programming Language. *J. Educ. Behav. Stat.* **2019**, *44*, 107699861983224. [CrossRef]
43. Hopkins, W.G.; Marshall, S.W.; Batterham, A.M.; Hanin, J. Progressive Statistics for Studies in Sports Medicine and Exercise Science. *Med. Sci. Sports Exerc.* **2009**, *41*, 3–12. [CrossRef]
44. Lehmann, J. Is There a Relationship between Executive Functions and Resilience in Youth Elite Soccer Players? *Brain Behav.* **2023**, *13*, e3122. [CrossRef]
45. Vestberg, T.; Gustafson, R.; Maurex, L.; Ingvar, M.; Petrovic, P. Executive Functions Predict the Success of Top-Soccer Players. *PLoS ONE* **2012**, *7*, e34731. [CrossRef] [PubMed]
46. Sabarit, A.; Morillo-Baro, J.P.; Reigal, R.E.; Vázquez-Diz, J.A.; Hernández-Mendo, A.; Morales-Sánchez, V. Analysis of Game Actions and Performance in Young Soccer Players: A Study Using Sequential Analysis. *Sustainability* **2022**, *14*, 13263. [CrossRef]
47. López-Segovia, M.; Dellal, A.; Chamari, K.; González-Badillo, J.J. Importance of Muscle Power Variables in Repeated and Single Sprint Performance in Soccer Players. *J. Hum. Kinet.* **2014**, *40*, 201–211. [CrossRef] [PubMed]
48. Duncan, M.J.; Eyre, E.L.J.; Noon, M.R.; Morris, R.; Thake, C.D.; Clarke, N.D.; Cunningham, A.J. Actual and Perceived Motor Competence Mediate the Relationship between Physical Fitness and Technical Skill Performance in Young Soccer Players. *Eur. J. Sport. Sci.* **2022**, *22*, 1196–1203. [CrossRef]
49. Wing, C.E.; Turner, A.N.; Bishop, C.J. Importance of strength and power on key performance indicators in elite youth soccer. *J. Strength Cond. Res.* **2020**, *34*, 2006–2014. [CrossRef]
50. Matsutake, T.; Nakata, H.; Matsuo, G.; Natsuhara, T.; Zippo, K.; Watanabe, K.; Sugo, T. Fast and Stable Responses during Decision Making Require Strong Inhibitory Processes in Soccer Players. *Brain Sci.* **2024**, *14*, 199. [CrossRef]
51. Iacono, A.D.; Beato, M.; Unnithan, V. Comparative Effects of Game Profile-Based Training and Small-Sided Games on Physical Performance of Elite Young Soccer Players. *J. Strength Cond. Res.* **2021**, *35*, 2810–2817. [CrossRef]
52. Lovecchio, N.; Manes, G.; Filipas, L.; Giuriato, M.; Torre, A.L.; Iaia, F.M.; Codella, R. Screening Youth Soccer Players by Means of Cognitive Function and Agility Testing. *Percept Mot. Ski.* **2021**, *128*, 2710–2724. [CrossRef]
53. Causer, J.; Smeeton, N.J.; Williams, A.M. Expertise Differences in Anticipatory Judgements during a Temporally and Spatially Occluded Task. *PLoS ONE* **2017**, *12*, e0171330. [CrossRef]
54. Teixeira, J.E.; Maio, E.; Afonso, P.; Encarnação, S.; Machado, G.F.; Morgans, R.; Barbosa, T.M.; Monteiro, A.M.; Forte, P.; Ferraz, R.; et al. Mapping Football Tactical Behavior and Collective Dynamics with Artificial Intelligence: A Systematic Review. *Front. Sports Act. Living* **2025**, *7*, 1569155. [CrossRef]

Journal of
Functional Morphology and Kinesiology

MDPI

Article

Sex-Specific Morphological and Neuromuscular Profiles of U-15 Colombian Basketball Players

Alex Ojeda-Aravena [1,2], María Alejandra Camacho-Villa [3,4], Fernando Millan-Domingo [5], Ronald Quintero-Bernal [3], Jeimy Andrea Merchán [6], Fabio Villafrades [7] and Adrián De la Rosa [3,5,7,*]

1 Departamento de Ciencias de la Actividad Física, Universidad de Los Lagos, Osorno 5290000, Chile; alex.ojeda@ulagos.cl
2 Programa de Investigación en Deporte, Sociedad y Buen Vivir (DSBv), Universidad de Los Lagos, Osorno 5290000, Chile
3 Harpeer Research Group, Yumbo 760001, Colombia; alejacvilla@gmail.com (M.A.C.-V.)
4 Pain Study Group (GED), Physical Therapy School, Universidad Industrial de Santander, Bucaramanga 680002, Colombia
5 Freshage Research Group, Department of Physiology, Faculty of Medicine, University of Valencia, Centro de Investigación Biomédica en Red Fragilidad y Envejecimiento Saludable (CIBERFES), Fundación Investigación Hospital Clínico Universitario/INCLIVA, 46010 Valencia, Spain; fernando.millan-domingo@uv.es
6 Physical Activity and Sport Program, Unidades Tecnológicas de Santander, Bucaramanga 680006, Colombia
7 Body, Physical Activity and Sport Study Group (GECAFD), Sports Department, Universidad Industrial de Santander, Bucaramanga 680002, Colombia; villafra@uis.edu.co
* Correspondence: adrian.rosa@uv.es

Abstract: Background: Basketball performance is highly dependent on morphological and neuromuscular traits, especially during adolescence, when rapid growth and maturation generate marked sex-based differences. However, limited data are available on Latin American players' performance. This study aimed to compare the anthropometric characteristics, body composition, somatotype, and neuromuscular performance of male and female Colombian U-15 national basketball players. **Methods:** The sample consisted of thirty-seven players (20 males, 17 females; mean age: 14.8 $\pm$ 0.4 years) during the preparatory phase of the 2022 South American U-15 Championships. Anthropometry and body composition were evaluated following ISAK standards, and somatotype was calculated using Carter and Heath's method. Neuromuscular performance included countermovement and squat jumps, bilateral handgrip strength, and isometric knee extensor and flexor peak torques. Between-sex differences were examined using *t*-tests, Welch's tests, or Mann–Whitney U tests. The effects of sex on body composition, somatotype, and neuromuscular outcomes were assessed using MANOVA. **Results:** Males had higher muscle mass, lower adipose mass, and greater limb lengths than females ($p < 0.01$). No sex differences were observed in BMI, waist or hip circumference, or quadriceps strength. Regarding neuromuscular performance, males exhibited higher handgrip strength, hamstring torque (absolute and relative), and jump performance than females. **Conclusions:** Males showed greater muscle mass, strength, and jump performance, whereas females displayed higher fat levels and endomorphy than males. These findings provide useful data for optimizing training load prescriptions, guiding targeted strength programs, and developing sex-specific strategies for injury prevention and talent identification in adolescents.

Keywords: team sports; adolescents; somatotype; body composition; athletic performance

1. Introduction

Basketball is an intermittent team sport characterized by a combination of high-intensity actions, such as accelerations, decelerations, changes in direction, jumps, lateral sliding, and static efforts, alternating with lower-intensity activities [1]. Studies have shown that junior male basketball players spend 22.57% of their total game time in high-moderate intensity activities and 63.28% in low-intensity zones [2,3]. Male and female elite basketball players change movement types every 1–3 s and perform an average of 44 ± 7 jumps per game [4]. Moreover, specific basketball movements (e.g., rebounding, sprinting, dribbling, shooting, and blocking) are performed at high intensities and are strongly associated with the development of strength, power, speed, and agility [5].

Adolescence is a critical period for athletic development and is marked by rapid growth and maturation processes that differ by sex. These biological differences affect physical performance, body composition, and neuromuscular capabilities, highlighting the importance of considering sex-specific characteristics in the assessment and training of young basketball players [6].

Anthropometric profiles and body composition play decisive roles in basketball performance. An adequate profile, characterized by greater height and lower fat mass, can provide significant advantages for the court [6]. Previous studies have demonstrated that a lower body fat percentage (BF%) is positively associated with competitive level and negatively related to the performance of explosive actions, such as sprinting, changes in direction, and vertical jumps [7–9]. Furthermore, dimensions such as body height, arm length, and upraised arm height are correlated with enhanced performance in basketball-specific tasks, such as obstacle dribbling and speed-related actions [10]. Furthermore, a correlation between morphology, physical capabilities, and game performance, through performance indices derived from match statistics, has been documented in elite female basketball [11].

A complementary approach to body composition assessment is the use of somatotypes, which provide a more comprehensive description of an athlete's physique by categorizing body structure into endomorphic, mesomorphic, and ectomorphic components [10]. These outcomes allow for a deeper understanding of how morphology influences basketball performance, offering valuable insights for designing tailored training programs and identifying morphological differences between male and female players [12].

In addition to morphology, basketball performance relies heavily on strength and power. Lower limb strength and explosive capacity are essential for game action, including jumping, rebounding, and blocking. The countermovement jump (CMJ) and squat jump (SJ) are widely used to assess explosive strength in basketball, providing insight into the contribution of the stretch-shortening cycle and pure concentric power, respectively [13,14]. Upper-body strength is also critical, with handgrip strength (HGS) serving as a practical indicator of muscular capacity that directly influences ball control, passing, and defensive actions. Upper and lower limb strength contribute decisively to overall basketball performance, with greater jumping and grip capacities consistently observed in players with higher competition levels [11].

A rigorous assessment of morphological and neuromuscular variables during adolescence is essential to underpin athletic development, guide training prescription, and optimize long-term performance trajectories. These variables are also significantly associated with basketball performance and talent progression. Anthropometric characteristics (e.g., stature, somatotype) and neuromuscular outputs (e.g., jump height, torque) discriminate selection levels and positional demands, and they explain variance in key performance indicators and game outcomes. These associations have been documented in elite youth cohorts and in meta-analytic syntheses on talent identification, reinforcing the utility of an integrated anthropometric–motor profile along developmental pathways [15].

Regarding the neuromuscular profile, and acknowledging its value, most evidence relies on vertical and horizontal jump tests to evaluate power performance. Measuring isometric knee torque adds joint specificity, in-field reliability, and the capacity to characterize strength components (peak and rate of torque development, RTD) that are not captured by jump tests or by the handgrip rate of torque development [16]. Integrating these measures with the squat jump (SJ) and countermovement jump (CMJ) yields a more robust neuromuscular profile that is useful for selection, individualized prescription, and decision-making in injury prevention and return to sport [17,18].

Despite its importance, sport and sex-specific normative references for youth basketball remain scarce in Latin America. Although age and sex-stratified reference values exist in other regions (e.g., federated Tunisian basketball) and in mixed cohorts from sports academies, Latin American datasets are fragmentary. This gap hinders benchmarking, talent identification, and load progression in the region, and justifies the development of local percentile charts and reference tables (by sex, age band, and playing position). This knowledge gap limits the ability to optimize training methodologies and to compare developmental trajectories with those observed in other populations.

Therefore, the present study aimed to compare anthropometric characteristics, body composition, and neuromuscular performance, including jump performance, HGS, and isometric lower limb strength, between young Colombian male and female basketball players.

2. Materials and Methods

2.1. Subjects

This cross-sectional analytical study was conducted during the preparatory phase for the 2022 South American U-15 Women's and Men's Championships. A total of 37 youth basketball players from Colombian U-15 national teams participated, comprising 17 females (mean age: 14.9 ± 0.24 years) and 20 males (mean age: 14.8 ± 0.44 years). All athletes who participated in the training microcycle were eligible for inclusion in the study.

Participants were healthy and free from any medical conditions that could affect performance, hand function, anthropometry, or daily activities. All procedures were explained in detail to the athletes and their legal guardians, and non-participation had no effect on team selection. Written informed consent was obtained from the parents or legal guardians, and assent was provided by all athletes. The study was conducted in accordance with the Declaration of Helsinki and approved by the University Ethics Committee (Approval No. 0010-2022; 2 May 2022).

2.2. Testing Procedures

Comprehensive details regarding the anthropometric parameters and neuromuscular performance are provided in the subsequent section. All assessments were performed before the training session in the laboratory of the University during morning hours (between 8:00 and 11:00 a.m.), on two separate days for females and males.

All the evaluations took place in a temperature-controlled environment and followed a predefined sequence: anthropometric measurements, Countermovement Jump (CMJ), Squat Jump (SJ), Handgrip Strength (HGS); and maximal isometric strength of the knee extensors and flexors.

Lower and upper limb dominance were determined by asking athletes which arm or leg they would use to throw/kick a ball [19,20].

Assessments were carried out by three researchers, each with nine years of experience in sports research. Anthropometric data were gathered by a level 2 anthropometrist, achieving an intra-rater intraclass correlation coefficient (ICC) between 0.91 and 0.96, indicating excellent reliability.

HGS, CMJ, and SJ were evaluated by a second researcher, trained in participant positioning and standardized verbal encouragement. Intra-rater ICC values for HGS were 0.98 for the dominant hand and 0.97 for the non-dominant hand. For CMJ and SJ, ICC values ranged from 0.94 to 0.95, indicating high reliability [21]. The maximal isometric strength test of the knee extensors and flexors was performed by a third researcher, who received specific training on participant positioning, device use and software protocols. The ICC values for these measures were 0.85 and 0.87, respectively.

2.3. Anthropometric Measurements and Body Composition

Anthropometric measurements were performed in accordance with the standards of the International Society for the Advancement of Kinanthropometry (ISAK). Participants wore minimal clothing and were barefoot to optimize measurement accuracy.

Stature was measured to the nearest 0.1 cm using a stadiometer (Seca® 274, Hamburg, Germany; Technical Error of Measurement = 0.019%), and body mass was measured to the nearest 0.1 kg using a TANITA BC 240 MA (Tanita Corporation, Arlington Heights, IL, USA). Eight skinfold thicknesses were assessed with a skinfold caliper (Cescorf, Porto Alegre, Brazil) at the following ISAK defined sites: triceps, bíceps, subscapular, supraspinale, iliac crest, abdominal, front thigh, and medial calf [22].

Five girths were measured with a metal tape (Cescorf, Porto Alegre, Brazil; measurement range of up to 100 cm and accuracy to 0.1 cm) at the following sites: relaxed arm, flexed arm, waist, hip, and calf. Bone breadths on both sides (humerus, bistyloid, and femur) were measured to the nearest 0.1 cm using a small bone anthropometer (Cescorf, Porto Alegre, Brazil). Upper limb lengths on both sides (arm and forearm length, hand breadth, hand length, and first-to-fifth finger distance) were recorded to the nearest 0.1 cm with a segmometer (Cescorf, Porto Alegre, Brazil), following previously described protocols [7,15].

Based on the measurements, body composition was estimated using the calculations proposed by De Rose and Guimaraes following their four-compartment model (fat mass, bone mass, muscle mass, and residual mass). BF% was calculated using the Yuhasz [23] equation adapted for adolescent athletes; fat mass was calculated as the product of BF% and body mass, divided by 100; bone mass from the bicondylar breadths of the humerus and femur [24]; residual mass as a fixed proportion of body mass [25]; and muscle mass by subtraction from total body mass. Likewise, the somatotype of these athletes was determined based on the model proposed by Heath and Carter, obtaining the value of the three components: endomorphic, mesomorphic, and ectomorphic. The analysis was performed using a Microsoft Excel spreadsheet.

The sum of six skinfold measurements (triceps, subscapular, supraspinale, abdominal, front thigh, and medial calf) and the sum of eight skinfold measurements (triceps, bíceps, subscapular, supraspinal, iliac crest, abdominal, front thigh, and medial calf) were also calculated. The estimated upper arm muscle area was calculated for the dominant side using the equation proposed by Frisancho [26].

All measurements were taken on the right side of the body by the same ISAK Level 2 anthropometrist to minimize inter-observer variability.

2.4. Neuromuscular Performance Assessment

Vertical Jump Tests

The CMJ and SJ were used to evaluate lower-limb neuromuscular performance using a contact platform (Chronojump Boscosystem, Barcelona, Spain) following standardized protocols described by Bosco et al. [27].

Before testing, athletes completed a five-minute warm-up consisting of light jogging, skipping, and dynamic exercises (half-squats, lunges, and leg swings). They were familiarized with the jumping technique with three submaximal practice trials of each type (CMJ and SJ) before performing maximal attempts.

For the CMJ, participants began in an upright standing position with hands placed on the hips to eliminate arm swing, performed a rapid downward movement to approximately 90° of knee flexion, and immediately executed a maximal vertical jump. For the SJ, participants started from a static squat position at ~90° of knee flexion, held the position for 2–3 s to minimize the contribution of the stretch-shortening cycle [28], and then jumped vertically without countermovement. Verbal encouragement was provided during testing to ensure maximal effort.

Three valid trials were performed for each jump type, separated by two minutes of passive recovery. Players were asked to jump as high as possible, and the highest value for each type of jump was used for analysis.

Handgrip Strength Assessment

Maximal HGS was measured bilaterally using a digital handheld dynamometer (Takei 5401; Tokyo, Japan) with a measurement accuracy of 0.1 kg. Participants were tested in a standing position, with the shoulder of the test arm adducted and the elbow flexed at 90°. The forearm and wrist were maintained in a neutral position to ensure proper alignment between the hand and forearm during grip assessment. The dynamometer was individually adjusted to each participant's hand size to ensure proper flexion of the metacarpophalangeal joints. Before testing, standardized verbal instructions were provided, and verbal encouragement was given throughout the procedure to elicit maximal effort [8].

Three maximal voluntary contractions were performed per hand, each lasting 3–5 s. A 60 s rest interval was implemented between trials to minimize potential fatigue. HGS values were recorded in kilograms (kg), and the highest value from the three trials, for each hand, was used for statistical analysis [22].

Maximal Isometric Lower Limb Strength

Maximal isometric strength of the quadriceps and hamstring muscles was assessed at 60° of knee flexion, with 0° representing full knee extension. Measurements were conducted using a portable hand-held dynamometer (Chronojump Boscosystem, Barcelona, Spain). This method has been demonstrated to be reliable for assessing lower-limb strength in both general and athletic populations [29].

Before testing, players completed a standardized, sport-specific warm-up as detailed elsewhere [30]. In addition, they were familiarized with the protocol by performing two submaximal contractions for both flexion and extension in both legs. After the warm-up, testing occurred with participants seated on an adjustable chair, with hips and knees stabilized by an adjustable and rigid strap around the thighs and hips to minimize trunk compensations (Figure 1). A knee flexion angle of 60° was verified using an electronic goniometer (K-force Sens, Kinvent, France). Figure 1 of the protocol was refined using the AI image generator Plus tool (Image Generator GPT (ChatGPT, version GPT-4o)) to enhance visual clarity.

For the quadriceps assessment, the dynamometer was attached to a rigid bar anchored to the back of the chair and connected to the tested leg's ankle via a strap positioned approximately 1–2 cm above the malleolus. The strap length was adjusted to achieve 60° of knee extension (Figure 1A). In contrast, during hamstring assessment, the dynamometer was attached to a rigid bar anchored to the wall in front of the chair, again using a malleolus level strap adjusted to ensure 60° of knee flexion (Figure 1B).

To avoid compensatory movements from the non-tested leg, participants were instructed to keep it relaxed (i.e., do not anchor the leg to the chair). Given that the original protocol does not allow support on the chair or any object, participants were asked to place their hands on the opposite shoulder (Figure 1). The dynamometer data were transmitted to an A/D converter and collected at 160 Hz, then filtered and smoothed using Chronojump software (v2.5.2-64, Chronojump Bosco System, Barcelona, Spain) according to manufacturer recommendations.

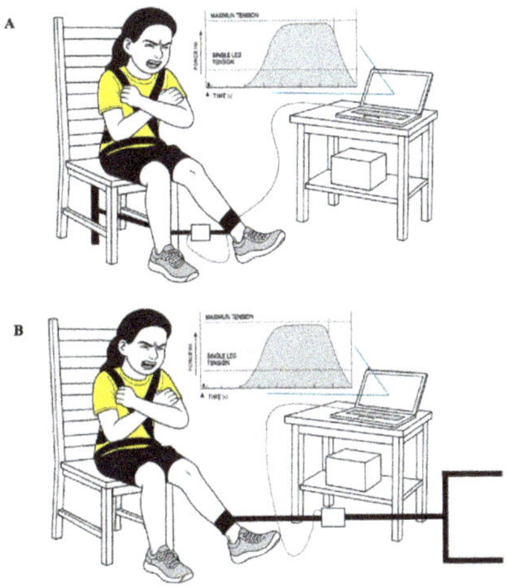

Figure 1. Set up of the maximal isometric lower limb strength. (**A**) Quadriceps and (**B**) hamstrings.

Participants performed two maximal isometric contractions per movement for both lower limbs, each lasting ~5 s, with 60 s rest intervals between trials. They were verbally encouraged to exert maximal effort throughout the test.

Torque was calculated by multiplying the measured force by the moment arm, which was defined as the linear distance from each participant's lateral epicondyle of the femur to the center of the ankle strap connected to the load cell. The highest peak torque value recorded for each muscle group was used for the analysis. Additionally, the relative peak torque was calculated by normalizing the absolute value to each participant's weight. Both values were included in the statistical analysis.

2.5. Statistical Analysis

All statistical analyses were performed using Jamovi software (v2.6, https://www.jamovi.org). Descriptive statistics were calculated for all variables and presented as mean ± standard deviation (SD) or median and interquartile range (IQR), as appropriate. The Shapiro–Wilk test was used to assess univariate normality, and Levene's test was applied to verify the homogeneity of variances.

Between-sex differences were examined using independent sample *t*-tests for normally distributed variables with homogeneous variances, and Mann–Whitney *U* tests for non-normally distributed variables. Effect sizes were reported as Cohen's d for *t* test and rank-biserial correlation (r) for the Mann–Whitney U test. For interpretation, we used conventional benchmarks: $d \approx 0.20$ (small), 0.50 (medium), 0.80 (large); and $r \approx 0.10$ (small), 0.30 (medium), 0.50 (large) [31].

Two separate multivariate analyses of variance (MANOVA) were performed to assess the effect of sex on (1) body composition variables and (2) neuromuscular performance variables. The assumptions for MANOVA were tested using Box's M test for homogeneity of covariance matrices and Shapiro–Wilk's test for multivariate normality. When multivariate normality was violated, Pillai's Trace was used as the multivariate test statistic due to its robustness to assumption violations.

Significant multivariate effects were followed by univariate analysis for each dependent variable. Partial eta squared (η^2p) was calculated as an estimate of the effect size for each univariate test and interpreted according to Richardson guidelines: small (0.01 to 0.05), medium (0.06 to 0.13), and large (≥0.14) [32]. The significance level was set at $p < 0.05$, with Bonferroni adjusted thresholds applied where appropriate.

3. Results

Table 1 summarizes the results of the analysis of anthropometric variables by sex. Multiple comparisons with the *t*-test and Mann-Whitney U test, with a Bonferroni correction, revealed significant sex differences in most variables. Males showed notably greater height ($p < 0.001$, d = 1.58), waist-to-hip ratio ($p < 0.001$, d = 1.64), and a wide range of bone breadths, including the humerus, femur, bistyloid, and hand widths ($p < 0.001$; $d = 1.53$–3.47) on both sides of the body (Table 1). Significant differences were also observed in limb lengths, with males exhibiting longer arms, forearms, hands, and finger lengths bilaterally ($p < 0.001$; d = 1.37–2.20 or r > 0.73).

On the other hand, female players exhibited significantly higher values in all skinfold thicknesses, including triceps, subscapular, supraspinale, abdominal, front thigh, and medial calf ($p < 0.001$; d = 1.21–2.54), as well as in the sum of 6 and 8 skinfolds ($p < 0.001$, $d = 2.26$ and 1.89, respectively), indicating greater subcutaneous fat (Table 2).

Table 1. Anthropometric profile of young male and female basketball players.

Variable	Male	Female	Statistic (t/U)	Effect Size (d/r)
Basic				
Height (m)	1.79 (0.08)	1.66 (0.07) ***	t = 5.02	d = 1.58
Weight (kg)	71.64 (9.85)	64.12 (7.49)	t = 2.58	d = 0.85
BMI (kg/m^2)	22.35 (2.34)	23.31 (2.22)	t = −1.28	d = 0.43
Waist-to-hip ratio	0.79 (0.03)	0.74 (0.03) ***	t = 4.98	d = 1.64
Skinfolds (mm)				
Triceps	8.80 (3.33)	14.88 (3.12) ***	t = −5.70	d = 1.91
Subscapular	8.70 (1.67)	12.21 (3.73) ***	t = −3.58	d = 1.21
Biceps	5.08 (1.49)	7.15 (1.52) ***	t = −4.18	d = 1.38
Ileocrestal	11.20 (4.05)	14.74 (6.28)	U = 114.0	r = 0.32
Supraspinal	7.58 (1.82)	13.18 (3.88) ***	t = −5.47	d = 1.86
Abdominal	11.45 (3.52)	18.03 (4.32) ***	t = −5.11	d = 1.67
Front thigh	10.88 (3.13)	20.21 (4.77) ***	t = −7.13	d = 2.35
Medial calf	7.55 (2.01)	15.32 (3.71) ***	t = −7.73	d = 2.54
Sum of 6 skinfolds	54.95 (13.87)	93.82 (19.86) ***	t = −6.78	d = 2.26
Sum of 8 skinfolds	71.23 (18.51)	115.71 (26.68) ***	t = −5.79	d = 1.89
Girths (cm)				
Waist	74.34 (3.54)	71.69 (4.91)	t = 1.90	d = 0.63
Hip	94.07 (5.16)	96.64 (5.19)	t = −1.51	d = 0.51
Arm (relaxed)	27.99 (2.42)	26.29 (1.98)	t = 2.30	d = 0.75
Arm (flexed and tensed)	30.63 (1.96)	27.33 (2.20) ***	t = 4.82	d = 1.59
Calf	37.95 (5.01)	35.64 (1.72)	t = 1.93	d = 0.64
Lengths (cm)				
Arm length	34.23 (1.94)	31.72 (1.50) ***	t = 4.34	d = 1.44
Forearm length	27.73 (2.61)	24.35 (1.52) ***	U = 22.5	r = 0.86
Hand length	20.30 (1.22)	18.85 (4.33) ***	U = 40.5	r = 0.76
First-to-fifth finger distance	22.75 (1.43)	20.73 (1.27) ***	U = 45.5	r = 0.73
Arm length (left)	34.05 (2.07)	31.44 (1.75) ***	t = 4.10	d = 1.37
Forearm length (left)	27.08 (2.08)	24.08 (1.37) ***	t = 5.24	d = 1.75
Hand length (left)	20.38 (1.23)	17.93 (1.00) ***	t = 6.53	d = 2.17
First-to-fifth finger dist. (left)	23.13 (1.64)	20.91 (1.07) ***	U = 32.0	r = 0.81

Table 1. *Cont.*

Variable	Male	Female	Statistic (t/U)	Effect Size (d/r)
Bone breadth (cm)				
Humerus	7.25 (0.38)	6.40 (0.33) ***	t = 7.15	d = 2.38
Femur	8.96 (1.19)	10.01 (0.58) ***	U = 54.5	r = 0.67
Bistyloid	6.67 (1.00)	5.58 (0.31) ***	t = 4.60	d = 1.53
Hand	9.72 (0.65)	7.98 (0.35) ***	t = 10.41	d = 3.47
Humerus (left)	7.32 (0.48)	6.48 (0.31) ***	t = 6.16	d = 2.05
Bistyloid (left)	6.19 (0.29)	5.58 (0.25) ***	t = 6.84	d = 2.27
Hand (left)	8.79 (0.43)	7.98 (0.35) ***	t = 6.64	d = 2.20

All data are presented as mean (standard deviation). *** denotes $p < 0.001$ vs. male; t/U: differences were assessed with *t*-test/Mann-Whithey U test; d/r: effect size was assessed with Cohen's d and rank-biserial correlation.

Table 2. Somatotype characteristics of young male and female basketball players.

Variable	Male	Female	F	Effect Size (η^2p)
Body Fat % (Yuhasz)	8.36 (1.46)	18.10 (3.07) ***	159.27	0.82
Adipose mass (kg)	6.09 (1.83)	11.76 (3.12) ***	47.19	0.57
Muscle mass (kg)	35.51 (4.49)	27.21 (2.79) ***	38.53	0.52
Residual mass (kg)	17.26 (2.37)	14.11 (1.75) ***	20.51	0.37
Bone mass (kg)	12.78 (1.83)	11.04 (1.33) **	11.37	0.24
Upper arm muscle area (cm^2)	50.82 (6.65)	37.38 (5.49) ***	43.85	0.56
Endomorphy	2.36 (0.64)	4.18 (0.90) ***	51.49	0.60
Mesomorphy	3.85 (1.31)	4.64 (1.21)	3.57	0.09
Ectomorphy	3.05 (1.24)	1.88 (1.09) **	9.13	0.21
SAM	1.64 (1.91)	1.31 (0.54)	1.65	0.04

All data are presented as mean (standard deviation). **/*** denotes $p < 0.01$/$p < 0.01$ vs. male.

Several variables did not show statistically significant differences after Bonferroni correction (Table 2), including body weight ($p = 0.014$), BMI ($p = 0.210$), waist circumference ($p = 0.066$), hip circumference ($p = 0.141$), relaxed arm girth ($p = 0.028$), calf girth ($p = 0.079$), and the ileocrestal skinfold ($p = 0.090$).

Overall, these findings highlight similar and substantial sex-based differences in body composition and structural dimensions among adolescent basketball players. Males generally displayed greater skeletal breadths and linear measurements, whereas females showed higher values in subcutaneous fat measures.

After studying the anthropometric profile, a MANOVA was conducted to assess the effect of sex on body composition, the somatotype components and the somatotype attitudinal mean (SAM). The overall multivariate effect was significant, Pillai's Trace = 0.97, $F(10,26) = 76.40$, $p < 0.001$, indicating marked sex-based differences in body composition profiles between male and female players (Table 2). The univariate analyses revealed that females presented significantly higher values for body fat percentage and adipose mass, whereas male showed greater muscle mass, residual mass, bone mass, and upper arm muscle area (Table 2).

Regarding somatotype components, females demonstrated markedly higher endomorphy values ($F = 51.49$, $p < 0.001$), while males scored significantly higher in ectomorphy. Mesomorphy tended to be higher in females, although this difference was not statistically significant ($F = 3.57$, $p = 0.067$). Thus, while male participants exhibited a meso-ectomorphic somatotype (2.36–3.85–3.05), with ranges of 1.40–3.94 for endomorphy, 1.50–6.90 for mesomorphy, and 0.86–5.93 for ectomorphy. Female participants demonstrated a meso-endomorphic somatotype (4.18–4.64–1.88), ranging from 2.78 to 5.75 for endomorphy, 2.58–6.39 for mesomorphy, and 0.56–4.42 for ectomorphy. To facilitate interpretation, individual somatotypical values are illustrated in Figure 2.

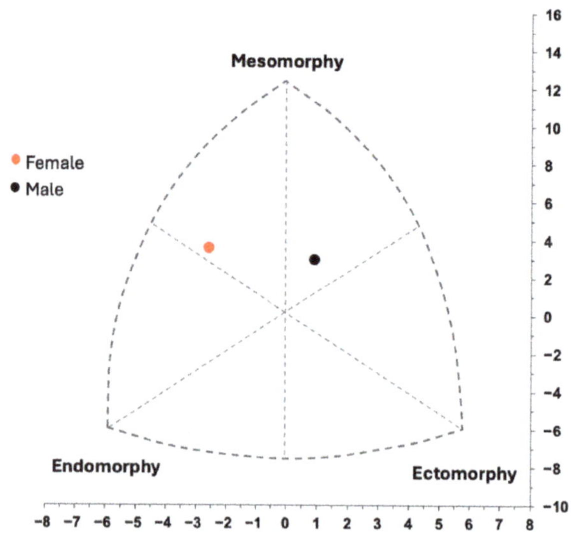

Figure 2. Somatotype distribution (somatochart) of young male and female basketball players.

Table 3 presents the results of the analysis of muscular strength and power output variables by sex. A MANOVA revealed a statistically significant overall multivariate effect (Pillai's Trace = 0.90, F(12,21) = 16,3, $p < 0.001$), indicating a strong influence of sex on physical performance variables.

Table 3. Sex-Based Differences in Neuromuscular Performance Variables.

Variable	Male	Female	F	η^2p
Handgrip Strength (kgf)				
Dominant	39.61 (7.13)	29.06 (5.24) ***	22.69	0.41
Non dominant	40.41 (6.63)	26.66 (4.77) ***	46.98	0.59
Quadriceps Performance				
Peak Torque $(N{\cdot}m)_{DS}$	177.01 (65.24)	145.17 (38.68)	2.48	0.07
Relative Peak Torque $(N{\cdot}m{\cdot}kg^{-1})_{DS}$	2.51 (0.90)	2.27 (0.58)	0.56	0.02
Peak Torque $(N{\cdot}m)_{NDS}$	186.66 (63.68)	155.09 (39.87)	3.27	0.09
Relative Peak Torque $(N{\cdot}m{\cdot}kg^{-1})_{NDS}$	2.64 (0.89)	2.40 (0.61)	0.93	0.03
Hamstring Performance				
Peak Torque $(N{\cdot}m)_{DS}$	155.11 (20.98)	121.14 (9.69) ***	33.89	0.51
Relative Peak Torque $(N{\cdot}m{\cdot}kg^{-1})_{DS}$	2.21 (0.27)	1.91 (0.23) ***	12.04	0.27
Peak Torque $(N{\cdot}m)_{NDS}$	161.53 (10.98)	121.14 (9.69) ***	19.58	0.09
Relative Peak Torque $(N{\cdot}m{\cdot}kg^{-1})_{NDS}$	2.30 (0.42)	1.90 (0.27) ***	9.43	0.23
Jump Performance (cm)				
CMJ height	34.80 (5.64)	23.28 (3.05) ***	55.62	0.63
SJ height	32.02 (5.95)	21.31 (3.04) ***	41.57	0.56

All data are presented as mean (standard deviation). *** denotes $p < 0.01$ vs. male, respectively.

Follow-up univariate analyses showed that males achieved higher HGS values than females in both ($F = 22.69$, $p < 0.001$, $\eta^2p = 0.41$) and non-dominant hand ($F = 46.98$, $p < 0.001$, $\eta^2p = 0.59$) hands, with very large effect sizes, reflecting superior upper body strength in male players.

In quadriceps performance, although males recorded higher peak torque values in both sides, the differences were not statistically significant ($p = 0.46$ and $p = 0.34$, respectively). In contrast, hamstring peak torque was significantly greater in males than females on both the dominant ($F = 12.04$, $p < 0.01$, $\eta^2 p = 0.27$) and non-dominant sides ($F = 9.43$, $p < 0.01$, $\eta^2 p = 0.23$.

The largest differences emerged in jump performance, where males outperformed females in both the CMJ and SJ height. These differences were statistically significant ($F = 55.62$ and 41.57, respectively; $p < 0.001$ for both) and accompanied by large effect sizes ($\eta^2 p = 0.63$ and 0.56), highlighting a clear disparity in explosive lower-limb power between sexes.

4. Discussion

4.1. Sex-Based Differences in Anthropometric and Body Composition Profiles

This study investigated sex-based differences in anthropometric characteristics, body composition, and neuromuscular performance among U-15 Colombian national basketball players. Here, we provide reference values of anthropometric and neuromuscular performance variables in highly trained players. Similarly, our findings showed significant sex-based differences in most variables assessed, highlighting the relevance of considering sex-specific profiles when evaluating youth athletes and designing development programs.

Basketball is highly dependent on physical and structural traits, particularly during adolescence [33,34], a critical period for growth and motor development. While several studies have documented anthropometric variations according to competitive level, to our knowledge, this is the first study to explore sex-specific differences in both anthropometric and neuromuscular characteristics among Colombian adolescent basketball players [35].

4.2. Anthropometric Profile and Body Composition

Significant sexual dimorphism arises during puberty, not only in the timing of pubertal milestones but also in changes in body composition [35,36]. Although both sexes experience a rapid increase in body fat during this period, the pattern and extent of fat accumulation differ between males and females. Changes during puberty include increased linear growth and muscle development in males, and greater fat accumulation and earlier skeletal maturity in females [37]. In boys, rising testosterone levels drive significant increase in bone length, height, and muscle mass, along with a decrease in limb fat [38]. In contrast, girls show a smaller increase in height and muscle mass, but a significant increase in body fat deposition [39].

Previous studies have reported significant differences in stature, arm span, leg length, and hand length among young basketball players across different levels of expertise [40]. Similarly, sex-specific differences in body fat percentage and body composition have also been documented in young athletes across different sports, including basketball [41].

Consistent with the above-mentioned reports, male players in our study exhibited significantly greater height, bone breadths, and limb lengths, especially in the arms, forearms, and hands, compared to females (Table 2). These morphological traits, especially longer arms, forearms, and hands, align with prior findings in elite young male basketball players [42–44]. Such advantages are critical in basketball, where reach, hand size, and stature directly influence shooting, passing, rebounding, defensive efficiency, and physical performance in a broad range of tests [45–47].

In terms of body composition, several studies have shown that male athletes typically display a more mesomorphic somatotype, characterized by greater muscularity and a more robust physique. In contrast, female athletes tend to present a higher endomorphic component, reflecting a greater proportion of body fat and a softer overall body composition [48,49]. In our study, female players had significantly higher values in skinfold thicknesses, endomorphy, body fat percentage, and adipose mass, indicating a greater accumulation of both subcutaneous and total fat. Conversely, male players exhibited more ectomorphic profiles and total muscle mass, denoting a leaner and more linear body shape (Table 3). This pattern aligns with typical sex-specific physiological development during adolescence, as documented in the studies of young athletes [50,51]. In basketball-specific contexts, it has been reported that female youth and professional players tend to display high endomorphy and low ectomorphy, which may affect performance and efficiency in specific actions within the sport [52,53]. On the other hand, the absence of significant differences in SAM values, which reflects the overall deviation from a balanced somatotype, suggests similar levels of somatotype dispersion in both sexes.

4.3. Neuromuscular Performance

It is well known that men's athletic performance exceeds that of women, especially in power sports, due to their greater strength, speed, and endurance [54,55]. This physical advantage arises during early adolescence, when male puberty begins. Afterwards, men acquire bigger muscle mass, greater strength, larger and stronger bones, and higher circulating hemoglobin. They also experience mental and/or psychological differences [56].

Muscle strength is a key component of athletic performance. In many sports, maximal repeated jumps are necessary, and it has been shown that there is a direct relationship between knee flexor and extensor muscle strength and jumping performance, subsequently leading to sports success [57,58]. In team sports, including basketball, lower limb strength is a critical determinant of performance, as it supports key actions such as sprinting, jumping, accelerations and decelerations, and physical duels during both offensive and defensive situations [59–62]. Athletes across all playing positions must perform shuffling movements at varying intensities during several match actions [63]. This demands the ability to perceive and react to opponents' movements rapidly, placing significant demands on their agility, lateral displacement, and acceleration capacities [64].

This study identified notable sex-based differences in almost all neuromuscular performance variables studied. Males exhibited higher hamstring isometric strength than females, whether measured as absolute or relative peak torque on the dominant or non-dominant side (Table 3). Interestingly, quadriceps strength values did not differ significantly between the sexes. This pattern aligns with previous studies indicating that sex-related differences in lower limb strength become more pronounced in the hamstring, widening with maturation, while quadriceps strength remains relatively comparable between the sexes during adolescence in isometric conditions [65].

Owoeye et al. [6] reported that although males generally achieve higher absolute strength in both muscle groups, the disparity between sexes is greater for the hamstrings, potentially due to sport-specific demands that preferentially stimulate hamstring development in males, hormonal influences, and differences in muscle fiber composition.

In the present study, the differences observed in the posterior chains may be linked to a greater reliance on hamstring-driven actions in boys during matches and training. Many studies have documented that male elite junior basketball players cover more high-intensity distance, perform more sprints and jumps, and engage in more deceleration activities than their female counterparts [66]. Because these actions typically require the primary participation of this muscle group, it is plausible to think that males have achieved greater adaptation. Moreover, differences could be explained, in part, by the contribution of maturation in our athletes. As documented, before, during, and after puberty, boys are on average stronger than girls [67]. The progression in physical performance during adolescence is typically slower and smaller in magnitude in girls compared to boys. This has been reported in both team and individual sport athletes [68], as well as in the general population [69]. This difference is largely attributed to a greater increase in fat mass in girls and muscle mass in boys during puberty and adolescence [70], as found in the present study.

Along with this, during adolescence, boys produce higher levels of circulating testosterone (about 15–20-fold more than females [71]), which has been associated with the increased growth velocity [72] and reduced adipose accumulation in males compared to females [73]. Higher testosterone levels, in turn, result in more muscle mass, which facilitates greater strength production and more advantageous ground reaction forces during high-intensity actions. In contrast, at this stage, girls produce higher levels of circulating estrogen, which are about 4-fold-higher among females than males [72]. This hormone has limited anabolic effects, and it is not a primary contributor to the large sex differences in athletic performance [66].

Regarding jumping performance, we found that male players significantly outperformed female players in both the CMJ and SJ jumps' height ($p < 0.001$), with moderate to large effect sizes (d = 0.63 and 0.57, respectively). These findings reflect sex differences in explosive power and stretch-shortening cycle utilization, which are critical for high-intensity basketball actions like jumping, rebounding, and quick take-offs [74]. Results in this study are consistent with those reported in adolescent and professional players of different team sports, including basketball. These differences may be attributed to morphological characteristics of the muscle between males and females, such as muscle thickness, pennation angle, and fascicle length, which tend to favor males in producing greater muscular strength [52]. Moreover, it has been described that the muscle area occupied by fast-twitch fibers is greater in males than in females [75], which is associated with greater strength [76].

On the other hand, in the present study, we also found significant sex-based differences in HGS for both the dominant and non-dominant side, with male players demonstrating higher values than their female counterparts ($p < 0.001$). The effect sizes were small to moderate (d = 0.41 and 0.59, respectively). However, no significant differences were found between the dominant and non-dominant sides in boys or girls.

While not a basketball-specific action, HGS has been associated with several movements that rely on the continuous use of wrist and digit flexors. These technical actions include catching, control, passing accuracy, shooting performance, speed dribbling, and time in an obstacle dribbling performance, in young and adult basketball players [77,78]. Results in our study align with prior research reporting superior upper-limb strength in adolescent male athletes, largely explained by greater muscle mass development and neuromuscular efficiency during puberty [79].

Our results were also higher than those reported in other male and female young and adult athletes across different sports, including basketball [80], but lower than those reported in Chilean professional players, and Italian and Greek adolescent basketball players [52,77,81]. The morphological differences in upper limbs have previously been described by our research group as a key factor in the HGS performance. Thus, athletes with larger hand breadth, hand length, upper arm length, and arm circumference display higher HGS [8]. Aligning with these previous reports, the results in the present study demonstrated that these anthropometric measures were larger in boys than in girls (Table 1). In addition, the estimated upper arm muscle area in our study was higher in boys than in girls (Table 2), which has also been related to greater HGS on both the dominant and non-dominant sides [22]. Thus, differences found in HGS could also be explained, in part, by the anthropometric characteristics of the sample.

In addition, as described above, greater testosterone production in males could modulate the force production in favor of boys. Recently, a study conducted in 641 Chinese male adolescents investigated the associations of sex steroids with muscle parameters [82]. The results indicated that the concentrations of serum testosterone and free testosterone were positively related to HGS. Researchers also found that this relationship was only observed in the late-post pubertal group (15.9 years), suggesting a potential threshold effect. Because the age of male players in our study is 14.9 years, these findings reinforce the idea that sex differences reported here could be strongly explained by the activating effects of sex hormones.

The present investigation offers a detailed characterization of the anthropometric, body composition, and neuromuscular performance profiles of a homogeneous sample of U-15 Colombian national basketball players. Our findings provide valuable insight into sex-specific physical and performance traits that may influence basketball development at this competitive stage. Nevertheless, caution should be exercised when extrapolating these results to athletes at different competitive levels, age categories, or training backgrounds, as the sample consisted exclusively of highly trained players engaged in national team preparation.

The sample size was determined by the availability of players selected for the preparatory phase of the South American U-15 Championships. Moreover, the cross-sectional nature of the study precludes any inferences about causality or longitudinal changes in these variables. While chronological age was controlled for, biological age or maturity status was not assessed, which could have provided an additional context for interpreting sex differences during adolescence. Additionally, the exclusive use of isometric strength assessment for knee muscles limits the ability to extrapolate to dynamic, sport-specific contexts.

Future research should employ longitudinal designs to monitor the evolution of anthropometric, body composition, and neuromuscular performance profiles during adolescence, as well as their relationship with performance metrics and injury risk in basketball. Incorporating dynamic strength tests, movement efficiency assessments, and match play performance analyses would provide a more holistic understanding of the sex-specific demands and adaptations in youth basketball.

5. Conclusions

This study provides novel, sex-specific data on anthropometric characteristics, body composition, and neuromuscular performance in U-15 Colombian national basketball players. Males presented greater height, skeletal breadths, and limb length, as well as higher HGS, hamstring strength, and vertical jump performance. The females showed higher subcutaneous fat levels and adiposity indicators. No significant sex-specific differences were found in quadriceps isometric strength; this may reflect similar developmental patterns for this muscle group in highly trained athletes during adolescence. These findings align with known growth and maturation processes, highlighting the influence of both biological factors and sport-specific demands on performance profiles. The reference values generated can inform coaches, strength and conditioning staff, and sports medicine professionals in designing targeted, sex-specific training and injury prevention strategies for highly trained young basketball players.

Author Contributions: Conceptualization, A.O.-A., A.D.l.R., M.A.C.-V., F.M.-D. and F.V.; methodology, A.O.-A., A.D.l.R., M.A.C.-V., R.Q.-B. and J.A.M.; software, A.D.l.R. and M.A.C.-V.; validation, R.Q.-B., J.A.M., F.M.-D. and F.V.; formal analysis, A.D.l.R., M.A.C.-V. and J.A.M.; investigation, A.D.l.R., M.A.C.-V. and F.M.-D.; resources, A.D.l.R. and A.O.-A.; data curation, R.Q.-B., F.M.-D. and J.A.M.; writing—original draft preparation, A.O.-A., A.D.l.R., M.A.C.-V., F.M.-D. and F.V.; writing—review and editing, A.D.l.R., M.A.C.-V. and A.O.-A.; visualization, M.A.C.-V., F.M.-D. and R.Q.-B.;

supervision, A.D.l.R. and M.A.C.-V.; project administration, A.D.l.R. and F.V.; funding acquisition, A.O.-A. and J.A.M. All authors have read and agreed to the published version of the manuscript.

Funding: This research received no external funding.

Institutional Review Board Statement: The study was conducted in accordance with the Declaration of Helsinki and approved by the Ethics Committee for Human Beings from the Unidades Tecnológicas de Santander (no. 0010-2022/2 May 2022).

Informed Consent Statement: Informed consent was obtained from all subjects involved in the study.

Data Availability Statement: The datasets used and/or analyzed during the current study are available from the corresponding author upon reasonable request.

Acknowledgments: The authors are thankful to the coaches and player participants.

Conflicts of Interest: The authors declare no conflicts of interest.

References

1. Scanlan, A.T.; Dascombe, B.J.; Kidcaff, A.P.; Peucker, J.L.; Dalbo, V.J. Gender-Specific Activity Demands Experienced during Semiprofessional Basketball Game Play. *Int. J. Sports Physiol. Perform.* **2015**, *10*, 618–625. [CrossRef]
2. Abian-Vicen, J.; Puente, C.; Salinero, J.J.; González-Millán, C.; Areces, F.; Muñoz, G.; Muñoz-Guerra, J.; Del Coso, J. A Caffeinated Energy Drink Improves Jump Performance in Adolescent Basketball Players. *Amino Acids* **2014**, *46*, 1333–1341. [CrossRef]
3. Sosa, C.; Lorenzo, A.; Trapero, J.; Ribas, C.; Alonso, E.; Jimenez, S.L. Specific Absolute Velocity Thresholds during Male Basketball Games Using Local Positional System; Differences between Age Categories. *Appl. Sci.* **2021**, *11*, 4390. [CrossRef]
4. Sugiyama, T.; Maeo, S.; Kurihara, T.; Kanehisa, H.; Isaka, T. Change of Direction Speed Tests in Basketball Players: A Brief Review of Test Varieties and Recent Trends. *Front. Sports Act. Living* **2021**, *3*, 645350. [CrossRef]
5. Spiteri, T.; Newton, R.U.; Binetti, M.; Hart, N.H.; Sheppard, J.M.; Nimphius, S. Mechanical Determinants of Faster Change of Direction and Agility Performance in Female Basketball Athletes. *J. Strength Cond. Res.* **2015**, *29*, 2205–2214. [CrossRef]
6. Owoeye, O.B.; Mulenga, D.; Kim, J.; Breitbach, A.; Neme, J.R. Normative Hamstrings and Quadriceps Isometric Strength Values and Hamstrings-Quadriceps Asymmetry in Healthy Collegiate Soccer and Basketball Players. *Int. J. Exerc. Sci.* **2024**, *17*, 768. [CrossRef]
7. Monterrosa Qiuntero, A.; De La Rosa, A.; Arc Chagnaud, C.; Gomez Qiuntero, J.M.; Pereira Moro, A.R. Morphology, Lower Limbs Performance and Baropodometric Characteristics of Elite Brazilian Jiu-Jitsu Athletes. *Ido Mov. Cult. J. Martial Arts Anthropol.* **2023**, *23*, 58–69. [CrossRef]
8. Camacho-Villa, M.A.; Hurtado-Alcoser, J.; Jerez, A.S.; Saavedra, J.C.; Paredes Prada, E.T.; Merchán, J.A.; Millan-Domingo, F.; Silva-Polanía, C.; De la Rosa, A. Handgrip Strength and Upper Limb Anthropometric Characteristics among Latin American Female Volleyball Players. *J. Funct. Morphol. Kinesiol.* **2024**, *9*, 168. [CrossRef]
9. Ojeda-Aravena, A.; Azocar-Gallardo, J.; Galle, F.; García-García, J.M. Relación Entre Las Características de La Composición Corporal y El Rendimiento Físico General y Específico En Competidores de Taekwondo Chilenos de Nivel Nacional de Ambos Sexos: Un Estudio Observacional. *Rev. Esp. Nutr. Humana Dietética* **2020**, *24*, 154–164. [CrossRef]
10. Apostolidis, N.; Zacharakis, E. The Influence of the Anthropometric Characteristics and Handgrip Strength on the Technical Skills of Young Basketball Players. *J. Phys. Educ. Sport* **2015**, *15*, 330–337. [CrossRef]
11. Garcia-Gil, M.; Torres-Unda, J.; Esain, I.; Duñabeitia, I.; Gil, S.M.; Gil, J.; Irazusta, J. Anthropometric Parameters, Age, and Agility as Performance Predictors in Elite Female Basketball Players. *J. Strength Cond. Res.* **2018**, *32*, 1723–1730. [CrossRef]
12. Han, M.; Gómez-Ruano, M.-A.; Calvo, A.L.; Calvo, J.L. Basketball Talent Identification: A Systematic Review and Meta-Analysis of the Anthropometric, Physiological and Physical Performance Factors. *Front. Sports Act. Living* **2023**, *5*, 1264872. [CrossRef]
13. Cui, J.; Liu, Y.; He, F.; Bu, Y. Systematic Review and Meta-Analysis on the Effect of Plyometric vs. Resistance Training on Lower Limb Explosive Power and Speed. *J. Sports Med. Phys. Fitness* **2025**, *65*, 69–79. [CrossRef]
14. Yamashita, D.; Henderson, F.J.; Ishida, Y. Assessing the Contribution of Arm Swing to Countermovement Jump Height Using Three Different Measurement Methods in Physically Active Men. *Biomechanics* **2025**, *5*, 45. [CrossRef]
15. Cengizel, Ç.Ö.; Cengizel, E.; Öz, E. Somatotype and Body Composition Profiles of Children and Adolescent Male Basketball Players. *Homo Int. Z. Vgl. Forsch. Am Menschen* **2024**, *75*, 41–49. [CrossRef]
16. Osawa, Y.; Studenski, S.A.; Ferrucci, L. Knee Extension Rate of Torque Development and Peak Torque: Associations with Lower Extremity Function. *J. Cachexia Sarcopenia Muscle* **2018**, *9*, 530–539. [CrossRef]
17. Hirano, M.; Katoh, M.; Gomi, M.; Arai, S. Validity and Reliability of Isometric Knee Extension Muscle Strength Measurements Using a Belt-Stabilized Hand-Held Dynamometer: A Comparison with the Measurement Using an Isokinetic Dynamometer in a Sitting Posture. *J. Phys. Ther. Sci.* **2020**, *32*, 120–124. [CrossRef]

18. Aedo-Muñoz, E.; Pérez-Contreras, J.; Bustamante-Garrido, A.; Arriagada-Tarifeño, D.; Cancino-Jiménez, J.; Retamal-Espinoza, M.; Argothy-Buchelli, R.; Brito, C.; Merino-Muñoz, P. Is Countermovement Jump an Indirect Marker of Neuromuscular Mechanism? Relationship with Isometric Knee Extension Test. *J. Funct. Morphol. Kinesiol.* **2024**, *9*, 242. [CrossRef]
19. Almeida, G.P.L.; das Neves Rodrigues, H.L.; de Freitas, B.W.; de Paula Lima, P.O. Reliability and Validity of the Hip Stability Isometric Test (HipSIT): A New Method to Assess Hip Posterolateral Muscle Strength. *J. Orthop. Sports Phys. Ther.* **2017**, *47*, 906–913. [CrossRef]
20. Mendonça, L.D.M.; Bittencourt, N.F.N.; Freire, R.L.; Campos, V.C.; Ferreira, T.V.; Silva, P.L. Hip External Rotation Isometric Torque for Soccer, Basketball, and Volleyball Athletes: Normative Data and Asymmetry Index. *Braz. J. Phys. Ther.* **2022**, *26*, 100391. [CrossRef]
21. Koo, T.K.; Li, M.Y. A Guideline of Selecting and Reporting Intraclass Correlation Coefficients for Reliability Research. *J. Chiropr. Med.* **2016**, *15*, 155–163, Erratum in *J. Chiropr. Med.* **2017**, *16*, 346. [CrossRef]
22. De la Rosa, A.; Camacho-Villa, M.A.; Millan-Domingo, F.; Saavedra, J.C.; Okoshi, M.P.; Urbano Pagan, L. Positional Profiling of Anthropometric, Baropodometric, and Grip Strength Traits in Male Volleyball Players: Insights from a National Colombian Study. *J. Funct. Morphol. Kinesiol.* **2025**, *10*, 197. [CrossRef]
23. Yuhasz, M.S. The Effects of Sports Training on Body Fat in Man with Predictions of Optimal Body Weight. Ph.D. Thesis, University of Illinois at Urbana-Champaign, Champaign, IL, USA, 1962.
24. Baglietto, N.; Albaladejo-Saura, M.; Esparza-Ros, F.; Vaquero-Cristóbal, R. Agreement and Differences between the Equations for Estimating Muscle and Bone Mass Using the Anthropometric Method in Recreational Strength Trainees. *PeerJ* **2024**, *12*, e17506. [CrossRef]
25. Geraldo, A.P.; Rincón, Y.G.; Vega, P.Á.; Villarreal, C. Selección y Análisis de Ecuaciones Antropométricas Para El Cálculo de La Composición Corporal En Adultos. *Rev. Ing. Matemáticas Cienc. Inf.* **2017**, *4*, 47–56. [CrossRef]
26. Frisancho, A.R. New Norms of Upper Limb Fat and Muscle Areas for Assessment of Nutritional Status. *Am. J. Clin. Nutr.* **1981**, *34*, 2540–2545. [CrossRef]
27. Bosco, C.; Luhtanen, P.; Komi, P.V. A Simple Method for Measurement of Mechanical Power in Jumping. *Eur. J. Appl. Physiol.* **1983**, *50*, 273–282. [CrossRef]
28. Bosco, C.; Belli, A.; Astrua, M.; Tihanyi, J.; Pozzo, R.; Kellis, S.; Tsarpela, O.; Foti, C.; Manno, R.; Tranquilli, C. A Dynamometer for Evaluation of Dynamic Muscle Work. *Eur. J. Appl. Physiol.* **1995**, *70*, 379–386. [CrossRef]
29. Chamorro, C.; Armijo-Olivo, S.; De la Fuente, C.; Fuentes, J.; Javier Chirosa, L. Absolute Reliability and Concurrent Validity of Hand Held Dynamometry and Isokinetic Dynamometry in the Hip, Knee and Ankle Joint: Systematic Review and Meta-Analysis. *Open Med.* **2017**, *12*, 359–375. [CrossRef]
30. Oranchuk, D.; Switaj, Z.; Zuleger, B. The Addition of a "Rapid Response" Neuromuscular Activation to a Standard Dynamic Warm-up Improves Isometric Force and Rate of Force Development. *J. Aust. Strength Cond.* **2017**, *25*, 19–24.
31. Cohen, J. *Statistical Power Analysis for the Behavioral Sciences*, 2nd ed.; Routledge: New York, NY, USA, 2013; ISBN 978-0-203-77158-7.
32. Richardson, J.T.E. Eta Squared and Partial Eta Squared as Measures of Effect Size in Educational Research. *Educ. Res. Rev.* **2011**, *6*, 135–147. [CrossRef]
33. Gryko, K.; Stastny, P.; Kopiczko, A.; Mikołajec, K.; Pecha, O.; Perkowski, K. Can Anthropometric Variables and Maturation Predict the Playing Position in Youth Basketball Players? *J. Hum. Kinet.* **2019**, *69*, 109–123. [CrossRef]
34. Karpowicz, K. Interrelation of Selected Factors Determining the Effectiveness of Training in Young Basketball Players. *HUMAN* **2006**, *7*, 130–146.
35. Ramos, S.; Volossovitch, A.; Ferreira, A.P.; Barrigas, C.; Fragoso, I.; Massuça, L. Differences in Maturity, Morphological, and Fitness Attributes Between the Better- and Lower-Ranked Male and Female U-14 Portuguese Elite Regional Basketball Teams. *J. Strength Cond. Res.* **2020**, *34*, 878–887. [CrossRef]
36. Maynard, L.M.; Wisemandle, W.; Roche, A.F.; Chumlea, W.C.; Guo, S.S.; Siervogel, R.M. Childhood Body Composition in Relation to Body Mass Index. *Pediatrics* **2001**, *107*, 344–350. [CrossRef]
37. Tanner, J.M.; Whitehouse, R.H.; Marshall, W.A.; Carter, B.S. Prediction of Adult Height from Height, Bone Age, and Occurrence of Menarche, at Ages 4 to 16 with Allowance for Midparent Height. *Arch. Dis. Child.* **1975**, *50*, 14–26. [CrossRef]
38. Rogol, A.D.; Roemmich, J.N.; Clark, P.A. Growth at Puberty. *J. Adolesc. Health* **2002**, *31*, 192–200. [CrossRef]
39. Loomba-Albrecht, L.A.; Styne, D.M. Effect of Puberty on Body Composition. *Curr. Opin. Endocrinol. Diabetes Obes.* **2009**, *16*, 10–15. [CrossRef]
40. Hoare, D.G. Predicting Success in Junior Elite Basketball Players—The Contribution of Anthropometic and Physiological Attributes. *J. Sci. Med. Sport* **2000**, *3*, 391–405. [CrossRef]
41. Tingelstad, L.M.; Raastad, T.; Myklebust, G.; Gjerstad Andersen, T.E.; Solstad, B.E.; Bugten, J.B.; Luteberget, L.S. Age and Sex Differences in Physical Performance Among Adolescent Team Sport Athletes. *Eur. J. Sport Sci.* **2025**, *25*, e12284. [CrossRef]
42. Ljubojevic, M.; Bojanic, D.; Nokic, A.; Malovic, P.; Bacovic, D. Anthropometric Characteristics and Body Composition of Elite Youth Male Basketball Players-Participants of Junior Euroleague. *Int. J. Morphol.* **2023**, *41*, 798–803. [CrossRef]

43. Ramos, S.; Massuça, L. Discriminant Attributes in U16 Basketball Selection in a National Elite Club. *Gymnasium* **2017**, *2*, 1–6.
44. Erol, E.; Özen, G.; Koç, H. General Anthropometric and Selected Motor Skills of Elite Young Male Basketball Players According to Position on the Court of Players. *J. Athl. Perform. Nutr.* **2016**, *1*, 1–9.
45. Gryko, K.; Kopiczko, A.; Mikołajec, K.; Stasny, P.; Musalek, M. Anthropometric Variables and Somatotype of Young and Professional Male Basketball Players. *Sports* **2018**, *6*, 9. [CrossRef]
46. Mikołajec, K.; Arede, J.; Gryko, K. Examining Physical and Technical Performance among Youth Basketball National Team Development Program Players: A Multidimensional Approach. *Sci. Rep.* **2025**, *15*, 3722. [CrossRef]
47. Canli, U. The Effect of Somatotype Profiles of Adolescent Basketball Players on Biomotoric Characteristics and Ability. *IOSR J. Sports Phys. Educ.* **2017**, *4*, 61–66. [CrossRef]
48. Stewar, A.D.; Benson, P.J.; Michanikou, E.G.; Tsiota, D.G.; Narli, M.K. Body Image Perception, Satisfaction and Somatotype in Male and Female Athletes and Non-Athletes: Results Using a Novel Morphing Technique. *J. Sports Sci.* **2003**, *21*, 815–823. [CrossRef]
49. Pettersson, S.; Kalén, A.; Gustafsson, M.; Grau, S.; Caspers, A. Off- to in-Season Body Composition Adaptations in Elite Male and Female Endurance and Power Event Athletics Competitors: An Observational Study. *BMC Sports Sci. Med. Rehabil.* **2024**, *16*, 90. [CrossRef]
50. Malina, R.M.; Geithner, C.A. Body Composition of Young Athletes. *Am. J. Lifestyle Med.* **2011**, *5*, 262–278. [CrossRef]
51. Santos, D.A.; Dawson, J.A.; Matias, C.N.; Rocha, P.M.; Minderico, C.S.; Allison, D.B.; Sardinha, L.B.; Silva, A.M. Reference Values for Body Composition and Anthropometric Measurements in Athletes. *PLoS ONE* **2014**, *9*, e97846. [CrossRef]
52. Hernandez-Martinez, J.; Perez-Carcamo, J.; Coñapi-Union, B.; Canales-Canales, S.; Negron-Molina, M.; Avila-Valencia, S.; Cid-Calfucura, I.; Herrera-Valenzuela, T.; Cisterna, D.; Branco, B.H.M.; et al. Relationship between Body Composition and Physical Performance by Sex in Professional Basketball Players. *Appl. Sci.* **2024**, *14*, 9165. [CrossRef]
53. Zaric, I.; Dopsaj, M.; Markovic, M.; Zaric, M.; Jakovljevic, S.; Beric, D.; Zaric, I.; Dopsaj, M.; Markovic, M.; Zaric, M. Body Composition Characteristics Measured by Multichannel Bioimpedance in Young Female Basketball Players: Relation with Match Performance. *Int. J. Morphol.* **2020**, *38*, 328–335. [CrossRef]
54. Ojeda-Aravena, A.P.; Azócar-Gallardo, J.; Hérnandez-Mosqueira, C.; Herrera-Valenzuela, T. Relación entre la prueba de agilidad específica en taekwondo (tsat), la fuerza explosiva y la velocidad líneal en 5-m atletas de taekwondo de ambos sexos (Relationship between the specific agility test in taekwondo (tsat), explosive strength and 5-m linea. *Retos* **2021**, *39*, 84–89. [CrossRef]
55. Quintero, A.M.; Ojeda-Aravena, A.; Arc-Chagnaud, C.; Quintero, J.M.G.; De la Rosa, A. Impact of COVID-19 Lockdown on Physical Activity and Psychological Well-Being in Taekwondo Athletes. *Idō Mov. Cult.* **2023**, *23*, 71.
56. Tønnessen, E.; Svendsen, I.S.; Olsen, I.C.; Guttormsen, A.; Haugen, T. Performance Development in Adolescent Track and Field Athletes According to Age, Sex and Sport Discipline. *PLoS ONE* **2015**, *10*, e0129014. [CrossRef]
57. Váczi, M.; Fazekas, G.; Pilissy, T.; Cselkó, A.; Trzaskoma, L.; Sebesi, B.; Tihanyi, J. The Effects of Eccentric Hamstring Exercise Training in Young Female Handball Players. *Eur. J. Appl. Physiol.* **2022**, *122*, 955–964. [CrossRef]
58. Križaj, J.; Rauter, S.; Vodičar, J.; Hadžić, V.; Šimenko, J. Predictors of Vertical Jumping Capacity in Soccer Players. *Isokinet. Exerc. Sci.* **2019**, *27*, 9–14. [CrossRef]
59. Spiteri, T.; Nimphius, S.; Hart, N.H.; Specos, C.; Sheppard, J.M.; Newton, R.U. Contribution of Strength Characteristics to Change of Direction and Agility Performance in Female Basketball Athletes. *J. Strength Cond. Res.* **2014**, *28*, 2415–2423. [CrossRef]
60. Beato, M.; Young, D.; Stiff, A.; Coratella, G. Lower-Limb Muscle Strength, Anterior-Posterior and Inter-Limb Asymmetry in Professional, Elite Academy and Amateur Soccer Players. *J. Hum. Kinet.* **2021**, *77*, 135–146. [CrossRef]
61. Ziv, G.; Lidor, R. Vertical Jump in Female and Male Basketball Players—A Review of Observational and Experimental Studies. *J. Sci. Med. Sport* **2010**, *13*, 332–339. [CrossRef]
62. Antoranz, Y.; Alonso-Pérez-Chao, E.; Tejero-González, C.M.; Salazar, H.; del Campo-Vecino, J.; Jiménez-Sáiz, S.L. Exploring External Peak Demands: The Influence of Contextual Factors on Male Basketball Players. *Appl. Sci.* **2024**, *14*, 10542. [CrossRef]
63. Stojanović, E.; Stojiljković, N.; Scanlan, A.T.; Dalbo, V.J.; Berkelmans, D.M.; Milanović, Z. The Activity Demands and Physiological Responses Encountered During Basketball Match-Play: A Systematic Review. *Sports Med.* **2018**, *48*, 111–135. [CrossRef]
64. Scanlan, A.; Humphries, B.; Tucker, P.S.; Dalbo, V. The Influence of Physical and Cognitive Factors on Reactive Agility Performance in Men Basketball Players. *J. Sports Sci.* **2014**, *32*, 367–374. [CrossRef] [PubMed]
65. Hewett, T.E.; Myer, G.D.; Zazulak, B.T. Hamstrings to Quadriceps Peak Torque Ratios Diverge between Sexes with Increasing Isokinetic Angular Velocity. *J. Sci. Med. Sport* **2008**, *11*, 452–459. [CrossRef] [PubMed]
66. Hunter, S.K.; Senefeld, J.W. Sex Differences in Human Performance. *J. Physiol.* **2024**, *602*, 4129–4156. [CrossRef] [PubMed]
67. Nuzzo, J.L.; Pinto, M.D. Sex Differences in Upper- and Lower-Limb Muscle Strength in Children and Adolescents: A Meta-Analysis. *Eur. J. Sport Sci.* **2025**, *25*, e12282. [CrossRef]
68. Tingelstad, L.M.; Raastad, T.; Till, K.; Luteberget, L.S. The Development of Physical Characteristics in Adolescent Team Sport Athletes: A Systematic Review. *PLoS ONE* **2023**, *18*, e0296181. [CrossRef]

69. Round, J.M.; Jones, D.A.; Honour, J.W.; Nevill, A.M. Hormonal Factors in the Development of Differences in Strength between Boys and Girls during Adolescence: A Longitudinal Study. *Ann. Hum. Biol.* **1999**, *26*, 49–62. [CrossRef]
70. Malina, R.M.; Bouchard, C.; Bar-Or, O. *Growth, Maturation, and Physical Activity*; Human Kinetics: Champaign, IL, USA, 2004.
71. Senefeld, J.W.; Lambelet Coleman, D.; Johnson, P.W.; Carter, R.E.; Clayburn, A.J.; Joyner, M.J. Divergence in Timing and Magnitude of Testosterone Levels Between Male and Female Youths. *JAMA* **2020**, *324*, 99–101. [CrossRef]
72. Hunter, S.K.; Angadi, S.S.; Bhargava, A.; Harper, J.; Hirschberg, A.L.; Levine, B.D.; Moreau, K.L.; Nokoff, N.J.; Stachenfeld, N.S.; Bermon, S. The Biological Basis of Sex Differences in Athletic Performance: Consensus Statement for the American College of Sports Medicine. *Transl. J. Am. Coll. Sports Med.* **2023**, *8*, 1–33, Erratum in *Transl. J. Am. Coll. Sports Med.* **2024**, *9*, e000240. [CrossRef]
73. Ma, H.; Sun, J.; Wu, X.; Mao, J.; Han, Q. Percent Body Fat Was Negatively Correlated with Testosterone Levels in Male. *PLoS ONE* **2024**, *19*, e0294567. [CrossRef]
74. Cao, S.; Li, Z.; Wang, Z.; Geok, S.K.; Liu, J. The Effects of High-Intensity Interval Training on Basketball Players: A Systematic Review and Meta-Analysis. *J. Sports Sci. Med.* **2025**, *24*, 31–51.
75. Alegre, L.M.; Lara, A.J.; Elvira, J.L.; Aguado, X. Muscle Morphology and Jump Performance: Gender and Intermuscular Variability. *J. Sports Med. Phys. Fitness* **2009**, *49*, 320.
76. Nuzzo, J.L. Narrative Review of Sex Differences in Muscle Strength, Endurance, Activation, Size, Fiber Type, and Strength Training Participation Rates, Preferences, Motivations, Injuries, and Neuromuscular Adaptations. *J. Strength Cond. Res.* **2023**, *37*, 494–536. [CrossRef]
77. Pizzigalli, L.; Micheletti Cremasco, M.; LA Torre, A.; Rainoldi, A.; Benis, R. Hand Grip Strength and Anthropometric Characteristics in Italian Female National Basketball Teams. *J. Sports Med. Phys. Fitness* **2017**, *57*, 521–528. [CrossRef] [PubMed]
78. Cortis, C.; Tessitore, A.; Lupo, C.; Pesce, C.; Fossile, E.; Figura, F.; Capranica, L. Inter-Limb Coordination, Strength, Jump, and Sprint Performances Following a Youth Men's Basketball Game. *J. Strength Cond. Res.* **2011**, *25*, 135–142. [CrossRef] [PubMed]
79. Gónzalez, I.P.; Javaloyes, A.; Moya-Ramón, M. The Effect of the Maturity Status on Strength Performance in Young Elite Basketball Players (El Efecto Del Estado Madurativo En El Rendimiento Físico de Jóvenes Jugadores de Baloncesto de Élite). *Retos* **2022**, *44*, 858–863. [CrossRef]
80. Guimarães, E.; Baxter-Jones, A.; Maia, J.; Fonseca, P.; Santos, A.; Santos, E.; Tavares, F.; Janeira, M. The Roles of Growth, Maturation, Physical Fitness, and Technical Skills on Selection for a Portuguese Under-14 Years Basketball Team. *Sports* **2019**, *7*, 61. [CrossRef] [PubMed]
81. Gerodimos, V. Reliability of Handgrip Strength Test in Basketball Players. *J. Hum. Kinet.* **2012**, *31*, 25–36. [CrossRef]
82. Xu, Y.; Wen, Z.; Deng, K.; Li, R.; Yu, Q.; Xiao, S.-M. Relationships of Sex Hormones with Muscle Mass and Muscle Strength in Male Adolescents at Different Stages of Puberty. *PLoS ONE* **2021**, *16*, e0260521. [CrossRef]

Journal of
Functional Morphology and Kinesiology

MDPI

Article

Predicting Sleep Quality Based on Metabolic, Body Composition, and Physical Fitness Variables in Aged People: Exploratory Analysis with a Conventional Machine Learning Model

Pedro Forte [1,2,3,4,*], Samuel G. Encarnação [2,3,5], José E. Teixeira [3,4,6,7,8], Luís Branquinho [9,10], Tiago M. Barbosa [3,4], António M. Monteiro [3] and Daniel Pecos-Martín [1]

1 Physiotherapy and Pain Group, Department of Physical Therapy, University of Alcala, 28801 Madrid, Spain; daniel.pecos@uah.es
2 Department of Sports, Higher Institute of Educational Sciences of the Douro, 4560-708 Penafiel, Portugal; samuel01.encarnacao@gmail.com
3 Research Centre for Active Living and Wellbeing (LiveWell), Instituto Politécnico de Bragança, 5300-253 Bragança, Portugal; jose.eduardo@ipg.pt (J.E.T.); barbosa@ipb.pt (T.M.B.); mmonteiro@ipb.pt (A.M.M.)
4 Department of Sports Sciences, Instituto Politécnico de Bragança, 5300-253 Bragança, Portugal
5 Department of Physical Activity and Sport Sciences, Universidad Autónoma de Madrid (UAM), 28049 Madrid, Spain
6 Department of Sports Sciences, Polytechnic of Guarda, 6300-559 Guarda, Portugal
7 Department of Sports Sciences, Polytechnic of Cávado and Ave., 4800-058 Guimarães, Portugal
8 SPRINT—Sport Physical Activity and Health Research & Inovation Center, 6300-559 Guarda, Portugal
9 Biosciences Higher School of Elvas, Polytechnic Institute of Portalegre, 7350-000 Portalegre, Portugal; luisbranquinho@ipportalegre.pt
10 Life Quality Research Center (LQRC-CIEQV), 2001-904 Santarém, Portugal
* Correspondence: pedromiguelforte@gmail.com

Abstract: Background: Sleep plays a crucial role in the health of older adults, and its quality is influenced by multiple physiological and functional factors. However, the relationship between sleep quality and physical fitness, body composition, and metabolic markers remains unclear. This exploratory study aimed to investigate the associations between sleep quality and physical, metabolic, and body composition variables in older adults, and to evaluate the preliminary performance of a logistic regression model in classifying sleep quality. **Methods**: A total of 32 subjects participated in this study, with a mean age of 69. The resting arterial pressure (systolic and diastolic), resting heart rate, anthropometrics (high waist girth), body composition (by bioimpedance), and physical fitness (Functional Fitness Test) and sleep quality (Pitsburg sleep-quality index) were evaluated. Group comparisons, associative analysis and logistic regression with 5-fold stratified cross-validation was used to classify sleep quality based on selected non-sleep-related predictors. **Results**: Individuals with good sleep quality showed significantly better back stretch (t = 2.592; p = 0.015; η^2 = 0.239), lower limb strength (5TSTS; t = 2.564; p = 0.016; η^2 = 0.476), and longer total sleep time (t = 6.882; p < 0.001; η^2 = 0.675). Exploratory correlations showed that poor sleep quality was moderately associated with reduced lower-limb strength and mobility. The logistic regression model including 5TSTS and TUG achieved a mean accuracy of 0.76 ± 0.15, precision of 0.79 ± 0.18, recall of 0.83 ± 0.21, and AUC of 0.74 ± 0.16 across cross-validation folds. **Conclusions**: These preliminary findings suggest that physical fitness and clinical variables significantly influence sleep quality in older adults. Sleep-quality-dependent patterns suggest that interventions to improve lower limb strength may promote better sleep outcomes.

Keywords: age; sleep quality; physical fitness; body composition

1. Introduction

Aging leads to significant changes in body composition and functional fitness [1]. These changes, notably the reduction in muscle mass and increase in adiposity, are associated with a higher risk of conditions like metabolic syndrome [1], and may lead to sarcopenia, contributing to frailty and loss of independence in older adults [2]. Physiologically, aging also affects hemodynamic parameters, including pulse pressure, arterial stiffness, and wave reflections, particularly in large arteries [3], leading to increased pulse pressure, especially in individuals over 60 years old [4].

Age-related alterations in body composition, functional fitness, resting heart rate, and arterial blood pressure are intricately linked to sleep quality [5]. In particular, obesity in older adults has been associated with reduced sleep duration, suggesting a potential interaction between metabolic and sleep-related factors [6]. Healthy behaviors may moderate this relationship, with evidence indicating that positive lifestyle factors contribute to improved sleep outcomes [7]. For example, Wu et al. [8] conducted a meta-analysis involving 197,906 participants, showing that obesity significantly shortens sleep duration. Moreover, overweight and obese individuals typically exhibit a pro-inflammatory profile with elevated levels of cytokines such as tumor necrosis factor α (TNF-α), interleukin 6 (IL-6), and C-reactive protein (CRP) [9,10], which disrupts sleep regulation by impairing hypothalamic control of non-rapid eye movement (Non-REM) sleep [11,12].

Physical exercise and improved fitness adaptations have also been shown to positively influence sleep regulation [13,14]. Several physiological mechanisms underlie this effect, including reductions in depression and anxiety [15], improved thermoregulation following physical exertion [16,17], enhanced muscle relaxation [18], and hormonal regulation involving melatonin, cortisol, growth hormone, adenosine, ghrelin, leptin orexin, prolactin, and serotonin [19,20]. These pathways support the hypothesis that maintaining or improving physical fitness may contribute to better sleep quality in older populations.

To assess physical fitness and function in elderly individuals, various tests have been employed, including the Eurofit battery [21], the Wii Fit Balance Board [22], and the Physical Activity Scale for Elderly (PASE) [23]. More commonly used assessments include the Timed Up and Go Test (TUG) and Tinetti Gait and Balance Test, which evaluate balance and fall risk [24,25], as well as handgrip strength, an indicator of motor function and overall health [26]. Among the most used protocols, the Fullerton Functional Fitness Test (FFFT) [27], developed by Rikli and Jones, comprehensively assesses strength, flexibility, coordination, and aerobic fitness [28]. This battery includes components such as strength, balance, coordination, flexibility, and aerobic fitness [29,30]. Many studies examining physical fitness in older adults also include body composition metrics to gain a more complete health profile [31–33].

Despite the known links between sleep quality, physical fitness, and metabolic health, the evidence base in older populations remains limited [34]. However, there are associations between metabolic rates and sleep quality, but research in aged adults is rare [35–39]. The only research found was the Schilling et al. [40] study. Additionally, Kohanmoo et al. [41] and Tan et al. [42] reported an inverse relationship between fat mass and sleep quality or duration. However, most prior studies examined these domains in isolation, without integrating physical, metabolic, and sleep-related variables into a unified predictive model.

Poor sleep quality in older adults (short duration and frequent awakenings) impairs recovery, increases cortisol and anxiety levels, and perpetuates fatigue and reduced quality of life [34]. It is also associated with worse mental health and diminished performance in daily activities [43]. Considering these challenges, predictive models that integrate multiple health domains may help identify older individuals at risk of poor sleep quality. Emerging tools such as machine learning (ML) have shown promise in modulating complex health outcomes. By processing multidimensional data, ML can uncover non-linear relationships and offer preliminary classification capabilities, even in exploratory settings. However, few studies have applied ML to predict sleep quality using physical fitness and body composition variables in older adults. Therefore, the present study aimed to compare clinical and functional fitness variables by sleep quality, and to assess the predictive value of these variables using machine learning algorithms. It was hypothesized that body composition, functional fitness, and metabolic variables would significantly predict sleep quality.

2. Materials and Methods

2.1. Study Design

This study followed a cross-sectional observational design, to explore associations between body composition, functional fitness, and sleep quality in older adults. A total of 32 community-dwelling individuals aged 60 years or older were assessed in a controlled, single-session setting. Participants were recruited from a local health and exercise program, using a convenient sampling approach. Therefore, all results should be interpreted as exploratory and hypothesis-generating, laying the groundwork for future studies with larger and more representative samples. Sleep quality was evaluated using the Pittsburgh Sleep Quality Index (PSQI), a validated subjective tool, while body composition was estimated using a bioimpedance scale (Tanita BC-601). The functional fitness test was used, and other parameters, including the sit-to-stand test, handgrip strength, and walking speed, were evaluated. All procedures were conducted by trained researchers following standardized protocols to ensure consistency across measurements. This study adhered to key elements of the STROBE checklist for cross-sectional studies, including clear reporting of variables, participant characteristics, statistical methods, and limitations.

2.2. Sample

Thirty-two subjects participated in this study; twenty-six were females, and six were males. The sample mean age was 69 years. The convenience sample was recruited in the Bragança Municipality. All the participants were aged community people. All the procedures were in agreement with Helsinki's declaration. The research project received approval by the Ethical Committee of the Instituto Politécnico de Bragança (number: 2576). The participants were instructed to maintain normal daily activities to prevent physical inactivity. The participants were asked to complete a sample characterization questionnaire during the first visit. The inclusion criteria were: (i) being aged 60 years or older, (ii) maintaining functional independence in daily activities; (iii) not having severe chronic diseases or taking sleep-related medication; (iv) not having any significant cardiovascular, musculoskeletal, metabolic, or joint conditions that could interfere with the assessments; (v) not having developed any new illness or begun any new medication during the study period that could affect sleep or physical function; (vi) being a non-smoker; (vii) not having undertaken long-distance travel during the study period that could cause jet lag and affect sleep quality. Given the small sample size, the results are exploratory in nature and not generalizable.

2.3. Procedures

2.3.1. Anthropometrics, Body Composition, and Metabolic Variables

Anthropometrics was evaluated by stature and body mass. Additionally, the body composition was estimated with a digital bioimpedance scale (Tanita BC-601, Arlington Heights, IL, USA), which is validated and used for research [44]; however, this equipment does not provide the estimation algorithms. The computed variables of body composition were lean mass, percentage of fat mass, bone mineral density, visceral fat, total body mass, muscle mass, fat mass, and bone mineral density. The Tanita BC-601 device uses proprietary algorithms to estimate body composition and does not provide raw impedance values. These estimates may not be fully validated for all populations, especially older adults with atypical body composition. Therefore, results should be interpreted with caution. The participants made the evaluations wearing light clothing and without shoes and socks during the morning and before breakfast. It is important to note that the use of BIA in elderly populations presents inherent limitations, as age-related alterations in fluid balance and body tissue conductivity may compromise measurement accuracy. The stature was evaluated standing with the head in the Frankfurt plane. Waist and hip circumferences were also evaluated. The cardiovascular measures variables were the arterial pressure (systolic and diastolic) and resting heart rate measured with an OMRON (M2 HEM-7143-E, Omron, Kyoto, Japan), which is also validated to be used in research [44]. The metabolic rate was estimated by bioimpedance with the Tanita.

2.3.2. Arterial Blood Pressure, Resting Heart Rate and Sleep Quality

Arterial systolic blood pressure (SBP), diastolic blood pressure (DBP), and resting heart rate (RHR) were measured following the 2018 European Society of Cardiology and the European Society of Hypertension (ESC/ESH) Guidelines for the management of arterial hypertension [45]. Two measurements were performed, and the average between metrics was calculated.

Sleep quality was verified through the use of the Pitsburg sleep-quality index (PSQI), a 19-item questionnaire [46], and validated for the Portuguese population [47], used in this research. The PSQI items are subdivided into the following components: (1) subjective sleep quality, (2) sleep latency, (3) sleep duration, (4) habitual sleep efficiency, (5) sleep disturbances, (6) use of sleeping medication, and (7) daytime dysfunction. Each component is scored from 0 to 3, with higher scores indicating poorer sleep quality. A global score greater than 5 indicates poor sleep quality [47].

2.3.3. Handgrip Strength

Handgrip strength was assessed using a digital palmar dynamometer (CAMRY®, Lisbon, Portugal), with the maximum kilograms-force (kgf) achieved using a palm grip as the measurement. The participants stood with their arms away from their bodies and, upon the researcher's signal, exerted maximum palm grip force on the dynamometer for four seconds [29]. Each participant was given two attempts, and the evaluator noted the highest recorded result.

2.3.4. Functional Fitness

The Functional Fitness Test by Rikli and Jones was used to assess the main physical parameters associated with functional mobility [28]. The battery was composed of the 2 min Step Test and the Seat To Stand, where each participant was positioned standing up in front of a 43 cm highchair. In the arm curl test, the participant was positioned in a chair 43 cm high, holding a 2 kg dumbbell. The Time-Up-and-Go Test was conducted with the participant seated in a chair 43 cm high, facing a cone at 2.44 m, and the time was recorded in seconds after two trials. Finally, the Sit and Reach and in the Back Scratch tests were applied.

2.3.5. Relative Lower Limb Muscle Power

The lower limb muscle power was measured through the five-time sit-to-stand (5TSTS) test. The test was performed in a standardized chair of 0.49 m in height. The evaluator encouraged the participants throughout the test to ensure they always perform the maximum movement speed and preserve the technique. Two attempts were performed with an interval of 60 s, and the shortest time was noted. Shorter completion time indicates greater lower-limb muscle power [48].

2.4. Statistical Analysis

The Kolmogorov–Smirnov test, kurtosis (<±3), and Skewness (−2 to +2 criteria) values allowed us to assess the normality of the distribution, and Levene's test assessed the homogeneity. Thus, the *t*-test allowed the comparison of variables' sleep quality, and Pearson's correlation test allowed the association between the variables. The test was carried out at a significance level of 5%. The effect size (eta square − η^2) was computed and interpreted as without effect (if $0 < \eta^2 \leq 0.04$), minimum (if $0.04 < \eta^2 \leq 0.25$), moderate (if $0.25 < \eta^2 \leq 0.64$), and strong (if $\eta^2 > 0.64$). The logistic regression Machine Learning algorithm was developed using two lower-body functional performance variables: the Five-Time Sit-to-Stand Test (5TSTS1) and the Timed Up and Go Test (TUG), with only two predictors to avoid overfitting (hence, the sample size still limits the power and generalizability of our results). The model was evaluated using 5-fold stratified cross-validation, with standardization of input variables applied within each fold. Performance metrics (accuracy, precision, recall, and ROC AUC) were computed for each fold and reported as mean ± standard deviation. Due to the limited sample size, default hyperparameters were used, and no optimization procedures (e.g., grid search) were performed to avoid overfitting. The 5-fold stratified cross-validation procedure was used to assess model performance. Evaluation metrics included accuracy, precision, recall, and the area under the receiver operating characteristic curve (ROC AUC), reported as mean ± standard deviation. The machine learning procedure was conducted in agreement with the literature for analysis over 20 participants [46]. All statistical analyses were performed using JASP version 0.18 (JASP Team, Amsterdam, The Netherlands), and the machine learning models were developed using Python 3.11 with the Scikit-learn library

3. Results

The analyses conducted in this study were structured to address two main objectives: (1) to compare clinical and functional variables according to sleep quality classification; and (2) to evaluate whether these variables could predict sleep quality using machine learning models. Descriptive statistics and group comparisons (*t*-tests and effect sizes) were used to explore differences by sleep quality status. Correlational analyses were performed to examine the relationships between physical fitness, body composition, and sleep quality components. Finally, supervised machine learning models were applied to assess the predictive value of selected variables for classifying sleep quality, providing a preliminary test of their potential in data-driven risk identification.

3.1. Descriptives

The study evaluated various health and fitness parameters among 32 individuals, distinguishing between those with good (15) and poor (17) sleep quality. The mean age was 69.28 years. Those with poor sleep demonstrated slightly better performance in the 5-time sit-to-stand test (6.75 s) than those with good sleep quality (7.69 s). These descriptive results provide the context for the subsequent analyses, beginning with comparisons between good and poor sleep quality.

3.2. Sleep Quality Comparisons

Descriptives (means and standard deviations) for good and poor sleep quality are presented in Table 1. Poor sleep quality was associated with lower total sleep (3.35 h) compared to good sleep quality (6.13 h). Individuals with poor sleep also had higher mean ages (73.24 years) and visceral fat (8.29) compared to those with good sleep (69.47 years and 7.47 visceral fat). Despite these differences, both groups had similar heart rates (72.38 vs. 71.82 bpm) and total fat percentages (29.93% vs. 32.13%). Comparing the variables between poor and good sleep quality, it is possible to find that the TUG (t = 2.564; $p = 0.016$; $\eta^2 = 0.476$), total fat (kg) (t = 2.592; $p = 0.015$; $\eta^2 = 0.239$), and total sleep (t = 6.882; $p < 0.001$; $\eta^2 = 0.675$) time significantly differed between groups, where the persons with good sleep quality presented higher scores.

Table 1. Comparisons between sleep quality.

Variables	Good Sleep Quality (n = 15)	Poor Sleep Quality (n = 17)	Sleep Quality Comparison		
	Mean (±SD)	Mean (±SD)	t	*p*	η^2
Age (yo)	69.47 (±5.99)	73.24 (±7.34)	−1.724	0.095	0.456
Mass (kg)	64.73 (±11.87)	67.44 (±9.25)	−0.724	0.474	0.002
Stature (cm)	157.49 (±4.30)	159.16 (±5.58)	−0.938	0.356	0.013
Rest Heart rate (Bpm)	71.82 (±6.26)	72.38 (±8.91)	1.742	0.197	0.411
Hand grip (kg)	21.67 (±8.61)	26.12 (±5.46)	0.411	0.526	0.002
Arm curl (Repetition)	21.29 (±4.75)	23.30 (±4.00)	−0.204	0.840	0.408
Waist circumference (cm)	88.00 (±10.23)	89.05 (±9.34)	−1.766	0.088	0.178
Hip circumference (cm)	101.53 (±9.68)	99.62 (±4.86)	−1.302	0.203	0.634
5TSTS (seconds)	7.69 (±1.00)	6.75 (±1.08)	−0.303	0.764	0.164
CS30 (repetitions)	21.14 (±4.05)	22.84 (±3.30)	0.691	0.497	0.138
TUG (seconds)	5.96 (±1.04)	5.36 (±0.67)	2.564	0.016 *	0.476
Seat and Reach (cm)	−2.14 (±7.39)	−0.15 (±9.92)	−1.309	0.200	0.733
Back Stretch (cm)	−5.21 (±8.40)	−14.49 (±11.40)	1.957	0.060	0.188
2MST (Repetitions)	180.52 (±22.38)	186.69 (±35.45)	−0.637	0.529	0.453
Total Fat (kg)	20.94 (±7.33)	20.05 (±5.27)	2.592	0.015 *	0.239
Total Fat (%)	32.13 (±6.79)	29.93 (±5.78)	−0.580	0.566	0.148
Lean Mass (kg)	41.23 (±5.60)	44.54 (±6.37)	0.990	0.330	0.337
Lean Mass (%)	63.73 (±7.54)	64.88 (±7.49)	−1.552	0.131	0.157
Body Water (%)	47.95 (±4.70)	49.33 (±3.83)	−0.430	0.670	0.431
Visceral Fat	7.47 (±2.23)	8.29 (±3.42)	−0.911	0.369	0.344
MET [KJ]	5405.20 (±709.26)	5648.75 (±599.06)	−0.792	0.435	0.000
MET [Kcal]	1411.87 (±521.07)	1477.87 (±463.82)	−0.379	0.707	0.438
Total Sleep	6.13 (±1.46)	3.35 (±0.61)	6.882	<0.001 *	0.675

Note: yo: years old; Bpm: beats per minute; 5TSTS: Five Times Sit to Stand Test; CS30: Sit to stand test 30 s; TUG: time up and go-test; 2MST: 2 min step test; Kcal: kilocalories; * $p < 0.05$.

Beyond group differences, we then examined associations between the studied variables to better understand the interrelationships underlying sleep quality.

3.3. Associative Analysis and Machine Learning

Based on the Pearsons correlation test, the total sleep scores presented significant associations with 5TSTS (r = 0.442; $p = 0.011$), TUG (r = 0.411; $p = 0.019$), and back stretch (r = 0.406; $p = 0.021$). Finally, to assess the potential predictive capacity of selected measures, we applied an exploratory logistic regression model with cross-validation.

Given the limited sample size (n = 32), the machine learning analyses were conducted in an exploratory manner to assess whether selected functional and physiological variables could provide preliminary classification of sleep quality. The logistic regression model using 5TSTS and TUG achieved a mean accuracy of 0.76 (±0.15), precision of 0.79 (±0.18), recall of 0.83 (±0.21), and ROC/AUC of 0.74 (±0.16) across 5-fold cross-validation. While the data showed a moderate association between functional lower-limb performance and sleep quality, the findings are tentative given the small and specific sample.

4. Discussion

This study aimed to understand the differences between sleep quality, physical fitness, body composition, cardiovascular measures, and their interplay with sleep quality. It was hypothesized that the variables of functional fitness, body composition, and cardiovascular measures explain sleep quality and its dependency. The results of the present study confirmed the hypothesis; however, while the findings revealed potential associations, they should be interpreted with caution, given the limited statistical power and small sample size. Different variables also explained the total sleep time.

4.1. Sleep Quality Comparisons

When comparing the participants between people with poor and good sleep quality, it is possible to find that the 5TSTS, Back Stretch, and total sleep time significantly differed between groups, where the people with good sleep quality presented higher scores of functional fitness. These exploratory results suggest a positive association between physical fitness levels, particularly in lower body strength and flexibility, and sleep quality, which may contribute to improved sleep patterns. The literature provides supportive information about the relationship between functional fitness and sleep quality. Studies have shown that improvements in functional fitness, including lower limb strength, have been associated with improved sleep outcomes [47,49,50]. Interventions that target lower limb strength, such as resistance training and specific exercises like yoga, have been proven to enhance sleep quality and overall wellbeing [51–53]. Lower limb strength plays an important role in mobility, and independence is highlighted as an important factor in supporting a healthy sleep pattern [47,54,55].

4.2. Associative Analysis and Machine Learning

The associative analysis allowed us to highlight the variables that primarily explain the total sleep time. For the total sleep, the Back Stretch, waist girth, 5TSTS, and Visceral Fat explained the sleep time. The sleep quality relationships with waist girth, visceral fat, and total fat can be explained by the body composition interplay with hormonal regulation, inflammation, and metabolic health [56,57]. Additionally, variables like 5TSTS, Back Stretch, and Seat and Reach test reflect aspects of functional fitness and flexibility, and again may influence comfort and relaxation [56,58,59].

The negative impact of body water is possibly explained by the increased urine output, resulting in the need for frequent urination during the night [60–62]. This negative effect on the circadian rhythm due to urine production can interfere with sleep continuity, resulting in poor sleep quality [62,63]. As for the 2MW, the literature presents evidence of the positive effects of aerobic exercise on arterial pressure and cardiovascular health [64,65]. Lately, aerobic exercise seems related to good sleep quality, time, efficiency, and latency [66–68].

The logistic regression model using 5TSTS and TUG achieved a mean accuracy of 0.76, precision of 0.79, recall of 0.83, and ROC/AUC of 0.74. These results suggest that functional fitness variables, particularly strength-related metrics, may serve as possible meaningful predictors of poor versus good sleep quality in older adults. This is aligned with the literature where body composition [56,57] and functional fitness [56,58,59] seem to be associated with good sleep quality and wellbeing. However, it was not possible to find studies with machine learning analysis to predict sleep quality based on anthropometrics, body composition, and functional fitness. For those reasons, comparisons with other studies regarding the algorithm's scores were difficult to perform. Anyway, other studies with accelerometers and electrodermal instruments reported that machine learning algorithms may predict sleep quality with a percentage of accuracy between 78% and 84% [69,70], aligned with the present study. The use of a limited number of predictors in the machine learning models allowed for greater interpretability and reduced the risk of overfitting given the small sample size [71]. However, this parsimony may overlook important variables and complex interactions, limiting predictive accuracy. Thus, the findings are preliminary and hypothesis-generating, requiring validation in larger, more diverse samples with broader variable sets.

4.3. Psychophysiological Remarks

Considering a psychophysiological approach, the interplay between lower limb strength and sleep quality can be explained by different mechanisms. Lower limb strength and consequent independence may promote physical activity and tiredness, resulting in relaxation and somnolence [72,73]. Second, the lower limbs' strength, relation to balance and stability may contribute to a perception of safety due to the reduced risk of falls and injuries during sleep, contributing to better sleep quality [72,74] because lower limb-related physical fitness is associated with reduced levels of pain and discomfort that may disrupt sleep [73]. Altogether, psychologically, these implications may also result in positive mental health and relaxation, resulting in better sleep quality [47,75]. Finally, physical activity and exercise regulate circadian rhythms and release endorphins, which are known to enhance sleep patterns [52,76].

4.4. Strengths, Limitations and Future Studies

This study presents a novel and exploratory approach to understanding the relationship between body composition, functional fitness, and sleep quality in older adults. A key strength lies in the integration of traditional statistical methods with a machine learning algorithm (logistic regression), which provides complementary perspectives on the data and enhances analytical robustness. The study uses validated tools such as the Pittsburgh Sleep Quality Index and BIA-based body composition estimates, alongside functional fitness tests, offering a multidimensional profile of the participants. Furthermore, the use of cross-validation and transparent model evaluation metrics (e.g., ROC AUC, accuracy, precision, recall) strengthens the internal consistency of the machine learning analysis and reflects a commitment to methodological rigor despite the exploratory nature of the work.

It is also possible to point out some of the limitations of this study: (i) this is not an interventional study comparing exercise base therapeutics; (ii) this study did not control biochemical, psychological, or physiological variables; (iii) the daily life routines including physical activity and nutritional habits were not evaluated or controlled; (iv) the sample size and the number of males and the exploratory characteristics of the study did not allow us to conduct sex comparisons with precise outputs. Although logistic regression modeling was performed with only two predictors to avoid overfitting, still limited the power and generalizability of our results, additionally no factor analysis were made to select important variables; (v) the reliance on estimated data from bioimpedance and subjective sleep measures (PSQI) introduces potential measurement bias, additionally, body composition estimates were derived from BIA using proprietary algorithms, which may lack validation across diverse populations, reducing reproducibility compared to reference methods such as DEXA; (vi) the machine learning algorithms were technically valid for application in this dataset. Further, future studies should (i) evaluate the effects of a regular training program on sleep quality; (ii) analyze biochemical, psychological, and physiological variables as well as body composition and physical fitness; (iii) assess the effects of active living and wellbeing lifestyles in sleep quality; (iiii) performing more robust approaches, such as classificatory machine learning approach, that can deal with several characteristics of predictors in the same set of analysis; (iv) recruit larger samples; (iv) mitigate the lack of data collection about bruxism incidence in the participants, which has been reported as a significant co-factor in sleep worsening; (v) employ precise instruments like DEXA and accelerometers to evaluate body composition and sleep quality.

5. Conclusions

This exploratory analysis provides early insight into potential functional predictors of sleep quality, which may inform future confirmatory research with larger samples that physical fitness and body composition may be important in sleep quality. The lower limbs' strength and upper limbs' flexibility seem to explain the sleep quality. This study suggests that improving muscle strength and managing body fat levels through regular physical activity may contribute to better sleep quality in older adults. Integrating simple functional fitness assessments (such as the 5TSTS) into routine geriatric evaluations may provide a practical pathway to identify older adults at higher risk of poor sleep quality and associated health outcomes.

Author Contributions: Conceptualization, P.F. and A.M.M.; methodology, T.M.B. and D.P.-M.; software, S.G.E.; validation, D.P.-M. and A.M.M.; formal analysis, P.F., J.E.T. and L.B.; investigation, P.F. and S.G.E.; resources, A.M.M.; data curation, J.E.T., L.B. and D.P.-M.; writing—original draft preparation, P.F.; writing—review and editing, J.E.T., L.B., T.M.B., S.G.E., A.M.M. and D.P.-M.; visualization, S.G.E.; supervision, T.M.B., A.M.M. and D.P.-M.; project administration, A.M.M.; funding acquisition, A.M.M. and T.M.B. All authors have read and agreed to the published version of the manuscript.

Funding: This research received no external funding.

Institutional Review Board Statement: The study was conducted in accordance with the Declaration of Helsinki, and approved by the Institutional Review Board and the research project received approval by the Ethical Committee of the Instituto Politécnico de Bragança (number: 2576, approved 25 October 2023).

Informed Consent Statement: Written informed consent was obtained from all subjects involved in the study.

Data Availability Statement: The data presented in this study are available on request from the corresponding author due to possible participants identification.

Conflicts of Interest: The authors declare no conflicts of interest.

References

1. Al-Sofiani, M.E.; Ganji, S.; Kalyani, R.R. Body Composition Changes in Diabetes and Aging. *J. Diabetes Complicat.* **2019**, *33*, 451–459. [CrossRef]
2. Picca, A.; Landi, F.; Cesari, M. Editorial: Age-Related Changes in Body Composition: Mechanisms, Clinical Implications and Possible Treatments. *Front. Med.* **2020**, *7*, 230. [CrossRef] [PubMed]
3. Steppan, J.; Barodka, V.; Berkowitz, D.E.; Nyhan, D. Vascular Stiffness and Increased Pulse Pressure in the Aging Cardiovascular System. *Cardiol. Res. Pract.* **2011**, *2011*, 263585. [CrossRef]
4. Uangpairoj, P. Shibata Evaluation of Vascular Wall Elasticity of Human Digital Arteries Using Alternating Current-Signal Photoplethysmography. *Vasc. Health Risk Manag.* **2013**, *9*, 283–295. [CrossRef]
5. Kawasaki, Y.; Kitamura, E.; Kasai, T. Impact of Body Composition on Sleep and Its Relationship with Sleep Disorders: Current Insights. *Nat. Sci. Sleep* **2023**, *15*, 375–388. [CrossRef]
6. De Moraes Coelho, V.; De Oliveira Sinhoroto, C.; Magnabosco, P.; Nasser Figueiredo, V.; Pereira de Almeida Neto, O. Sleep Quality and Body Composition in a Nursing Team. *Biosci. J.* **2022**, *38*, e38082. [CrossRef]

7. Casas, R.S.; Pettee Gabriel, K.K.; Kriska, A.M.; Kuller, L.H.; Conroy, M.B. Association of Leisure Physical Activity and Sleep with Cardiovascular Risk Factors in Postmenopausal Women. *Menopause* **2012**, *19*, 413–419. [CrossRef] [PubMed]
8. Wu, Y.; Zhai, L.; Zhang, D. Sleep Duration and Obesity among Adults: A Meta-Analysis of Prospective Studies. *Sleep Med.* **2014**, *15*, 1456–1462. [CrossRef]
9. Wang, M.; Wei, J.; Yang, X.; Ni, P.; Wang, Y.; Zhao, L.; Deng, W.; Guo, W.; Wang, Q.; Li, T.; et al. The Level of IL-6 Was Associated with Sleep Disturbances in Patients with Major Depressive Disorder. *Neuropsychiatr. Dis. Treat.* **2019**, *15*, 1695–1700. [CrossRef]
10. Dolsen, E.A.; Harvey, A.G. IL-6, sTNF-R2, and CRP in the Context of Sleep, Circadian Preference, and Health in Adolescents with Eveningness Chronotype: Cross-Sectional and Longitudinal Treatment Effects. *Psychoneuroendocrinology* **2021**, *129*, 105241. [CrossRef]
11. Krueger, J.M.; Obál, F.J.; Fang, J.; Kubota, T.; Taishi, P. The Role of Cytokines in Physiological Sleep Regulation. *Ann. N. Y. Acad. Sci.* **2001**, *933*, 211–221. [CrossRef] [PubMed]
12. Majde, J.A.; Krueger, J.M. Links between the Innate Immune System and Sleep. *J. Allergy Clin. Immunol.* **2005**, *116*, 1188–1198. [CrossRef]
13. De Nys, L.; Anderson, K.; Ofosu, E.F.; Ryde, G.C.; Connelly, J.; Whittaker, A.C. The Effects of Physical Activity on Cortisol and Sleep: A Systematic Review and Meta-Analysis. *Psychoneuroendocrinology* **2022**, *143*, 105843. [CrossRef] [PubMed]
14. Gao, X.; Qiao, Y.; Chen, Q.; Wang, C.; Zhang, P. Effects of Different Types of Exercise on Sleep Quality Based on Pittsburgh Sleep Quality Index in Middle-Aged and Older Adults: A Network Meta-Analysis. *J. Clin. Sleep Med. JCSM Off. Publ. Am. Acad. Sleep Med.* **2024**, *20*, 1193–1204. [CrossRef]
15. Piva, T.; Masotti, S.; Raisi, A.; Zerbini, V.; Grazzi, G.; Mazzoni, G.; Belvederi Murri, M.; Mandini, S. Exercise Program for the Management of Anxiety and Depression in Adults and Elderly Subjects: Is It Applicable to Patients with Post-COVID-19 Condition? A Systematic Review and Meta-Analysis. *J. Affect. Disord.* **2023**, *325*, 273–281. [CrossRef]
16. Caldwell, H.G.; Coombs, G.B.; Howe, C.A.; Hoiland, R.L.; Patrician, A.; Lucas, S.J.E.; Ainslie, P.N. Evidence for Temperature-Mediated Regional Increases in Cerebral Blood Flow during Exercise. *J. Physiol.* **2020**, *598*, 1459–1473. [CrossRef]
17. Aritake-Okada, S.; Tanabe, K.; Mochizuki, Y.; Ochiai, R.; Hibi, M.; Kozuma, K.; Katsuragi, Y.; Ganeko, M.; Takeda, N.; Uchida, S. Diurnal Repeated Exercise Promotes Slow-Wave Activity and Fast-Sigma Power during Sleep with Increase in Body Temperature: A Human Crossover Trial. *J. Appl. Physiol.* **2019**, *127*, 168–177. [CrossRef]
18. Alparslan, G.B.; Orsal, Ö.; Unsal, A. Assessment of Sleep Quality and Effects of Relaxation Exercise on Sleep Quality in Patients Hospitalized in Internal Medicine Services in a University Hospital: The Effect of Relaxation Exercises in Patients Hospitalized. *Holist. Nurs. Pract.* **2016**, *30*, 155–165. [CrossRef]
19. Stokes, K.A.; Sykes, D.; Gilbert, K.L.; Chen, J.-W.; Frystyk, J. Brief, High Intensity Exercise Alters Serum Ghrelin and Growth Hormone Concentrations but Not IGF-I, IGF-II or IGF-I Bioactivity. *Growth Horm. IGF Res. Off. J. Growth Horm. Res. Soc. Int. IGF Res. Soc.* **2010**, *20*, 289–294. [CrossRef] [PubMed]
20. Athanasiou, N.; Bogdanis, G.C.; Mastorakos, G. Endocrine Responses of the Stress System to Different Types of Exercise. *Rev. Endocr. Metab. Disord.* **2023**, *24*, 251–266. [CrossRef]
21. Kreivėnaitė, L.; Streckis, V.; Visagurskienė, K.; Buliuolis, A.; Lileikienė, A. Dynamics of Variation in Physical Capacity of Elderly People. *Balt. J. Sport Health Sci.* **2018**, *4*. [CrossRef]
22. Chang, W.-D.; Chang, W.-Y.; Lee, C.; Feng, C.-Y. Validity and Reliability of Wii Fit Balance Board for the Assessment of Balance of Healthy Young Adults and the Elderly. *J. Phys. Ther. Sci.* **2013**, *25*, 1251–1253. [CrossRef]
23. Jayawardena, R.; Wickramawardhane, P.; Dalpadadu, C.; Hills, A.P.; Ranasignhe, P. The Impact of an Oral Nutritional Supplement on Body Weight Gain in Older Adults With Malnutrition: An Open-Label Randomized Controlled Trial. *Trials* **2023**, *24*, 625. [CrossRef]
24. Jiang, S.; Li, P. Current Development in Elderly Comprehensive Assessment and Research Methods. *BioMed Res. Int.* **2016**, *2016*, 1–10. [CrossRef]
25. Rodrigues, F.; Teixeira, J.E.; Forte, P. The Reliability of the Timed Up and Go Test among Portuguese Elderly. *Healthcare* **2023**, *11*, 928. [CrossRef]
26. Nakagaichi, M.; Anan, Y.; Hikiji, Y.; Uratani, S. Developing an Assessment Based on Physical Fitness Age to Evaluate Motor Function in Frail and Healthy Elderly Women. *Clin. Interv. Aging* **2018**, *13*, 179–184. [CrossRef] [PubMed]
27. Honório, S.; Batista, M.; Paulo, R.; Mendes, P.; Serrano, J.; Petrica, J.; Faustino, A.; Santos, J.; Martins, J. Functional Fitness and Nutritional Status of Institutionalized Elderly. *Med. Sport* **2017**, *70*, 200–211. [CrossRef]
28. Rikli, R.; Jones, J. *Senior Fitness Test Manual*, 2nd ed.; Human Kinetics: Champaign, IL, USA, 2013; 200p.
29. Marques, E.A.; Baptista, F.; Santos, R.; Vale, S.; Santos, D.A.; Silva, A.M.; Mota, J.; Sardinha, L.B. Normative Functional Fitness Standards and Trends of Portuguese Older Adults: Cross-Cultural Comparisons. *J. Aging Phys. Act.* **2014**, *22*, 126–137. [CrossRef] [PubMed]
30. Monteiro, A.M.; Bartolomeu, R.F.; Forte, P.; Carvalho, J. The Effects of Three Different Types of Training in Functional Fitness and Body Composition in Older Women. *J. Sport Health Res.* **2019**, *11*, 289–304.

31. Monteiro, A.M.; Rodrigues, S.; Matos, S.; Teixeira, J.E.; Barbosa, T.M.; Forte, P. The Effects of 32 Weeks of Multicomponent Training with Different Exercises Order in Elderly Women's Functional Fitness and Body Composition. *Medicina* **2022**, *58*, 628. [CrossRef]
32. Zhuang, M.; Shi, J.; Liu, J.; He, X.; Chen, N. Comparing the Efficacy of Low-Load Resistance Exercise Combined with Blood Flow Restriction versus Conventional-Load Resistance Exercise in Chinese Community-Dwelling Older People with Sarcopenic Obesity: A Study Protocol for a Randomised Controlled Trial. *BMC Geriatr.* **2023**, *23*, 874. [CrossRef]
33. Jiang, Y.; Tan, S.; Wang, Z.; Guo, Z.; Li, Q.; Wang, J. Aerobic Exercise Training at Maximal Fat Oxidation Intensity Improves Body Composition, Glycemic Control, and Physical Capacity in Older People with Type 2 Diabetes. *J. Exerc. Sci. Fit.* **2020**, *18*, 7–13. [CrossRef]
34. Alnawwar, M.A.; Alraddadi, M.I.; Algethmi, R.A.; Salem, G.A.; Salem, M.A.; Alharbi, A.A. The Effect of Physical Activity on Sleep Quality and Sleep Disorder: A Systematic Review. *Cureus* **2023**, *15*, e43595. [CrossRef]
35. Aggio, D.A.; Sartini, C.; Papacosta, O.; Lennon, L.T.; Ash, S.; Whincup, P.H.; Wannamethee, S.G.; Jefferis, B.J. Cross-Sectional Associations of Objectively Measured Physical Activity and Sedentary Time with Sarcopenia and Sarcopenic Obesity in Older Men. *Prev. Med.* **2016**, *91*, 264–272. [CrossRef]
36. Hernandez-Martinez, J.; González-Castillo, C.; Herrera-Valenzuela, T.; Muñoz-Vásquez, C.; Magnani Branco, B.H.; Valdés-Badilla, P. Association between Physical Activity Habits with Cardiometabolic Variables, Body Composition, and Physical Performance in Chilean Older Women. *Int. J. Environ. Res. Public Health* **2023**, *20*, 6688. [CrossRef] [PubMed]
37. Buckinx, F.; Peyrusqué, É.; Granet, J.; Aubertin-Leheudre, M. Impact of Current or Past Physical Activity Level on Functional Capacities and Body Composition among Elderly People: A Cross-Sectional Analysis from the YMCA Study. *Arch. Public Health* **2021**, *79*, 50. [CrossRef] [PubMed]
38. Pedrero-Chamizo, R.; Gómez-Cabello, A.; Mélendez, A.; Vila-Maldonado, S.; Espino, L.; Gusi, N.; Villa, G.; Casajús, J.A.; González-Gross, M.; Ara, I. Higher Levels of Physical Fitness Are Associated with a Reduced Risk of Suffering Sarcopenic Obesity and Better Perceived Health among the Elderly. The EXERNET Multi-Center Study. *J. Nutr. Health Aging* **2015**, *19*, 211–217. [CrossRef] [PubMed]
39. Shimoda, T.; Suzuki, T.; Tsutsumi, K.; Samukawa, M.; Yoshimura, S.; Ogasawara, K. Association between Physical Activity Levels and Body Composition among Healthy Older Japanese Adults during a Snowy Winter: A Cross-Sectional Study. *Int. J. Environ. Res. Public. Health* **2020**, *17*, 5316. [CrossRef]
40. Schilling, R.; Schmidt, S.C.E.; Fiedler, J.; Woll, A. Associations between Physical Activity, Physical Fitness, and Body Composition in Adults Living in Germany: A Cross-Sectional Study. *PLoS ONE* **2023**, *18*, e0293555. [CrossRef]
41. Kohanmoo, A.; Kazemi, A.; Zare, M.; Akhlaghi, M. Gender-Specific Link between Sleep Quality and Body Composition Components: A Cross-Sectional Study on the Elderly. *Sci. Rep.* **2024**, *14*, 8113. [CrossRef]
42. Tan, X.; Titova, O.E.; Lindberg, E.; Elmståhl, S.; Lind, L.; Schiöth, H.B.; Benedict, C. Association Between Self-Reported Sleep Duration and Body Composition in Middle-Aged and Older Adults. *J. Clin. Sleep Med.* **2019**, *15*, 431–435. [CrossRef] [PubMed]
43. Xu, X.; Li, W.; Zou, S.; Li, Y.; Wang, H.; Yan, X.; Du, X.; Zhang, L.; Zhang, Q.; Cheung, T.; et al. Sleep Disturbances and Their Association With Quality of Life in Older Psychiatric Patients During the COVID-19 Pandemic. *J. Geriatr. Psychiatry Neurol.* **2022**, *35*, 229–236. [CrossRef] [PubMed]
44. Topouchian, J.; Agnoletti, D.; Blacher, J.; Youssef, A.; Ibanez, I.; Khabouth, J.; Khawaja, S.; Beaino, L.; Asmar, R. Validation of Four Automatic Devices for Self-Measurement of Blood Pressure According to the International Protocol of the European Society of Hypertension. *Vasc. Health Risk Manag.* **2011**, *7*, 709–717. [CrossRef]
45. Williams, B.; Mancia, G.; Spiering, W.; Rosei, E.A.; Azizi, M.; Burnier, M.; Clement, D.L.; Coca, A.; De Simone, G.; Dominiczak, A.; et al. 2018 ESC/ESH Guidelines for the Management of Arterial Hypertension. *Eur. Heart J.* **2018**, *39*, 3021–3104. [CrossRef]
46. Rajput, D.; Wang, W.-J.; Chen, C.-C. Evaluation of a Decided Sample Size in Machine Learning Applications. *BMC Bioinform.* **2023**, *24*, 48. [CrossRef]
47. Fraga, A.; Rodrigues, L.; Catarino, B. Correlation of Lower Limb Muscle Strength with Functional Mobility and Quality of Life in Patients with Multiple Sclerosis. *Med. Res. Arch.* **2022**, *10*. [CrossRef]
48. Alcazar, J.; Losa-Reyna, J.; Rodriguez-Lopez, C.; Alfaro-Acha, A.; Rodriguez-Mañas, L.; Ara, I.; García-García, F.J.; Alegre, L.M. The Sit-to-Stand Muscle Power Test: An Easy, Inexpensive and Portable Procedure to Assess Muscle Power in Older People. *Exp. Gerontol.* **2018**, *112*, 38–43. [CrossRef]
49. Zhang, H.; Chen, X.; Han, P.; Ma, W.; Zhang, Y.; Song, P.; Wu, Y.; Zhu, Y.; Jiang, Z.; Cai, M.; et al. Mediating Effects of Lower Extremity Function on the Relationship between Night Sleep Duration and Cardiovascular Disease Risk: A Cross-Sectional Study in Elderly Chinese without Cardiovascular Diseases. *BMJ Open* **2021**, *11*, e046015. [CrossRef]
50. Antônio Gomes de Resende, N.; Matheus Amarante do, N.; Danilo Rodrigues Pereira da, S.; Raquel Simões Mendes, N.; Josimari Melo de, S.; Marzo Edir Da Silva, G. Effects of Multicomponent Training on Functional Fitness and Quality of Life in Older Women: A Randomized Controlled Trial. *Int. J. Sports Exerc. Med.* **2019**, *5*, 126. [CrossRef]

51. Sivaramakrishnan, D.; Fitzsimons, C.; Kelly, P.; Ludwig, K.; Mutrie, N.; Saunders, D.H.; Baker, G. The Effects of Yoga Compared to Active and Inactive Controls on Physical Function and Health Related Quality of Life in Older Adults- Systematic Review and Meta-Analysis of Randomised Controlled Trials. *Int. J. Behav. Nutr. Phys. Act.* **2019**, *16*, 33. [CrossRef]
52. Oshima-Saeki, C.; Taniho, Y.; Arita, H.; Fujimoto, E. Lower-Limb Warming Improves Sleep Quality in Elderly People Living in Nursing Homes. *Sleep Sci.* **2017**, *10*, 87–91. [CrossRef]
53. Corrêa, H.L.; Moura, S.R.G.; Neves, R.V.P.; Tzanno-Martins, C.; Souza, M.K.; Haro, A.S.; Costa, F.; Silva, J.A.B.; Stone, W.; Honorato, F.S.; et al. Resistance Training Improves Sleep Quality, Redox Balance and Inflammatory Profile in Maintenance Hemodialysis Patients: A Randomized Controlled Trial. *Sci. Rep.* **2020**, *10*, 11708. [CrossRef]
54. Azimizadeh, M.J.; Tabatabai Asl, S.M.; Hoseini, S.H. The Effects of an Eight-Week Cawthorne-Cooksey Training Program on Balance and Lower Limb Strength in the Elderly. *J. Sport Biomech.* **2021**, *7*, 68–77. [CrossRef]
55. Kato, T.; Ikezoe, T.; Tabara, Y.; Matsuda, F.; Tsuboyama, T.; Ichihashi, N. Differences in Lower Limb Muscle Strength and Balance Ability between Sarcopenia Stages Depend on Sex in Community-Dwelling Older Adults. *Aging Clin. Exp. Res.* **2022**, *34*, 527–534. [CrossRef] [PubMed]
56. McCoy, T.; Sochan, A.J.; Spaeth, A.M. The Relationship between Sleep and Physical Activity by Age, Race, and Gender. *Rev. Cardiovasc. Med.* **2024**, *25*, 378. [CrossRef]
57. Knobbe, T.J.; Kremer, D.; Eisenga, M.F.; van Londen, M.; Annema, C.; Bültmann, U.; Kema, I.P.; Navis, G.J.; Berger, S.P.; Bakker, S.J.L.; et al. Sleep Quality, Fatigue, Societal Participation and Health-Related Quality of Life in Kidney Transplant Recipients: A Cross-Sectional and Longitudinal Cohort Study. *Nephrol. Dial. Transplant.* **2023**, *39*, 74–83. [CrossRef]
58. Li, C.-H.; Chung, M.-H.; Liao, C.-H.; Su, C.-C.; Lin, Y.; Liao, Y.-M. Urinary Incontinence and Sleep Quality in Older Women with Type 2 Diabetes: A Cross-Sectional Study. *Int. J. Environ. Res. Public. Health* **2022**, *19*, 15642. [CrossRef]
59. Zhang, J.; Wang, C.; Gong, W.; Peng, H.; Tang, Y.; Li, C.C.; Zhao, W.; Ye, Z.; Lou, T. Association between Sleep Quality and Cardiovascular Damage in Pre-Dialysis Patients with Chronic Kidney Disease. *BMC Nephrol.* **2014**, *15*, 131. [CrossRef] [PubMed]
60. Vu, H.M.; Tran, V.T.H.; Hoang, H.Q.; Han, B.; Hoang, B.X. Efficacy and Tolerability of Ich Nieu Khang Dietary Supplement for Overactive Bladder. *J. Med. Food* **2023**, *26*, 262–269. [CrossRef] [PubMed]
61. Torimoto, K.; Uchimura, N.; Roitmann, E.; Marumoto, M.; Hirakata, T.; Burtea, T. A Large Survey of Nocturia Related to Sleep Quality and Daytime Quality of Life in a Young Japanese Population: NOCTURNE Study. *Neurourol. Urodyn.* **2021**, *40*, 340–347. [CrossRef]
62. Kanammit, P.; Boonchan, T.; Sirisreetreerux, P.; Viseshsindh, W.; Kochakarn, W. Nocturia and Effect on the Quality of Life. A Study at Ramathibodi Hospital. *Insight Urol.* **2021**, *42*, 144–153. [CrossRef]
63. Udo, Y.; Nakao, M.; Honjo, H.; Ukimura, O.; Kitakoji, H.; Miki, T. Sleep Duration Is an Independent Factor in Nocturia: Analysis of Bladder Diaries. *BJU Int.* **2009**, *104*, 75–79. [CrossRef]
64. Akazawa, N.; Ra, S.; Sugawara, J.; Maeda, S. Influence of Aerobic Exercise Training on Post-Exercise Responses of Aortic Pulse Pressure and Augmentation Pressure in Postmenopausal Women. *Front. Physiol.* **2015**, *6*, 268. [CrossRef] [PubMed]
65. Schroeder, E.C.; Ranadive, S.M.; Yan, H.; Lane-Cordova, A.D.; Kappus, R.M.; Cook, M.D.; Fernhall, B. Effect of Acute Maximal Exercise on Vasodilatory Function and Arterial Stiffness in African-American and White Adults. *J. Hypertens.* **2019**, *37*, 1262–1268. [CrossRef] [PubMed]
66. Ezati, M.; Keshavarz, M.; Barandouzi, Z.A.; Montazeri, A. The Effect of Regular Aerobic Exercise on Sleep Quality and Fatigue among Female Student Dormitory Residents. *BMC Sports Sci. Med. Rehabil.* **2020**, *12*, 44. [CrossRef]
67. Brupbacher, G.; Zander-Schellenberg, T.; Straus, D.; Porschke, H.; Infanger, D.; Gerber, M.; von Känel, R.; Schmidt-Trucksäss, A. The Acute Effects of Aerobic Exercise on Sleep in Patients with Unipolar Depression: A Randomized Controlled Trial. *Sleep* **2021**, *44*, zsab177. [CrossRef]
68. Abd El-Kader, S.M.; Al-Jiffri, O.H. Aerobic Exercise Modulates Cytokine Profile and Sleep Quality in Elderly. *Afr. Health Sci.* **2019**, *19*, 2198–2207. [CrossRef]
69. Piccini, J.; August, E.; Óskarsdóttir, M.; Arnardóttir, E.S. Using the Electrodermal Activity Signal and Machine Learning for Diagnosing Sleep. *Front. Sleep* **2023**, *2*, 1127697. [CrossRef]
70. Sundararajan, K.; Georgievska, S.; te Lindert, B.H.W.; Gehrman, P.R.; Ramautar, J.; Mazzotti, D.R.; Sabia, S.; Weedon, M.N.; van Someren, E.J.W.; Ridder, L.; et al. Sleep Classification from Wrist-Worn Accelerometer Data Using Random Forests. *Sci. Rep.* **2021**, *11*, 24. [CrossRef] [PubMed]
71. Han, M.; Cai, D.; Huo, Z.; Shen, Z.; Tang, L.; Yang, S.; Wang, C. Reducing Overfitting Risk in Small-Sample Learning with ANN: A Case of Predicting Graduate Admission Probability. In *Communications in Computer and Information Science, Proceedings of the Artificial Intelligence and Machine Learning, Nanjing, China, 25–27 November 2023*; Jin, H., Pan, Y., Lu, J., Eds.; Springer Nature: Singapore, 2024; pp. 404–419.
72. Zhang, Y.; Zhao, M.; Zhang, B.; Zhang, K.; Zhou, Z. Acupuncture as an Adjunctive Treatment for Post-Stroke Epilepsy: Protocol for a Randomized Controlled Trial. *Front. Neurol.* **2021**, *12*, 711390. [CrossRef]

73. Szlejf, C.; Suemoto, C.K.; Drager, L.F.; Griep, R.H.; Fonseca, M.J.M.; Diniz, M.F.H.S.; Lotufo, P.A.; Benseãor, I.M. Association of Sleep Disturbances with Sarcopenia and Its Defining Components: The ELSA-Brasil Study. *Braz. J. Med. Biol. Res.* **2021**, *54*, e11539. [CrossRef] [PubMed]
74. Cauley, J.A.; Hovey, K.M.; Stone, K.L.; Andrews, C.A.; Barbour, K.E.; Hale, L.; Jackson, R.D.; Johnson, K.C.; LeBlanc, E.S.; Li, W.; et al. Characteristics of Self-Reported Sleep and the Risk of Falls and Fractures: The Women's Health Initiative (WHI). *J. Bone Miner. Res.* **2019**, *34*, 464–474. [CrossRef] [PubMed]
75. Fábrega-Cuadros, R.; Aibar-Almazán, A.; Martínez-Amat, A.; Hita-Contreras, F. Impact of Psychological Distress and Sleep Quality on Balance Confidence, Muscle Strength, and Functional Balance in Community-Dwelling Middle-Aged and Older People. *J. Clin. Med.* **2020**, *9*, 3059. [CrossRef]
76. Delgado-Floody, P.; Caamaño Navarrete, F.; Chirosa-Ríos, L.; Martínez-Salazar, C.; Vargas, C.A.; Guzmán-Guzmán, I.P. Exercise Training Program Improves Subjective Sleep Quality and Physical Fitness in Severely Obese Bad Sleepers. *Int. J. Environ. Res. Public. Health* **2022**, *19*, 13732. [CrossRef] [PubMed]

Journal of
Functional Morphology and Kinesiology

MDPI

Article

A National Study of Somatotypes in Mexican Athletes Across 43 Sports

Ximena Martínez-Mireles [1], José Omar Lagunes-Carrasco [2], Vianney Curiel-Cervantes [3], Ximena Ortega-Salinas [4], Mauro E. Valencia [5], Ricardo López-García [2], Silvia García [6,7,8], Cristina Bouzas [6,7,8], Rogelio Salas-García [1], Erik Ramírez [1] and Josep A. Tur [6,7,8,*]

1 Facultad de Salud Pública y Nutrición, Universidad Autónoma de Nuevo León UANL, Av. Universidad S/N Ciudad Universitaria, San Nicolás de los Garza 66451, Mexico; ximenamtznutricion@gmail.com (X.M.-M.); erik.ramirezlp@uanl.edu.mx (E.R.)

2 Facultad de Organización Deportiva, Universidad Autónoma de Nuevo León UANL, Av. Universidad S/N Ciudad Universitaria, San Nicolás de los Garza 66451, Mexico

3 Secretaría de Ciencia, Humanidades, Tecnología e Innovación SECIHTI, Universidad Nacional Autónoma de México, UNAM, Campus Juriquilla, Blvd. Juriquilla 3001, Juriquilla, Querétaro 76230, Mexico; viana.ccjv@gmail.com

4 Instituto del Deporte y la Recreación del Estado de Querétaro INDEREQ, Blvd. Bernardo Quintana, S/N, Col. Villas del Parque, Querétaro 76090, Mexico

5 Centro de Investigación en Alimentación y Desarrollo A.C. CIAD, Departamento de Nutrición, Carretera Gustavo Enrique Astiazarán Rosas 46, La Victoria, Hermosillo 83304, Mexico

6 Research Group on Community Nutrition & Oxidative Stress, University of Balearic Islands-IUNICS, 07122 Palma de Mallorca, Spain; cristina.bouzas@uib.es (C.B.)

7 Health Research Institute of the Balearic Islands (IdISBa), 07120 Palma de Mallorca, Spain

8 CIBER Fisiopatología de la Obesidad y Nutrición (CIBEROBN), Instituto de Salud Carlos III (ISCIII), 28029 Madrid, Spain

* Correspondence: pep.tur@uib.es; Tel.: +34-971173146; Fax: +34-971173184

Abstract: Background: In Mexico, research on somatotypes in athletes has primarily focused on team sports, taekwondo, climbing, and triathlon. However, the available evidence remains limited. Therefore, the purpose of this study was to determine the somatotype of Mexican athletes by sex, and to compare somatotype and body composition across sport macro-categories in 43 disciplines. **Methods**: Anthropometric measurements were conducted according to the International Society for the Advancement of Kinanthropometry (ISAK) protocol. Athletes who participated in regional, national, or international competitions between 2008 and 2024 were included. **Results**: A total of 889 Mexican athletes (477 males and 412 females) across 43 disciplines were evaluated. Among male athletes, the predominant somatotype was endomorphic mesomorph (52.4%), followed by balanced mesomorph (17.6%) and ectomorphic mesomorph (13.6%). Among female athletes, the most reported somatotypes were endomorphic mesomorph (24.5%), mesomorphic endomorph (24.0%), and mesomorph-endomorph (21.4%). Athletes in endurance sports showed significant differences for both sexes compared to those in power and skill-based sports for both sexes ($p < 0.05$). Among males, team sports showed the highest values for body mass, height, and body fat percentage (%BF), while mesomorphy was greatest in track and field and ectomorphy in sprint events. Among females, track and field athletes presented the highest values for body mass, height, %BF, mesomorphy, and endomorphy, whereas endurance athletes exhibited the highest ectomorphy values. **Conclusions**: The findings suggest that, compared to international athletes, Mexican athletes exhibited a higher endomorphic component. It is recommended that somatotype assessments should be incorporated into regular monitoring protocols at national sports centers and considered in physical training programs to optimize performance and reduce the risk of injury.

https://doi.org/10.3390/jfmk10030329

Keywords: anthropometric profile; physical status; somatochart; body composition; Mexican athletes

1. Introduction

Physical status refers to the physical characteristics of the body, including size, shape, and body composition [1]. Genetic factors, diet, training, and the specific demands of each sport influence this configuration. Somatotype, in turn, is a classification system that describes the shape and composition of the human body, applicable to both athletes and non-athletes. It was initially proposed by Sheldon et al. [2] and later adapted by Heath and Carter as the anthropometric somatotype, which adds measurements of skinfolds, girths, and bone diameters [3,4]. This model considers three main components: endomorphy, mesomorphy, and ectomorphy. Endomorphy refers to a body type with a higher proportion of body fat and a softer body with curves; mesomorphy describes a muscular and well-developed body structure; and ectomorphy characterizes individuals with a thin, linear physique, low muscle mass, and fat, giving them a slender appearance with long limbs. Although most individuals exhibit a combination of these components, one typically predominates, influencing their physical capabilities and athletic performance [5–8].

Among male athletes, a predominant mesomorphic profile has been observed, characterized by a strong body structure and higher muscle mass. In contrast, female athletes tend to exhibit a more endomorphic profile, marked by a higher accumulation of adipose tissue [9,10]. The application of somatotype analysis contributes to optimizing athletic performance, assessing the balance between fat and muscle mass, determining nutritional status, identifying physical profiles prone to injury, and adapting training to the specific demands of each sport [6,11–18]. Moreover, determining somatotypes across sports macro-categories enhances the identification of morphological patterns and the design of targeted training programs [19,20]. Genetically, mesomorphic and ectomorphic configurations are primarily inherited, while environmental factors such as diet and physical activity exert a higher influence on endomorphy [5,21,22]. From a functional perspective, mesomorphy is particularly favorable for the development of strength and power, whereas ectomorphy is associated with improved performance in endurance disciplines [17,23].

The classification of somatotypes among athletes from Latin America and Spain reveals a predominance of the endomorphic component, often in combination with mesomorphic traits, depending on the sport and playing position. For instance, Spanish futsal players [24] present endomorphy values of 3.8, which are lower compared to Brazilian players with 4.7 [25] and differ from Mexican soccer players with 4.3 [26]; this reflects a higher proportion of body fat in sports involving Latin ethnic populations. Additionally, other studies have reported differences in skinfold thickness across Latin American countries among female soccer players [27]. The field of sports science is still developing and consolidating in Latin America. The quality of research is affected by limited resources, language barriers (given that English dominates international research), and the need for greater interdisciplinary integration [28,29] which may contribute to the relatively lower number of studies addressing somatotype and body composition in athletes compared to other regions. Regarding body composition, evidence suggests that Mexican adults, particularly women, tend to have higher levels of adiposity compared to other ethnic groups [30]. Specifically, among Mexican American athletes, a higher accumulation of fat in the limbs has been observed compared to the trunk [31].

In Mexico, research on somatotype in athletes was primarily focused on team sports [26,32–35], as well as on individual disciplines such as taekwondo, climbing, and triathlon [36–38]. Although these studies have contributed valuable insights into the somatotype of Mexican athletes, the available data remain limited and are concentrated in a small number of sports disciplines. Nationally, the lack of updated and representative morphological references hinders the establishment of practical standards for coaches, selectors, nutritionists, and institutional sports programs. Understanding the somatotype of Mexican athletes would provide essential reference information on body composition, physical characteristics, and nutritional status, which are key elements for monitoring and optimizing athletic performance. Moreover, the absence of specific information on sex-based and discipline-specific differences limits the development of more effective strategies within the national sports context. Therefore, the aim of this study was (1) to determine the somatotype of Mexican athletes by sex and (2) to compare somatotype and body composition across sport macro-categories in 43 disciplines.

2. Materials and Methods

2.1. Study Design

This observational, cross-sectional, and descriptive-analytical study was conducted using data previously collected from Mexican athletes. The study followed the guidelines of the Strengthening the Reporting of Observational Studies in Epidemiology (STROBE) statement for cross-sectional studies [39].

2.2. Setting

Athletes from northern, central, and southern Mexico were included in the study. All participants competed at regional, national, or international levels between 2008 and 2024, and the recruitment and data collection were conducted. The data originated from athletes affiliated with public and private institutions, as well as from individual sources, all of whom provided informed consent for the use of their data for research purposes. The study protocol was approved by the Ethics and Research Committee of the Facultad de Salud Pública y Nutrición at the Universidad Autónoma de Nuevo León (UANL) (registration number 24-FaSPyN-SA-04; 11 June 2024). All athletes included in the study signed a written informed consent form, which explained the purpose of the research, the procedures involved, potential benefits and risks, and the confidentiality of their data. In the case of underage athletes, assent was obtained from the participants along with written consent from their parents or legal guardians [40].

To ensure confidentiality, each athlete was identified using a unique code to guarantee the anonymous handling of information. The protocol complied with the guidelines established by the NOM-012-SSA3-2012, "*Que establece los criterios para la ejecución de proyectos de investigación para la salud en seres humanos*" (translation: which sets the criteria for conducting health research projects involving human subjects) [41] in Mexico, as well as the Declaration of Helsinki [42], to ensure adherence to ethical principles in research.

2.3. Participants

Athletes from 43 sports across various regions of Mexico were included. For the analysis, sport macro-categories were considered within each discipline, including differences by event type, playing position, and competition format. Athletes participated in regional, national, and international competitions between 2008 and 2024, during which recruitment and data collection took place. Participants were selected through purposive sampling. From a total of 889 Mexican athletes, 477 male and 412 female athletes were analyzed, all of whom met the following inclusion criteria: (i) assessment conducted during pre-competition, general, specific, or pre-season phases; (ii) measurements taken between 2008 and 2024; (iii) age range between 14 and 35 years; and (iv) active participation in regional, national, or international competitions. To be classified as an athlete, individuals were required to be actively engaged in official competitions, to be part of a systematic training program, and to be formally registered with a recognized sports organization at the regional, national, or international level [43,44]. This definition ensured that all included participants met formal criteria for athletic representation and competition.

The athletes included in this study participated in one or more of the following competitions: Universiada Nacional, ranked among the top 10 in the Comisión Nacional de Cultura Física y Deporte (CONADE) Games; Central American and Caribbean Games, Junior Pan American Games, Pan American Games, World Championships, Junior World Championships, Youth Olympic Games, and the World Series of Team Roping. They were also members of or participants in events organized by the Liga Mexicana de Powerlifting, the Comisión Nacional Deportiva Estudiantil de Instituciones Privadas (CONADEIP), the Organización Nacional Estudiantil de Fútbol Americano (ONEFA), the Mexican National American Football Team, the Mexican National Rugby Team, and the Federación Mexicana de Rodeo.

Exclusion criteria were as follows: (i) inactive athletes, (ii) injured athletes, (iii) paralympic-level athletes, preschool- and school-aged individuals, and (iv) athletes participating in winter or extreme sports. Elimination criteria included (i) incomplete anthropometric measurements, (ii) body composition assessed during non-designated macrocycle phases, (iii) voluntary withdrawal from extreme somatotype characteristic participation, and (iv) atypical physiological conditions at the time of assessment (e.g., dehydration, illness).

2.4. Variables

The variables assessed included sex, body mass (kg), height (m), age (years), body fat percentage (%BF) (calculated using Equation (5) from Lean et al. [45] for females and Equation (2) for males; based on triceps skinfold, age, and body mass index), triceps skinfold (mm), subscapular skinfold (mm), suprailiac skinfold (mm), thigh skinfold (mm), flexed arm girth (cm), thigh girth (mm), humerus breadth (cm), femur breadth (cm), and the three somatotype components: endomorphy, mesomorphy, and ectomorphy [3].

2.5. Measurements

Body weight was measured using a SECA® 813bt scale (Seca GmbH & Co. KG, Hamburg, Germany) with a precision of ±0.1 kg. Height was measured with a SECA® 213 stadiometer (Seca GmbH & Co. KG, Hamburg, Germany) (±0.1 cm). Flexed arm and calf girths were measured using a Lufkin® anthropometric tape (±0.1 mm; Cooper Industries, Houston, TX, USA). Triceps, subscapular, supraspinal, and thigh skinfolds were measured with a Slim Guide® caliper (±1.0 mm; Creative Health Products, Ann Arbor, MI, USA). Humerus and femur bone breadths were assessed using a Lenart® anthropometer (Lenart Instruments®; ±0.1 mm).

2.6. Bias

All measurements were taken in duplicate, and a third measurement was performed if the intra-evaluator technical error of measurement (TEM) threshold was exceeded [46]. Assessments were conducted following the measurement protocol of the International Society for the Advancement of Kinanthropometry (ISAK) [47] and were carried out by certified anthropometrists at Levels 1, 2, and 3. The intra-evaluator TEM was 2.60% for skinfolds, 1.18% for breadths, and 0.85% for girths. Anthropometric assessments were conducted prior to each athlete's training session and after a minimum fasting period of 4 h.

2.7. Somatotype Calculation

Somatotype and its three components—endomorphy, mesomorphy, and ectomorphy—were calculated in a Microsoft® Excel® spreadsheet (Microsoft 365, Version 2508, Build 19127.20134, Microsoft Corporation, Redmond, WA, USA) using the equations proposed by Heath–Carter [3]; Appendix A.1. The classification into 13 somatotype categories [3] was performed using the NutriSolver® software, version 1.0.0, Monterrey, N.L., Mexico [48].

Somatochart and Tables by Sports

The somatochart is a graphical tool used to visually represent and classify somatotypes based on established anthropometric measurements [3] within a two-dimensional plane [49]. The location of the somatotype on the somatochart was determined by two coordinates derived from the endomorphic, mesomorphic, and ectomorphic components. These were calculated using the X and Y axes, where the X-axis represented the difference between ectomorphy and endomorphy, and the Y-axis reflected mesomorphic predominance relative to the other two components. The equations can be found in Appendix A.2.

2.8. Macro-Categories by Sports

To facilitate a better understanding of the somatotype distribution among the analyzed athletes, sports disciplines were grouped into six functional macro-categories: team sports, combat sports, individual sports, track and field, endurance events, and sprint events. This categorization was structured based on functional and morphophysiological criteria, with the aim of identifying potential common somatotype patterns within groups sharing similar competitive characteristics [20]. Therefore, they were considered equivalent for somatotype characterization. The sports included in each macro-category are detailed and visually represented in the corresponding somatochart in the Supplementary Material (Figures S1 and S2).

Tables were created showing the mean and standard deviation for age, body mass (kg), height (cm), body mass index (BMI) (kg/m^2), body fat percentage (%BF), and the three somatotype components (endomorphy, mesomorphy, and ectomorphy), stratified by sport discipline, position, and category. Data were organized separately for male and female athletes.

2.9. Statistics

Descriptive statistics (mean $\pm$ standard deviation) were used for continuous variables. Data normality was assessed using the D'Agostino–Pearson test and further verified through histogram distributions. To analyze the distribution of the three most frequent somatotypes by sex, the chi-square test (χ^2) was applied along with the Marascuilo procedure to identify group differences.

As the assumption of normal distribution was not met, the Kruskal–Wallis test and Dunn's post hoc test with Bonferroni correction were used to compare sport macro-categories. Results were reported as medians and interquartile ranges (Q1–Q3). Outliers were identified using Tukey's method [50]. A value was classified as outside if it was lower than the first quartile minus 1.5 times the interquartile range (IQR) or higher than the third quartile plus 1.5 $\times$ IQR (inner fences). Far-out values were defined as those lower than the first quartile minus 3 $\times$ IQR or higher than the third quartile plus 3 $\times$ IQR (outer fences). Quantitative variables were treated as continuous for both descriptive and inferential statistical analyses. No recategorization or dichotomization of these variables was performed.

As no comparative hypotheses were formulated, the results are presented without adjustments and represent the main estimates required by the STROBE guidelines [39].

For all statistical tests, the significance level was set at $p < 0.05$. Statistical analyses were performed using NCSS 8 software (version 8.0.24, Kaysville, UT, USA) [51].

3. Results

3.1. Selected Athletes

Out of 43 sports disciplines, 1224 Mexican athletes agreed to participate and completed the anthropometric assessment. Two disciplines were excluded: chess (n = 11 male players; n = 8 female players) and esports (n = 27 male players), as they do not involve physical-athletic demands suitable for morphological evaluation. Consequently, 1178 athletes proceeded to the selection process; of these, 42 were excluded due to incomplete records for the analyzed variables, and 119 were removed due to measurement errors confirmed through dispersion analysis. A total of 302 outliers were identified using the interquartile range (IQR) method described in the Materials and Methods section, of which 174 were retained, as they represented extreme somatotypes characteristic of their respective sports disciplines (American football, powerlifting, and volleyball). The final sample consisted of 889 Mexican athletes (477 males and 412 females) (Figure 1).

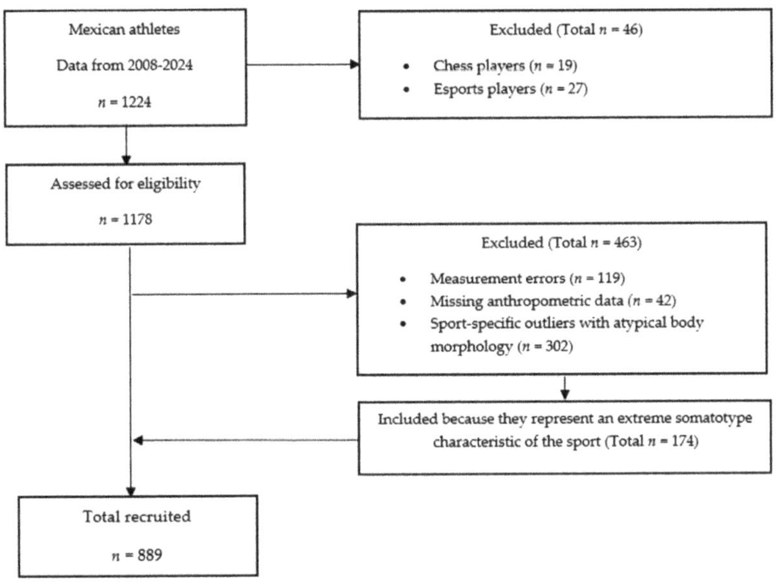

Figure 1. Flow diagram for Mexican athlete eligibility criteria.

3.2. Descriptive Data and Main Results

A total of 889 Mexican athletes (477 males and 412 females) from 43 sports disciplines across northern, central, and southern Mexico were evaluated. Table 1 shows the mean and standard deviation of body composition parameters and somatotype components by discipline and, when applicable, by playing position for male athletes. Similarly, Table 2 shows the corresponding data for female athletes.

Table 1. Descriptive characteristics and somatotype ratings of male athletes by sport.

Sport	*n*	Age	Body Mass (kg)	Height (cm)	BMI (kg/m^2)	%BF (Lean et al.) [45]	ENDO	MESO	ECTO
American football, defensive back	1	22.7	76.4	182.0	23.1	9.9	2.2	4.3	2.8
American football, defensive end	14	22.8 ± 1.6	110.4 ± 16.2	184.8 ± 3.5	32.4 ± 5.3	21.6 ± 10.9	4.9 ± 1.9	7.0 ± 1.6	0.7 ± 0.6
American football, linemen	31	22.3 ± 1.6	94.3 ± 22.7	178.6 ± 5.8	29.4 ± 5.8	16.6 ± 7.4	4.1 ± 1.9	6.4 ± 1.5	1.1 ± 0.8
American football, linebacker	15	22.2 ± 2.1	93.7 ± 6.1	178.9 ± 4.5	29.2 ± 1.4	15.7 ± 4.8	4.2 ± 0.8	6.8 ± 1.0	0.6 ± 0.4
American football, quarterback	5	22.5 ± 1.9	93.6 ± 7.5	184.8 ± 4.0	27.4 ± 1.4	15.0 ± 4.4	3.3 ± 0.8	6.2 ± 0.3	1.3 ± 0.3
American football, running back	11	22.0 ± 1.9	80.6 ± 9.2	171.2 ± 6.4	27.5 ± 2.3	12.2 ± 3.0	3.2 ± 0.7	7.0 ± 0.9	0.8 ± 0.6
American football, safety	1	20.8	87.4	177.0	27.9	17.1	4.4	5.3	0.8
American football, wide receiver	11	23.3 ± 1.6	85.0 ± 16.1	178.7 ± 5.5	26.6 ± 4.4	13.7 ± 5.7	3.1 ± 1.5	5.7 ± 1.5	1.5 ± 0.9
Archery	1	21.0	89.1	180.0	27.5	14.8	3.5	6.2	1.0
Baseball	1	21.0	71.5	170.0	24.7	15.2	3.2	6.6	1.4
Baseball, catcher	1	21.0	92.7	182.0	28.0	17.2	4.3	5.6	1.0
Baseball, center fielder	1	19.0	59.7	168.0	21.2	7.3	1.8	4.8	2.9
Baseball, infielder	1	17.0	64.8	175.0	21.2	6.7	1.5	4.6	3.3
Baseball, pitcher	8	19.6 ± 1.9	77.9 ± 9.5	177.9 ± 7.5	24.6 ± 2.7	15.9 ± 4.9	3.5 ± 1.0	4.6 ± 1.0	2.0 ± 1.2
Baseball, second baseman	1	23.0	76.5	173.0	25.6	13.4	2.9	6.4	1.3
Basketball	25	20.9 ± 1.2	87.0 ± 10.1	186.5 ± 6.9	25.0 ± 2.2	11.5 ± 3.9	2.6 ± 0.9	5.1 ± 0.9	2.3 ± 0.9
Basketball, center	2	20.5 ± 0.7	83.9 ± 1.3	190.0 ± 1.4	23.2 ± 0.0	7.5 ± 0.7	1.8 ± 0.1	4.3 ± 0.4	3.1 ± 0.1
Basketball, forward	4	19.5 ± 0.6	79.2 ± 6.4	185.0 ± 3.5	23.2 ± 1.3	9.4 ± 2.2	2.3 ± 0.7	4.4 ± 0.8	3.0 ± 0.6
Basketball, point guard	3	19.7 ± 3.8	74.1 ± 3.0	177.0 ± 4.4	23.6 ± 0.6	9.5 ± 0.4	2.1 ± 0.1	4.9 ± 0.8	2.3 ± 0.5
Beach volleyball	3	20.3 ± 1.5	77.5 ± 14.8	184.0 ± 6.1	22.8 ± 2.9	11.7 ± 6.0	2.8 ± 1.3	4.0 ± 0.9	3.1 ± 1.0
Boxing	14	20.4 ± 2.0	70.4 ± 11.9	171.4 ± 6.9	23.9 ± 3.3	12.2 ± 3.1	2.9 ± 1.0	5.6 ± 1.6	2.0 ± 1.3
Boxing < 63 kg	1	22.0	64.3	174.0	21.2	11.7	2.1	4.6	3.2
Boxing < 69 kg	1	19.0	70.2	174.0	23.2	9.4	2.0	4.5	2.3
Boxing < 75 kg	1	18.0	76.9	175.0	25.1	14.7	3.3	5.8	1.5
Boxing > 91 kg	1	19.0	94.3	184.0	27.9	15.5	4.4	6.1	1.1

Table 1. *Cont.*

Sport	n	Age	Body Mass (kg)	Height (cm)	BMI (kg/m^2)	%BF (Lean et al.) [45]	ENDO	MESO	ECTO
Fencing, épée	2	18.5 ± 0.7	73.4 ± 32.7	178.0 ± 12.7	22.6 ± 7.1	15.6 ± 14.8	3.5 ± 3.2	4.0 ± 1.8	3.1 ± 2.5
Fencing, foil	3	20.7 ± 2.1	62.0 ± 5.5	167.7 ± 2.1	22.1 ± 2.4	12.6 ± 1.7	3.1 ± 0.4	4.8 ± 1.1	2.5 ± 1.2
Flag football	4	22.2 ± 1.5	81.8 ± 5.2	182.0 ± 4.1	24.6 ± 0.9	13.6 ± 2.4	2.8 ± 0.4	4.8 ± 0.4	2.1 ± 0.4
Freestyle wrestling	1	21.0	91.8	182.0	27.7	13.1	4.1	5.9	1.0
Freestyle wrestling < 74 kg	1	24.0	75.7	168.0	26.8	12.3	3.0	6.2	0.8
Greco-Roman wrestling	1	17.0	68.1	167.0	24.4	9.7	2.7	5.4	1.4
Greco-Roman wrestling < 60 kg	1	21.0	63.2	162.0	24.1	10.1	2.3	5.7	1.2
Greco-Roman wrestling < 63 kg	1	21.0	64.7	167.0	23.2	7.6	1.9	5.7	1.9
Greco-Roman wrestling < 82 kg	1	23.0	83.7	169.0	29.3	14.1	3.9	6.9	0.3
Gymnastics	1	21.0	62.5	166.0	22.7	7.3	2.1	6.1	2.0
Half marathon, 21 km	2	21.0 ± 1.4	72.7 ± 12.2	177.0 ± 5.7	23.1 ± 2.4	9.2 ± 2.5	2.3 ± 0.6	4.5 ± 0.3	2.5 ± 0.8
Handball	14	20.1 ± 1.1	76.9 ± 10.9	175.5 ± 6.8	24.9 ± 2.9	10.7 ± 2.8	2.9 ± 0.9	6.0 ± 1.6	1.8 ± 1.1
Handball, back	1	23.0	93.2	184.0	27.5	17.0	3.1	6.2	1.2
Handball, center	3	19.7 ± 1.5	70.8 ± 11.0	170.3 ± 3.8	24.3 ± 2.9	13.0 ± 2.8	2.9 ± 0.7	5.5 ± 2.0	1.7 ± 0.9
Handball, goalkeeper	4	21.2 ± 1.0	81.4 ± 12.4	180.8 ± 5.2	24.9 ± 3.0	16.3 ± 3.0	3.4 ± 0.6	5.0 ± 1.1	2.1 ± 1.2
Handball, left back	1	18.0	78.6	186.0	22.7	9.3	2.2	4.7	3.2
Handball, left wing	1	21.0	70.2	164.0	26.1	6.8	1.7	7.1	0.8
Handball, line player	1	23.0	93.5	174.0	30.9	16.6	4.3	8.5	0.1
Handball, right back	2	20.5 ± 3.5	88.0 ± 0.0	182.5 ± 0.7	26.6 ± 0.0	16.8 ± 2.0	3.4 ± 0.2	6.0 ± 0.3	1.4 ± 0.1
Handball, right wing	2	21.5 ± 0.7	69.8 ± 2.0	173.5 ± 0.7	23.2 ± 0.4	11.6 ± 0.9	2.7 ± 0.0	5.0 ± 0.1	2.3 ± 0.1
Handball, wing	1	22.0	71.2	165.0	26.2	8.4	1.9	6.7	0.8
High jump	1	20.4	70.0	178.4	22.1	12.1	2.3	4.1	3.1
Indoor soccer	5	20.4 ± 2.6	71.3 ± 17.8	170.1 ± 9.9	24.3 ± 3.9	14.0 ± 5.5	3.5 ± 1.5	5.3 ± 0.9	1.8 ± 1.0
Indoor soccer, defender	5	19.8 ± 3.0	79.8 ± 14.0	176.0 ± 4.3	25.8 ± 4.6	15.4 ± 8.4	3.8 ± 1.8	5.0 ± 1.7	1.7 ± 1.6
Indoor soccer, forward	1	18.0	83.8	175.0	27.4	12.2	4.0	5.7	0.9
Indoor soccer, goalkeeper	4	21.0 ± 1.8	69.7 ± 9.5	174.2 ± 7.8	22.9 ± 2.8	16.1 ± 8.6	3.6 ± 1.8	4.2 ± 1.4	2.5 ± 1.2

Table 1. *Cont.*

Sport	*n*	Age	Body Mass (kg)	Height (cm)	BMI (kg/m^2)	%BF (Lean et al.) [45]	ENDO	MESO	ECTO
Indoor soccer, midfielder	5	20.2 ± 1.6	70.2 ± 5.2	170.4 ± 5.2	24.2 ± 2.3	14.8 ± 4.5	3.5 ± 1.1	5.6 ± 1.1	1.8 ± 1.0
Javelin throw	2	19.0 ± 1.4	79.2 ± 8.8	175.0 ± 0.0	25.9 ± 2.9	11.7 ± 1.8	2.5 ± 0.3	6.2 ± 0.7	1.4 ± 0.9
Judo	7	19.6 ± 2.1	65.2 ± 8.5	167.0 ± 6.3	23.4 ± 2.3	11.6 ± 5.3	2.9 ± 1.4	5.7 ± 1.2	1.9 ± 0.8
Judo < 100 kg	1	18.0	97.2	170.0	33.6	17.3	5.5	8.8	0.1
Judo < 55 kg	1	20.0	53.5	164.0	19.9	11.2	2.6	4.3	3.3
Judo < 73 kg	1	19.0	73.6	167.0	26.4	15.9	4.7	5.8	0.8
Judo < 81 kg	1	17.0	81.2	164.0	30.2	10.3	3.2	8.1	0.1
Karate	4	19.5 ± 1.7	69.5 ± 14.9	174.8 ± 6.0	22.6 ± 4.1	12.2 ± 4.6	2.8 ± 0.7	4.5 ± 1.5	2.8 ± 1.6
Karate, kata	2	17.0 ± 0.0	56.4 ± 3.8	162.5 ± 0.7	21.4 ± 1.2	7.2 ± 2.4	1.9 ± 0.3	5.1 ± 0.1	2.5 ± 0.6
Karate, kumite	6	19.8 ± 1.5	72.3 ± 7.0	174.8 ± 5.2	23.7 ± 2.3	14.1 ± 3.9	3.2 ± 0.7	4.9 ± 1.6	2.2 ± 1.1
Kickboxing, low kick	6	19.7 ± 2.0	64.6 ± 7.6	171.2 ± 4.0	22.1 ± 2.4	10.9 ± 3.4	2.6 ± 0.9	4.4 ± 0.5	2.7 ± 1.1
Kickboxing, point fighting	2	18.0 ± 1.4	76.6 ± 10.6	175.0 ± 7.1	24.9 ± 1.5	15.9 ± 7.8	4.5 ± 0.9	5.5 ± 0.1	1.6 ± 0.1
Long jump	1	20.0	70.1	173.0	23.4	10.4	2.5	5.3	2.1
Olympic wrestling	19	20.9 ± 1.4	73.0 ± 13.9	170.0 ± 6.7	25.1 ± 3.4	11.9 ± 4.4	3.2 ± 1.3	5.9 ± 1.2	1.5 ± 0.9
Olympic wrestling < 65 kg	1	19.0	66.3	168.0	23.5	9.9	3.1	5.4	1.8
Padel	4	21.2 ± 1.7	74.2 ± 10.0	175.0 ± 2.5	24.2 ± 3.2	15.3 ± 1.0	3.5 ± 0.4	4.3 ± 1.3	2.1 ± 1.1
Padel, doubles	1	19.0	77.6	176.0	25.1	18.1	4.0	5.7	1.6
Pentathlon	1	23.5	76.2	184.5	22.5	6.8	1.6	4.9	3.3
Powerlifting < 90 kg	1	25.0	80.1	170.0	27.7	17.8	4.7	6.5	0.6
Powerlifting < 125 kg	1	32.0	128.4	179.0	40.1	34.7	5.6	10.2	0.1
Powerlifting < 100 kg	1	31.0	100.3	166.0	36.4	26.4	6.3	8.7	0.1
Powerlifting < 140 kg	1	18.0	131.6	190.0	36.5	27.7	7.3	7.5	0.1
Racewalking	1	19.0	59.5	168.0	21.1	7.4	2.0	3.6	2.9
Rugby	16	19.9 ± 1.9	78.6 ± 9.4	172.6 ± 4.8	26.5 ± 3.2	14.5 ± 6.5	3.6 ± 1.2	6.5 ± 1.5	1.3 ± 1.1
Rugby, center	2	21.0 ± 1.4	77.5 ± 3.6	174.0 ± 2.8	25.6 ± 0.3	16.9 ± 0.9	3.4 ± 0.1	6.0 ± 0.3	1.3 ± 0.0
Rugby, fly-half	1	20.0	63.1	170.0	21.8	12.1	3.0	4.3	2.7

Table 1. *Cont.*

Sport	*n*	Age	Body Mass (kg)	Height (cm)	BMI (kg/m^2)	%BF (Lean et al.) [45]	ENDO	MESO	ECTO
Rugby, hooker	3	20.0 ± 1.0	84.6 ± 7.3	168.3 ± 2.1	29.8 ± 1.9	12.6 ± 4.2	3.6 ± 1.1	8.7 ± 0.9	0.2 ± 0.2
Rugby, prop	2	22.0 ± 1.4	83.8 ± 5.2	176.2 ± 3.9	27.1 ± 0.3	13.2 ± 4.8	2.9 ± 0.3	7.0 ± 0.8	1.0 ± 0.0
Rugby, scrum-half	1	19.0	74.0	180.0	22.8	5.3	1.5	4.0	2.8
Rugby, wing	1	21.0	66.2	171.0	22.6	5.1	1.4	5.8	2.4
Soccer	43	21.2 ± 1.7	73.6 ± 7.2	175.9 ± 5.3	23.8 ± 1.6	11.6 ± 3.5	2.6 ± 0.8	5.2 ± 0.8	2.2 ± 0.7
Soccer, defender	6	21.0 ± 2.0	72.8 ± 5.0	173.7 ± 5.7	24.2 ± 2.2	11.0 ± 3.1	2.5 ± 0.5	5.7 ± 1.2	1.9 ± 1.1
Soccer, forward	7	20.3 ± 1.8	68.0 ± 11.0	170.3 ± 8.0	23.3 ± 1.9	9.1 ± 3.2	2.5 ± 0.7	5.3 ± 0.7	2.1 ± 0.5
Soccer, goalkeeper	1	20.0	81.7	173.0	27.3	12.7	3.7	7.4	0.8
Soccer, midfielder	6	21.0 ± 1.3	68.0 ± 9.7	171.9 ± 7.9	22.9 ± 1.7	10.3 ± 2.9	2.4 ± 0.6	5.3 ± 0.9	2.3 ± 0.7
Sport climbing	13	19.5 ± 1.3	62.9 ± 7.4	170.8 ± 6.7	21.5 ± 2.1	9.6 ± 4.6	2.3 ± 0.8	4.5 ± 1.2	2.9 ± 1.2
Sprint	9	20.8 ± 1.7	70.6 ± 9.4	179.5 ± 8.3	21.8 ± 1.4	5.6 ± 0.9	1.3 ± 0.3	4.4 ± 0.5	3.3 ± 0.7
Sprint, 100 m	3	22.3 ± 0.6	76.0 ± 9.9	173.3 ± 2.9	25.2 ± 2.5	8.1 ± 0.4	2.0 ± 0.2	6.3 ± 1.4	1.5 ± 0.8
Sprint, 200 m	1	23.0	73.5	181.0	22.4	5.6	1.1	4.5	3.0
Sprint, 300 m hurdles	1	22.0	59.7	174.0	19.7	6.7	1.4	3.9	4.0
Sprint, 400 m	2	20.0 ± 2.8	77.3 ± 6.7	186.0 ± 2.8	22.4 ± 2.6	8.8 ± 0.6	2.3 ± 0.0	4.2 ± 0.9	3.4 ± 1.4
Table tennis	9	19.8 ± 1.4	69.5 ± 5.9	173.8 ± 4.4	23.1 ± 2.4	12.0 ± 4.6	3.1 ± 1.0	4.2 ± 1.4	2.5 ± 1.1
Taekwondo	6	18.3 ± 1.4	67.4 ± 8.4	174.2 ± 4.8	22.1 ± 1.7	9.6 ± 3.5	2.4 ± 0.8	4.6 ± 0.7	2.8 ± 0.6
Taekwondo < 74 kg	1	24.0	71.9	174.0	23.7	14.0	2.5	4.9	2.1
Team roping, heeler	1	21.0	82.4	184.5	24.3	26.1	5.5	4.6	2.5
Track and field, 4 × 100 m relay	2	20.5 ± 2.1	72.8 ± 5.6	175.0 ± 1.4	23.8 ± 1.5	8.9 ± 0.8	2.0 ± 0.3	4.8 ± 0.5	2.1 ± 0.6
Track and field, 4 × 400 m relay	3	22.0 ± 2.6	74.0 ± 7.1	182.7 ± 7.1	22.3 ± 3.7	8.5 ± 1.2	1.8 ± 0.2	3.9 ± 1.9	3.4 ± 2.0
Track and field long-distance	3	22.4 ± 2.9	58.9 ± 4.4	171.1 ± 2.5	20.2 ± 2.1	7.8 ± 2.4	1.6 ± 0.3	3.9 ± 1.0	3.6 ± 1.2
Track and field middle-distance	1	21.0	59.1	164.0	22.0	7.3	1.8	5.6	2.2
Track cycling	2	20.0 ± 1.4	80.2 ± 10.0	175.1 ± 9.8	26.1 ± 0.3	11.2 ± 8.7	2.6 ± 2.5	5.3 ± 0.1	1.2 ± 0.3
Triathlon	6	19.8 ± 1.3	69.5 ± 6.8	171.7 ± 6.5	23.6 ± 2.1	15.6 ± 5.2	3.5 ± 1.1	4.9 ± 1.1	2.1 ± 0.9

Table 1. *Cont.*

Sport	n	Age	Body Mass (kg)	Height (cm)	BMI (kg/m^2)	%BF (Lean et al.) [45]	ENDO	MESO	ECTO
Triple jump	2	22.5 ± 0.7	79.0 ± 1.8	183.9 ± 4.1	23.3 ± 0.6	10.6 ± 2.1	1.7 ± 0.3	4.8 ± 0.9	2.8 ± 0.4
Volleyball	12	20.7 ± 1.8	75.7 ± 10.2	186.3 ± 10.7	22.0 ± 3.5	9.7 ± 4.1	2.3 ± 1.3	3.8 ± 1.9	3.8 ± 2.1
Volleyball, center	1	19.0	73.5	200.0	18.4	6.1	1.2	1.3	6.4
Volleyball, libero	2	21.0 ± 2.8	69.2 ± 7.0	169.5 ± 3.5	24.0 ± 1.4	8.9 ± 3.7	2.4 ± 0.9	5.8 ± 0.9	1.6 ± 0.3
Volleyball, middle blocker	4	21.0 ± 0.8	91.6 ± 17.4	190.8 ± 4.9	25.1 ± 4.7	14.6 ± 2.9	3.4 ± 1.1	4.6 ± 1.9	2.6 ± 2.0
Volleyball, opposite hitter	1	23.0	85.9	200.0	21.5	12.9	2.2	2.3	4.6
Volleyball, outside hitter	5	22.4 ± 1.1	81.5 ± 9.6	179.7 ± 2.6	25.2 ± 2.5	10.7 ± 3.4	2.1 ± 0.7	6.2 ± 0.9	1.9 ± 1.0
Volleyball, setter	1	21.0	71.3	171.0	24.4	5.2	1.9	5.5	1.6
Weightlifting	2	20.0 ± 0.0	83.7 ± 8.0	171.5 ± 7.8	28.4 ± 0.1	13.1 ± 6.2	3.6 ± 1.3	7.0 ± 1.7	0.6 ± 0.2
Weightlifting < 67 kg	1	18.0	68.3	169.0	23.9	6.6	2.1	5.4	1.7
Weightlifting < 81 kg	1	19.0	80.8	175.0	26.4	12.2	3.4	6.3	1.1
Weightlifting < 89 kg	1	20.0	89.4	167.0	32.1	9.9	4.2	8.9	0.1
TOTAL	477								

Note. This table shows the sample size (n), the mean and standard deviation of the somatotype components (endomorphy, mesomorphy, and ectomorphy) according to the Heath–Carter method, body mass (kg), height (cm), body mass index (BMI, kg/m^2), and body fat percentage estimated using Equation (2) proposed by Lean et al. [45].

Table 2. Descriptive characteristics and somatotype ratings of female athletes by sport.

Sport	n	Age	Body Mass (kg)	Height (cm)	BMI (kg/m^2)	%BF (Lean et al.) [45]	ENDO	MESO	ECTO
Aerobic gymnastics	11	20.2 ± 2.4	56.4 ± 4.2	158.7 ± 5.3	22.4 ± 1.7	23.0 ± 2.9	3.5 ± 1.1	4.9 ± 0.5	1.8 ± 0.8
American football, left guard	1	22.0	89.2	168.1	31.6	37.9	7.9	6.6	0.1
American football, quarterback	1	24.0	62.4	169.6	21.8	23.1	2.6	4.0	2.7
Archery	1	17.0	44.0	159.0	17.4	16.1	2.5	2.7	4.4
Basketball	22	20.4 ± 1.8	69.0 ± 12.6	172.6 ± 7.6	22.8 ± 2.9	24.4 ± 4.0	3.8 ± 1.3	3.6 ± 1.0	2.4 ± 1.0
Basketball, center	2	22.0 ± 0.0	78.8 ± 0.8	183.0 ± 0.0	23.6 ± 0.2	27.8 ± 0.5	4.5 ± 0.2	2.4 ± 0.4	2.6 ± 0.1
Basketball, forward	1	25.0	66.3	167.0	23.8	27.0	3.6	4.0	1.6

Table 2. *Cont.*

Sport	n	Age	Body Mass (kg)	Height (cm)	BMI (kg/m^2)	%BF (Lean et al.) [45]	ENDO	MESO	ECTO
Basketball, point guard	1	20.0	56.2	160.0	22.0	22.8	3.6	3.6	2.0
Beach volleyball	4	20.5 ± 1.7	61.2 ± 7.1	166.8 ± 2.7	22.1 ± 2.6	22.8 ± 3.4	3.8 ± 1.7	3.4 ± 0.8	2.5 ± 1.3
Beach volleyball, all-round player	1	22.0	67.8	162.0	25.8	27.1	4.6	3.9	0.8
Beach volleyball, blocker	2	21.0 ± 0.0	67.2 ± 0.3	173.0 ± 0.0	22.5 ± 0.1	24.2 ± 0.1	3.6 ± 0.0	3.1 ± 0.6	2.5 ± 0.1
Beach volleyball, defender	3	21.3 ± 1.1	58.5 ± 4.1	163.3 ± 1.1	21.9 ± 1.2	23.4 ± 1.9	4.7 ± 0.4	3.2 ± 0.3	2.2 ± 0.5
Boxing	7	21.0 ± 1.6	60.2 ± 6.6	161.5 ± 4.0	23.0 ± 2.2	26.4 ± 3.9	4.8 ± 1.4	4.2 ± 1.2	1.7 ± 1.0
Discus throw	1	21.0	80.0	165.0	29.4	30.5	5.2	6.4	0.1
Fencing, foil	1	22.0	63.1	163.3	23.7	31.1	6.2	4.7	1.4
Fencing, sabre	1	19.0	55.5	161.0	21.4	24.3	4.3	3.5	2.3
Fencing, épée	1	18.0	67.9	178.0	21.4	25.8	5.0	2.3	3.4
Flag football	11	20.0 ± 2.0	58.1 ± 7.3	160.2 ± 6.5	22.7 ± 1.9	24.7 ± 2.5	4.1 ± 0.9	4.3 ± 1.2	1.8 ± 0.8
Flag football, cornerback	3	17.3 ± 1.5	50.3 ± 0.1	159.7 ± 2.3	19.8 ± 0.6	19.4 ± 1.1	2.8 ± 0.0	3.2 ± 0.9	3.1 ± 0.4
Flag football, quarterback	3	23.0 ± 0.0	67.6 ± 14.3	164.7 ± 8.1	24.7 ± 2.9	30.3 ± 3.6	5.6 ± 0.6	5.1 ± 0.5	1.3 ± 0.6
Flag football, safety	2	22.5 ± 0.7	70.8 ± 8.0	162.2 ± 3.2	26.9 ± 2.0	29.8 ± 4.7	5.0 ± 0.3	6.5 ± 1.0	0.6 ± 0.3
Flag football, wide receiver	10	19.7 ± 1.3	59.2 ± 4.4	157.9 ± 6.1	23.8 ± 1.9	25.8 ± 2.7	4.3 ± 0.5	5.1 ± 1.0	1.3 ± 0.8
Gymnastics	1	21.0	60.4	162.0	23.0	23.5	3.7	4.9	1.6
Half marathon, 21 km	2	20.5 ± 3.5	49.3 ± 3.2	157.9 ± 0.8	19.9 ± 1.5	21.4 ± 1.1	3.3 ± 0.4	3.4 ± 1.7	3.0 ± 0.8
Hammer throw	1	19.0	96.7	165.0	35.5	43.1	8.8	8.6	0.1
Handball	13	20.6 ± 2.0	64.2 ± 6.5	163.2 ± 5.7	24.1 ± 2.1	26.7 ± 4.2	4.7 ± 1.3	4.8 ± 1.4	1.4 ± 0.7
Handball, back	3	20.3 ± 0.6	60.9 ± 10.3	162.0 ± 11.4	23.0 ± 0.6	24.5 ± 1.2	4.1 ± 0.5	4.3 ± 0.6	1.6 ± 0.5
Handball, center	3	20.3 ± 0.6	56.7 ± 1.1	153.3 ± 4.9	24.2 ± 1.2	27.2 ± 3.1	4.8 ± 1.4	5.3 ± 0.3	0.9 ± 0.6
Handball, goalkeeper	2	21.5 ± 2.1	66.5 ± 11.0	164.5 ± 2.1	24.6 ± 4.7	29.8 ± 6.0	6.2 ± 1.9	4.2 ± 1.5	1.5 ± 1.7
Handball, lateral	1	19.0	69.2	165.0	25.4	26.0	3.7	5.5	1.0
Handball, left back	1	20.0	70.0	165.0	25.7	28.3	4.2	4.7	0.9
Handball, left wing	3	20.3 ± 2.1	63.8 ± 2.8	158.3 ± 4.6	25.4 ± 1.0	26.7 ± 1.3	5.0 ± 0.8	5.1 ± 0.6	0.7 ± 0.4

Table 2. *Cont.*

Sport	*n*	Age	Body Mass (kg)	Height (cm)	BMI (kg/m^2)	%BF (Lean et al.) [45]	ENDO	MESO	ECTO
Handball, pivot	2	20.5 ± 0.7	73.0 ± 8.6	161.0 ± 1.4	28.1 ± 2.8	32.4 ± 4.2	5.7 ± 1.7	7.4 ± 1.0	0.3 ± 0.3
Handball, right wing	1	20.0	58.8	160.0	23.0	25.7	4.5	4.7	1.5
Handball, wing	1	17.0	63.6	162.7	23.9	27.7	5.6	5.3	1.3
Heptathlon	1	22.0	54.1	156.5	22.2	22.2	2.5	3.9	1.7
Indoor soccer	4	20.5 ± 1.0	59.4 ± 2.9	161.5 ± 1.3	22.8 ± 1.1	26.6 ± 3.1	4.6 ± 0.9	4.3 ± 0.8	1.7 ± 0.5
Indoor soccer, defender	4	20.8 ± 2.2	56.5 ± 3.4	161.5 ± 6.8	21.8 ± 2.3	23.9 ± 4.0	4.2 ± 1.6	4.0 ± 1.2	2.3 ± 1.3
Indoor soccer, forward	5	19.6 ± 1.5	53.3 ± 3.6	159.8 ± 3.8	20.8 ± 1.2	22.4 ± 2.7	3.6 ± 0.9	3.7 ± 1.0	2.5 ± 0.7
Indoor soccer, goalkeeper	4	20.0 ± 1.8	64.3 ± 1.6	158.8 ± 5.7	25.6 ± 1.4	30.2 ± 1.4	5.9 ± 0.7	4.9 ± 0.9	0.8 ± 0.7
Indoor soccer, midfielder	5	20.8 ± 2.2	53.9 ± 4.6	154.8 ± 4.1	22.6 ± 2.6	24.4 ± 3.5	4.9 ± 1.0	4.2 ± 1.2	1.6 ± 1.1
Javelin throw	1	22.0	76.2	163.5	28.7	29.2	4.9	5.7	0.2
Judo	6	19.8 ± 2.6	55.4 ± 5.7	161.2 ± 6.2	21.3 ± 1.7	21.9 ± 4.0	3.2 ± 1.1	4.3 ± 0.8	2.4 ± 1.0
Judo < 44 kg	1	20.0	44.1	145.0	21.0	23.6	5.2	4.5	1.5
Judo < 48 kg	1	20.0	48.7	152.0	21.1	23.5	4.5	4.6	1.9
Judo < 57 kg	2	20.5 ± 0.7	56.9 ± 3.5	160.5 ± 2.1	22.1 ± 2.0	23.0 ± 3.1	3.2 ± 0.5	4.1 ± 0.4	2.0 ± 1.0
Karate	8	18.5 ± 0.9	58.8 ± 11.4	158.9 ± 8.3	23.1 ± 2.6	27.0 ± 4.2	5.0 ± 1.2	4.4 ± 0.8	1.6 ± 0.8
Karate, kata	2	17.5 ± 0.7	51.0 ± 2.9	157.5 ± 5.0	20.6 ± 0.1	21.4 ± 1.4	3.8 ± 0.6	3.2 ± 0.1	2.5 ± 0.4
Karate, kumite	4	18.5 ± 1.0	59.1 ± 4.9	158.2 ± 4.6	23.6 ± 1.2	25.3 ± 1.4	4.3 ± 1.2	4.5 ± 1.0	1.3 ± 0.5
Kickboxing, low kick	3	19.3 ± 1.1	57.8 ± 3.0	163.0 ± 5.3	21.8 ± 0.4	25.7 ± 0.6	4.6 ± 0.5	3.4 ± 0.7	2.3 ± 0.5
Long jump	4	21.4 ± 1.4	62.4 ± 5.7	170.7 ± 7.9	21.4 ± 0.5	23.6 ± 1.3	3.4 ± 0.7	3.1 ± 0.7	2.9 ± 0.6
Olympic wrestling	17	18.9 ± 1.5	57.2 ± 8.5	156.7 ± 6.1	23.2 ± 2.3	23.9 ± 3.8	4.4 ± 1.5	4.8 ± 0.9	1.4 ± 0.7
Olympic wrestling < 53 kg	1	19.0	53.0	157.0	21.5	21.6	3.0	4.6	2.0
Padel, doubles, backhand player	1	21.0	59.2	154.0	25.0	27.7	4.3	5.2	0.7
Padel, doubles, right-handed player	1	23.0	57.2	164.0	21.3	25.5	3.5	3.4	2.6
Pole vault	1	23.0	60.9	164.0	22.6	23.6	2.6	4.8	1.9
Powerlifting < 44 kg	1	26.0	40.3	150.9	17.9	20.7	3.4	2.0	3.6

Table 2. *Cont.*

Sport	n	Age	Body Mass (kg)	Height (cm)	BMI (kg/m^2)	%BF (Lean et al.) [45]	ENDO	MESO	ECTO
Powerlifting < 90 kg	1	17.0	88.6	167.0	31.8	37.8	7.5	6.6	0.1
Racewalking, 20 km	1	18.0	48.0	157.0	19.5	22.8	5.5	3.3	3.0
Road cycling	1	17.0	67.1	164.0	24.9	24.6	4.0	4.1	1.1
Rugby	6	19.5 ± 1.4	59.4 ± 9.0	158.5 ± 3.4	23.6 ± 2.8	26.3 ± 4.5	4.8 ± 1.5	4.2 ± 1.1	1.4 ± 1.0
Rugby 7s, prop	1	18.0	73.7	165.2	27.1	26.9	5.5	5.2	0.6
Rugby, center	2	19.0 ± 1.4	54.6 ± 2.9	156.5 ± 6.4	22.4 ± 0.6	24.9 ± 2.3	4.8 ± 1.5	4.0 ± 0.7	1.6 ± 0.6
Rugby, front row	1	27.0	74.5	160.8	29.1	33.6	4.6	6.4	0.1
Rugby, inside center	1	21.0	57.9	161.0	22.3	22.9	3.1	3.6	1.9
Rugby, prop	3	22.0 ± 2.6	66.2 ± 1.1	161.7 ± 2.3	25.3 ± 0.8	28.1 ± 1.5	4.5 ± 1.4	4.8 ± 0.2	0.9 ± 0.2
Rugby, scrum-half	2	19.5 ± 0.7	56.1 ± 3.0	156.8 ± 6.7	23.0 ± 3.1	25.8 ± 8.8	4.6 ± 3.1	4.2 ± 1.5	1.6 ± 1.6
Rugby, wing	7	20.6 ± 1.0	51.7 ± 3.6	158.6 ± 6.3	20.6 ± 1.6	22.9 ± 2.9	3.6 ± 0.9	3.6 ± 1.2	2.6 ± 1.1
Shot put	1	21.0	82.5	164.6	30.3	33.4	4.7	7.7	0.1
Soccer	30	20.8 ± 1.7	56.4 ± 6.3	159.2 ± 6.4	22.3 ± 2.0	23.9 ± 2.8	3.9 ± 1.0	4.3 ± 1.4	1.9 ± 0.9
Soccer, defender	20	19.2 ± 1.3	55.4 ± 7.8	161.3 ± 5.7	21.2 ± 2.4	22.4 ± 3.8	3.9 ± 1.3	3.7 ± 0.9	2.5 ± 1.1
Soccer, forward	13	20.5 ± 2.1	59.0 ± 6.3	161.7 ± 4.2	22.6 ± 2.4	24.5 ± 3.4	4.1 ± 0.8	4.1 ± 1.4	2.0 ± 1.0
Soccer, goalkeeper	7	19.6 ± 1.0	65.1 ± 8.7	162.5 ± 8.1	24.6 ± 1.6	27.9 ± 3.7	5.3 ± 0.8	4.5 ± 0.9	1.2 ± 0.6
Soccer, midfielder	10	20.1 ± 2.4	51.0 ± 7.8	155.7 ± 6.4	20.9 ± 2.3	22.1 ± 4.0	3.4 ± 0.9	4.1 ± 1.4	2.3 ± 1.0
Softball	11	19.8 ± 1.8	60.3 ± 10.9	161.8 ± 7.1	22.9 ± 2.7	26.3 ± 5.6	4.5 ± 1.3	4.0 ± 1.5	1.9 ± 0.8
Softball, fielder	3	18.0 ± 1.0	59.5 ± 5.8	164.0 ± 6.9	22.1 ± 1.8	24.3 ± 4.8	4.6 ± 1.4	3.6 ± 0.7	2.2 ± 1.0
Softball, second base	2	19.5 ± 0.7	53.2 ± 2.0	158.0 ± 0.0	21.3 ± 0.8	21.3 ± 1.1	3.1 ± 0.1	3.7 ± 0.0	2.1 ± 0.3
Softball, shortstop	1	20.0	60.5	163.0	22.8	26.1	4.1	4.5	1.8
Sport climbing	3	22.0 ± 1.0	44.4 ± 3.7	154.0 ± 5.3	18.7 ± 1.7	20.0 ± 1.6	3.3 ± 0.7	2.9 ± 0.6	3.3 ± 1.2
Sprint	7	19.2 ± 0.8	54.9 ± 10.8	162.4 ± 9.6	20.8 ± 2.7	20.5 ± 4.1	2.8 ± 1.0	3.6 ± 1.2	2.9 ± 1.4
Sprint, 100 m	1	21.0	57.7	162.0	22.0	22.7	3.4	3.4	2.1
Sprint, 200 m	3	23.0 ± 2.6	60.5 ± 1.9	164.2 ± 2.6	22.5 ± 0.5	23.6 ± 1.1	3.8 ± 0.4	3.6 ± 0.8	2.0 ± 0.3
Sprint, 400 m	2	23.0 ± 1.4	51.9 ± 4.9	162.0 ± 1.4	19.8 ± 1.5	20.9 ± 1.0	3.0 ± 1.1	2.8 ± 0.7	3.2 ± 0.8

Table 2. *Cont.*

Sport	*n*	Age	Body Mass (kg)	Height (cm)	BMI (kg/m²)	%BF (Lean et al.) [45]	ENDO	MESO	ECTO
Sprint, 400 m hurdles	1	18.4	63.8	163.7	23.7	21.9	2.6	4.6	1.4
Table tennis	6	20.8 ± 1.4	57.4 ± 7.5	160.5 ± 10.3	22.4 ± 3.8	25.6 ± 4.4	4.7 ± 1.3	4.0 ± 1.9	2.2 ± 1.9
Taekwondo	8	20.0 ± 1.1	60.0 ± 10.2	164.7 ± 5.4	22.0 ± 2.8	24.0 ± 4.7	4.2 ± 1.5	3.8 ± 1.2	2.4 ± 1.3
Taekwondo < 46 kg	1	22.0	48.4	162.0	18.4	18.1	2.4	2.0	4.0
Taekwondo < 49 kg	1	22.0	51.9	165.0	19.1	20.5	2.7	2.6	3.8
Track and field long-distance	4	20.7 ± 1.4	53.5 ± 7.4	158.7 ± 7.2	21.3 ± 2.9	23.6 ± 3.8	4.2 ± 0.9	3.5 ± 0.9	2.4 ± 1.5
Track and field middle-distance	1	23.0	51.3	156.0	21.1	21.9	3.4	4.1	2.2
Track and field, 1500 m	1	20.0	53.9	158.0	21.6	24.0	3.8	2.7	2.0
Track and field, 10,000 m	2	20.5 ± 0.7	50.8 ± 0.0	157.0 ± 5.7	20.6 ± 1.5	24.9 ± 0.8	4.2 ± 0.2	3.6 ± 1.0	2.5 ± 1.1
Track and field, 3000 m steeplechase	1	20.0	47.1	155.0	19.6	18.5	3.1	4.9	2.8
Track and field, 4 × 100 m relay	1	20.0	59.0	162.0	22.5	22.8	3.1	3.9	1.9
Track and field, 5000 m	1	22.0	49.9	156.4	20.5	20.6	2.6	4.8	2.5
Track and field, 800 m	3	19.0 ± 0.0	53.4 ± 4.0	157.1 ± 1.9	21.6 ± 1.1	21.2 ± 2.0	3.0 ± 0.9	4.1 ± 0.2	2.0 ± 0.4
Track cycling	1	21.0	56.9	158.0	22.8	21.9	2.5	5.0	1.5
Triathlon	6	19.2 ± 1.3	58.1 ± 4.7	160.7 ± 7.2	22.5 ± 1.4	24.8 ± 2.7	4.4 ± 1.0	4.1 ± 0.6	1.8 ± 0.9
Volleyball	21	20.7 ± 1.8	70.0 ± 11.0	172.4 ± 7.8	23.5 ± 2.6	26.0 ± 3.7	4.3 ± 1.3	3.6 ± 1.1	2.2 ± 1.1
Volleyball, center	1	21.0	76.2	183.0	22.8	27.3	5.9	3.1	3.0
Volleyball, outside hitter	1	21.0	70.4	166.0	25.5	29.5	6.0	6.4	1.0
Volleyball, setter	1	19.0	66.1	167.0	23.7	24.5	3.3	4.9	1.7
Weightlifting	4	20.8 ± 2.6	55.6 ± 8.4	154.8 ± 4.1	23.2 ± 3.5	24.0 ± 4.0	3.7 ± 0.9	4.8 ± 1.6	1.4 ± 1.0
Weightlifting < 45 kg	1	22.0	47.6	155.0	19.8	18.7	2.1	4.5	2.7
Weightlifting < 55 kg	1	21.0	55.8	154.0	23.5	23.6	3.8	5.4	1.0
Weightlifting < 59 kg	2	18.5 ± 0.7	62.8 ± 2.6	152.0 ± 0.0	27.2 ± 1.1	28.6 ± 3.0	5.2 ± 1.1	6.8 ± 0.1	0.2 ± 0.1
Weightlifting < 64 kg	1	24.0	66.1	153.0	28.2	29.3	5.7	6.1	0.1
TOTAL	412								

Note. This table shows the sample size (*n*), means of somatotype components (endomorphy, mesomorphy, and ectomorphy) according to the Heath–Carter method, body mass (kg), height (cm), body mass index (BMI, kg/m²), and body fat percentage estimated using Equation 5 proposed by Lean et al. [45].

Among male athletes, the predominant somatotype was endomorphic mesomorph (52.4%), followed by balanced mesomorph (17.6%) and ectomorphic mesomorph (13.6%). Differences were found between somatotype categories ($p < 0.001$). However, no difference was observed between the proportions of balanced mesomorph and ectomorphic mesomorph ($p = 0.236$). Other somatotypes included mesomorph-endomorph (4.6%), mesomorphic ectomorph (4.2%), mesomorph-ectomorph (3.1%), central (1.9%), mesomorphic endomorph (1.3%), balanced ectomorph (1.0%), and balanced endomorph (0.2%) (Figure 2).

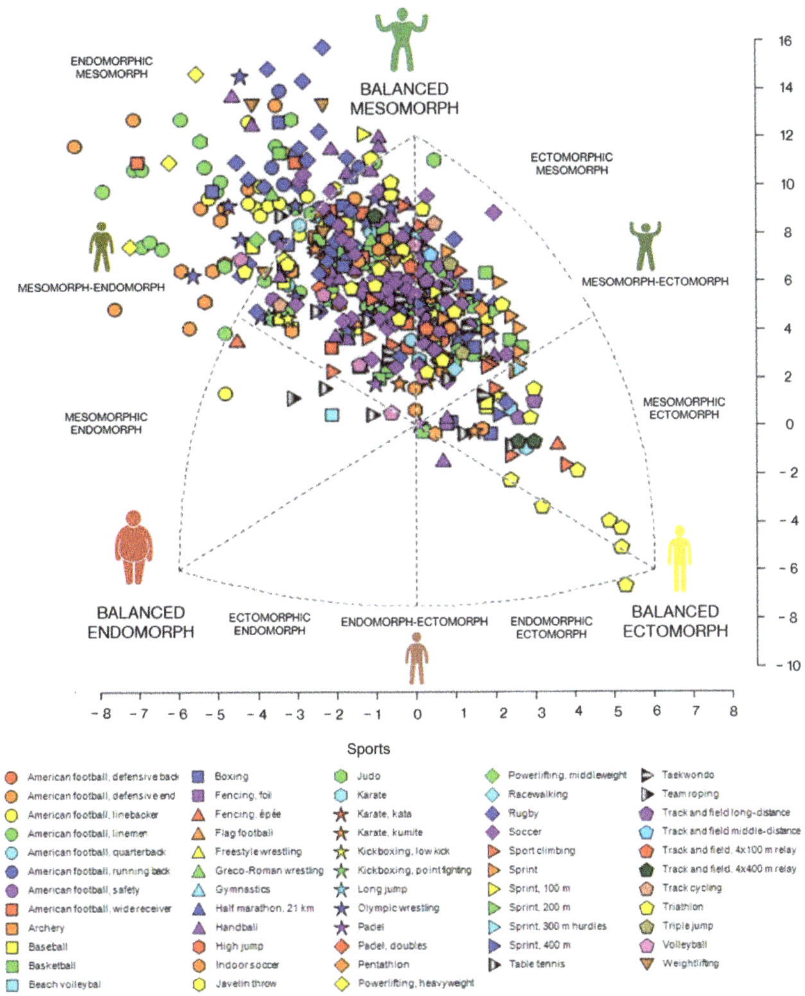

Figure 2. Somatotype of Mexican male athletes by sport ($n = 477$).

Among female athletes, no differences were found between somatotype categories ($p = 0.514$). The most frequently reported somatotypes were endomorphic mesomorph (24.5%), mesomorphic endomorph (24.0%), and mesomorph-endomorph (21.4%). Other observed somatotypes included central (8.0%), balanced endomorph (6.1%), balanced mesomorph (5.6%), balanced ectomorph (3.6%), endomorphic ectomorph (2.4%), endomorph-ectomorph (1.5%), mesomorphic ectomorph (1.2%), mesomorph-ectomorph (1.0%), ectomorphic mesomorph (0.5%), and ectomorphic endomorph (0.2%) (Figure 3).

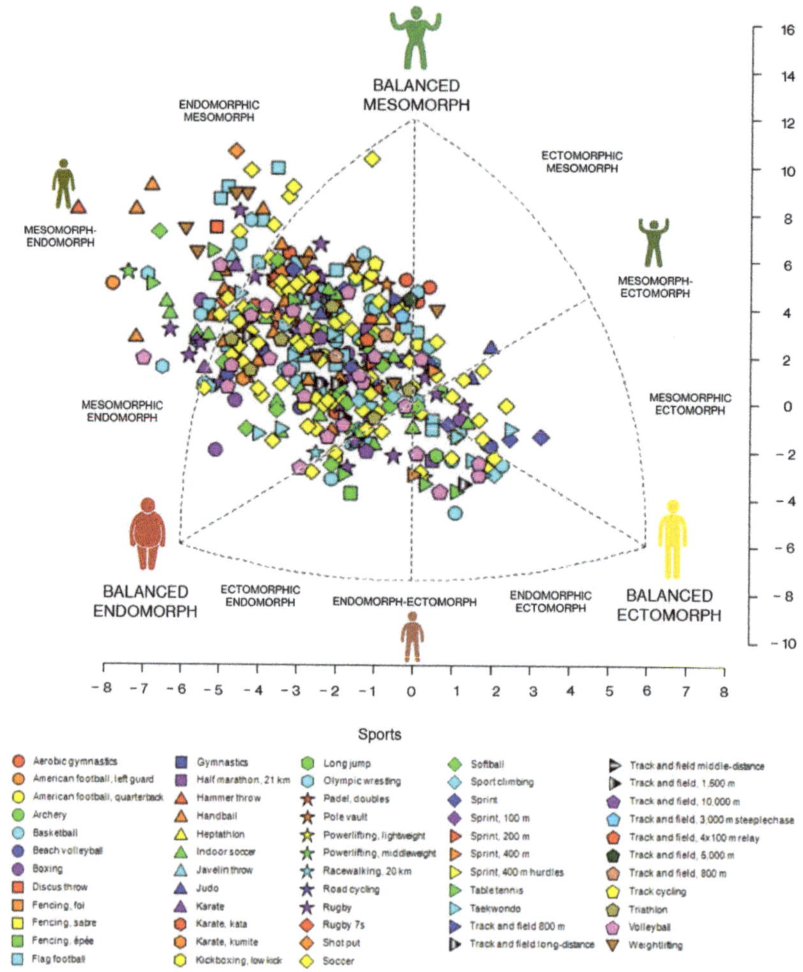

Figure 3. Somatotype of Mexican female athletes by sport ($n = 412$).

3.3. Other Analyses

Athletes were divided into six macro-categories to group sports by sex (Tables 3 and 4). Among male athletes (Table 3), differences were observed across all analyzed variables. Team sports exhibited the highest values for body mass (79.6 kg), height (178.0 cm), and body fat percentage (12.5%) compared to the other macro-categories ($p < 0.001$). The highest mesomorphy values were recorded in track and field (6.2), while the highest ectomorphy values were observed in sprint events (3.0). Endomorphy was greater in team sports (2.9) and combat sports (2.9) ($p < 0.001$).

Table 3. Descriptive table of body composition and somatotype characteristics in male athletes by sport macro-category.

Sport Macro-Category	Combat Sports (n = 87)	Endurance Events (n = 13)	Individual Sports (n = 44)	Sprint Events (n = 21)	Team Sports (n = 310)	Track and Field (n = 2)	p Value
Age	20.0 ^{a} (18.2–22.0)	20.0 (19.0–22.0)	20.0 (19.0–21.0)	22.0 (19.3–23.0)	21.0 ^{b} (20.0–22.5)	19.0 (18.0–20.0)	<0.001 **
Body mass (kg)	68.5 ^{a} (62.5–75.6)	63.9 ^{a} (59.4–74.4)	72.3 ^{a} (65.9–79.0)	73.6 (64.5–79.4)	79.6 ^{b} (71.2–88.9)	79.2 (73.0–85.4)	<0.001 **
Height (cm)	170.0 ^{a,b,c} (167.0–175.0)	170.0 ^{a,b,e} (168.0–174.0)	173.2 ^{a,b,c,e} (168.5–178.7)	176.2 ^{b,c,d,e} (174.0–185.4)	178.0 ^{d,e} (172.0–183.0)	175.0 ^{a,c,d,e} (175.0–175.0)	<0.001 **
BMI (kg/m^2)	23.5 ^{a} (21.7–25.9)	22.5 ^{a} (21.0–23.8)	23.3 (22.2–26.1)	22.4 ^{a} (20.5–24.2)	25.2 ^{b} (23.1–27.4)	25.8 (23.8–27.9)	<0.001 **
%BF (Lean et al.) [45]	11.4 ^{a} (8.8–14.6)	9.7 (7.3–17.5)	10.5 ^{a} (7.6–16.7)	6.9 ^{b} (5.6–8.4)	12.5 ^{a} (9.3–16.1)	11.6 (10.4–12.9)	<0.001 **
Endomorphy	2.8 ^{a} (2.1–3.6)	2.0 (1.8–3.7)	2.7 ^{a} (1.9–3.9)	1.6 ^{b} (1.1–2.0)	2.9 ^{a} (2.2–3.8)	2.5 (2.3–2.7)	<0.001 **
Mesomorphy	5.4 (4.5–6.2)	4.3 (3.6–5.4)	5.0 (4.0–6.1)	4.5 ^{a} (3.9–5.1)	5.5 ^{b} (4.7–6.5)	6.2 (5.7–6.7)	<0.001 **
Ectomorphy	1.9 ^{a} (1.1–2.8)	2.2 (1.9–3.1)	2.1 (1.3–3.0)	3.0 ^{b} (2.2–4.0)	1.7 ^{a} (1.0–2.5)	1.4 (0.8–2.1)	<0.001 **

Note. Data are shown as median and interquartile range (Q1–Q3). Different superscript letters (^{a,b,c,d,e}) within the same row indicate statistically significant differences between sport macro-categories according to Dunn's post hoc test with Bonferroni correction after a Kruskal–Wallis analysis ($p < 0.05$). %BF = body fat percentage; BMI = body mass index. ** $p < 0.001$.

Table 4. Descriptive table of body composition and somatotype characteristics in female athletes by sports macro-category.

Sport Macro-category	Combat Sports (n = 65)	Endurance Events (n = 22)	Individual Sports (n = 41)	Sprint Events (n = 15)	Team Sports (n = 265)	Track and Field (n = 4)	p Value
Age	19.0 ^{b} (18.0–20.2)	20.0 (19.0–21.0)	21.0 ^{a} (19.0–22.2)	20.0 (19.0–21.7)	20.0 ^{a} (19.0–21.5)	21.0 (20.0–21.5)	0.011 *
Body mass (kg)	56.1 ^{a,b,c,d,e} (51.7–62.2)	51.5 ^{a,b,c,d} (49.2–57.1)	58.1 ^{a,b,c,d,e} (52.8–60.9)	57.7 ^{a,b,c,d,e} (49.2–60.1)	59.3 ^{a,c,d,e} (54.1–66.0)	81.2 (78.1–89.6)	<0.001 **
Height (cm)	160.0 (154.8–164.2)	157.2 ^{a} (155.3–159.0)	156.5 ^{a} (154.0–164.0)	162.8 (161.2–164.2)	161.6 ^{b} (158.0–167.0)	164.8 (164.0–165.0)	<0.001 **
BMI (kg/m^2)	22.2 ^{a,b,c,d,e} (20.9–23.7)	21.0 ^{a,b,c,d} (20.2–22.2)	22.2 ^{a,b,c,d,e} (20.2–24.2)	22.0 ^{a,b,c,d,e} (19.4–22.8)	22.8 ^{a,c,d,e} (21.1–24.3)	29.8 (29.0–32.9)	<0.001 **
%BF (Lean et al.) [45]	24.8 ^{a,b,c,d,e} (21.9–27.0)	22.9 ^{a,b,c,d,e} (20.6–24.4)	23.5 ^{a,b,c,d,e} (21.1–25.4)	21.9 ^{a,b,c,d} (20.2–23.1)	25.0 ^{a,b,c,e,f} (22.4–27.4)	31.9 ^{e,f} (29.8–38.2)	<0.001 **
Endomorphy	4.3 ^{a} (3.2–5.2)	3.8 (3.1–4.5)	3.7 (2.7–4.5)	3.2 ^{b} (2.5–3.7)	4.2 ^{a} (3.3–5.0)	5.0 ^{a} (4.8–7.0)	<0.001 **
Mesomorphy	4.2 ^{a} (3.5–4.9)	4.1 ^{a} (3.3–4.4)	4.8 (3.2–5.4)	3.6 ^{a} (2.6–4.2)	4.1 ^{a} (3.3–4.9)	7.0 ^{b} (6.0–8.1)	0.002 **
Ectomorphy	1.8 ^{a} (1.2–2.3)	2.3 ^{a} (1.6–2.9)	1.7 ^{a} (0.9–2.7)	2.2 ^{a} (1.6–3.3)	1.8 ^{a} (1.1–2.6)	0.1 ^{b} (0.1–0.1)	0.002 **

Note. Data are shown as median and interquartile range (Q1–Q3). Different superscript letters (^{a,b,c,d,e}) within the same row indicate statistically significant differences between sport macro-categories according to Dunn's post hoc test with Bonferroni correction after a Kruskal–Wallis analysis ($p < 0.05$). %BF = body fat percentage; BMI = body mass index. * $p < 0.05$; ** $p < 0.001$.

Among female athletes (Table 4), differences between sport macro-categories were also identified. Athletes in track and field showed higher values for body mass (81.2 kg), height (164.8 cm), and body fat percentage (31.9%) ($p < 0.001$). The highest ectomorphy values were recorded in endurance events (2.3) ($p = 0.002$), while both mesomorphy and endomorphy were greater in track and field ($p = 0.002$; 7.0 and 5.0, respectively).

Figures S1 and S2 (Supplementary Materials) present somatocharts with athletes grouped into six macro-categories structured to facilitate the interpretation of the most predominant somatotypes among athletes. Figure S1 displays the somatochart for male athletes. In Figure S1a, corresponding to team sports, a predominance of the endomorphic mesomorph somatotype is observed ($p < 0.001$). The same somatotype also predominates in combat sports ($p < 0.001$) (Figure S1b). However, no predominant somatotype was found in individual sports ($p = 0.056$) (Figure S1c), endurance events ($p = 0.718$) (Figure S1d), or sprint events ($p = 0.544$) (Figure S1e). In track and field, only two athletes were reported, with no significance ($p = 1.000$) (Figure S1f).

Figure S2 illustrates the somatochart for female athletes, organized by macro-category and further subdivided by sport. No differences were found across any of the macro-categories among female athletes ($p = 0.050$).

4. Discussion

4.1. Key Results

The primary objective of the current study was to determine the somatotype of Mexican athletes by sex. Among male athletes, the predominant somatotype was endomorphic mesomorph (52.4%), followed by balanced mesomorph (17.6%) and ectomorphic mesomorph (13.6%), with differences between somatotype classifications. Among female athletes, the most frequently reported somatotypes were endomorphic mesomorph (24.5%), mesomorphic endomorph (24.0%), and mesomorph-endomorph (21.4%), with no differences between somatotype classifications. Male athletes exhibited greater morphological variability, whereas female athletes showed a more homogeneous distribution of morphological characteristics. Similar patterns have been reported in male international athletes [20,52,53]. However, studies on female athletes from other countries have reported a more centrally distributed somatotype with greater variability across disciplines [54–56]. This difference suggests that Mexican female athletes may exhibit lower morphological variability across disciplines, potentially influenced by ethnic and sociocultural factors or a generalized approach to their athletic training.

The second objective was to compare somatotype and body composition among sport macro-categories. In male athletes, differences were observed among all variables analyzed; team sports exhibited higher values for body mass (79.6 kg), height (178.0 cm), and %BF (12.5%). The highest mesomorphy was recorded in track and field (6.2), while the highest ectomorphy appeared in sprint events (3.0). Among female athletes, differences were also observed ($p < 0.05$); track and field athletes displayed higher body mass (81.2 kg), height (164.8 cm), and %BF (31.9%). Ectomorphy was highest in endurance events (2.3), while both mesomorphy and endomorphy reached their highest values in track and field (7.0 and 5.0, respectively). Sport macro-categories reflected specific morphological profiles: higher mesomorphy and endomorphy in team and combat sports, and higher ectomorphy in sprint and endurance events. In female athletes, high %BF and mesomorphy were observed in track and field. Grouping athletes by macro-categories facilitates the interpretation of somatotype patterns by highlighting the most suitable body prototypes for performance in each category. For example, Baranauskas et al. [6] reported higher endomorphy and mesomorphy in athletes from combat sports, while Gutnik et al. [57] and Campa et al. [58] described distinct profiles in team sports with higher mesomorphy in basketball players and higher ectomorphy in soccer players. These findings support the idea that somatotype reflects both training adaptations and the morphological demands inherent to each sport modality, a pattern clearly observed in the context of Mexican athletes.

Overall, the results of this study demonstrate that somatotype varies according to sex and type of sport, reflecting both functional adaptations and structural requirements related to body physique. These findings may be helpful for talent identification, designing individualized training programs, and implementing sport-specific nutritional strategies. Moreover, they provide a solid scientific foundation for the monitoring and physical development of high-performance athletes in Mexico. Higher mesomorphy is consistently associated with superior strength, power, and explosive performance, as seen in exercises such as the bench press, back squat, vertical jump, and sprinting. In contrast, higher ectomorphy tends to favor flexibility, aerobic capacity, and endurance, but may negatively impact strength and power outputs. Conversely, higher endomorphy generally predicts poorer performance in explosive and aerobic tasks, but may be advantageous in some strength-related activities. These relationships are evident across a range of sports and age groups, and somatotype can explain a significant portion of variance in physical fitness and sport-specific skills, making it a valuable consideration for talent identification and individualized training programs [7,17,23,59,60].

4.2. Strengths and Limitations

This study provides valuable insights into the somatotypes of Mexican athletes across various sports disciplines; however, it also reveals several limitations. First, the sample was obtained through purposive sampling and included only athletes who voluntarily agreed to participate, with representation limited to specific regions of Mexico. Second, for some sports disciplines, data were available for only a single subject. This limited representation constrains the generalizability of the findings within those particular disciplines. This limitation is primarily due to the logistical and structural challenges of recruiting athletes across multiple competitive modalities in a national context. Consequently, results in these disciplines should be interpreted with caution. These constraints reinforce the exploratory nature of the study and highlight the need for further research with larger and more balanced samples. Moreover, the sample encompassed a heterogeneous range of competitive levels, including athletes from regional competitions to world-class performers. Such variability may have attenuated the somatotype differences typically observed at specific elite levels, thereby introducing a bias toward intermediate and less sport-specific profiles. The magnitude of this potential bias may be moderate, particularly in disciplines where performance is closely linked to body morphology, as elite athletes tend to exhibit superior proprioceptive capacities, initiate training earlier, and engage in more intensive training regimens compared to their lower-level counterparts [61,62].

4.3. Interpretation

The findings of this study provide a robust characterization of somatotype and body composition among Mexican athletes, identifying relevant differences by sex and sports disciplines. These differences reflect not only physiological adaptations to training but also specific morphological demands inherent to each sport modality, consistent with findings reported in international studies, particularly among male athletes [20,52,53]. The data obtained represent a valuable foundation for understanding the predominant body profiles among Mexican athletes. They may contribute to the development of tailored strategies for assessment, selection, and planning within the national sports context.

Carter [16] demonstrated differences in somatotype components, particularly endomorphy, among male athletes practicing the same sport, such as weightlifters and wrestlers, from different nationalities. Similarly, the visual trends observed in the somatocharts of both male and female Mexican athletes align with those reported by Carter and Heath [63]. Ethnic group differences are evident, suggesting that the use of global averages based on international competitions or events [20,64] may introduce bias into data interpretation. Although regional or national variations are occasionally observed, current evidence indicates that somatotype distribution is more strongly influenced by factors such as genetics, biological maturation, training, environment, and nutrition. At the same time, nationality acts as an indirect and not necessarily determinant factor [6,57,65,66].

To the best of our knowledge, only one study conducted among Lithuanian athletes has compared somatotypes across multiple sports disciplines at the national level. The findings revealed that each sport favored a distinct body type: kayakers were predominantly endomorphic, basketball players presented an endomorphic-mesomorphic profile, and soccer players exhibited a more ectomorphic build [57]. It is important to note that somatotype is influenced by both training-induced adaptations and self-selection into sports. Individuals may be drawn to disciplines that match their natural physique, such as leaner athletes choosing endurance running or more muscular individuals gravitating toward strength and power sports. Therefore, somatotype should be understood as the result of a dynamic interaction between biological predisposition and the specific demands of sport participation.

Although somatotype may serve as a valuable tool for identifying athletic potential, it should not be considered in isolation from other factors. Successful athletes are also distinguished by psychological attributes such as self-confidence, motivation, and resilience, as well as by their sport-specific experience, technical skills, and environmental support [67–69]. These elements, when considered alongside physical profile, are critical for the long-term development and maintenance of athletic performance.

Current research strongly supports emphasizing in the discussion that anthropometric data, such as somatotype, should not be used as standalone criteria for talent identification. While anthropometric and physical performance measures can help distinguish between competitive levels and contribute to early talent identification, their predictive power is limited when used in isolation. Multiple studies highlight that talent identification is a complex, multidimensional process influenced by technical, tactical, psychological, and sociological factors in addition to physical attributes. Relying solely on anthropometric data risks overlooking late-maturing or otherwise talented individuals who may excel in other critical domains. Therefore, integrating anthropometric data within a broader, holistic assessment framework is necessary to improve the accuracy and fairness of talent identification and development programs [70–74].

Moreover, it is essential to recognize that anthropometric and somatotype profiles can be influenced by training phases within the competitive cycle. Recent evidence highlights that variations in preparation and competition periods can affect both body composition and performance-related parameters, underscoring the dynamic nature of morphological characteristics [75].

4.4. Generalizability

A wide variety of disciplines were included in the study. However, representation by sport was limited in some cases. This was partly influenced by the popularity of certain sports practiced in Mexico and those that receive greater institutional support. Such limitations may reduce the applicability of the findings to disciplines with lower representation. Therefore, the results may be beneficial for national sports contexts or countries with similar competitive structures and anthropometric characteristics. Future multicenter studies at the national level are recommended to strengthen the external validity of these findings.

5. Conclusions

This study provided a detailed characterization of somatotypes in Mexican athletes, revealing specific patterns by sex and sport macro-categories. In a sample of 477 male and 412 female athletes, the most frequent somatotypes among males were endomorphic mesomorph, balanced mesomorph, and ectomorphic mesomorph, with differences among somatotype classifications, reflecting a predominance of traits associated with strength and power. Among females, the reported somatotypes were endomorphic mesomorph, mesomorphic endomorph, and mesomorph-endomorph, with no differences among categories, suggesting a relatively higher proportion of adiposity. Compared to international athletes, Mexican female athletes exhibited a more pronounced endomorphic component. Additionally, the macro-categorical groupings revealed somatotype differences in both males and females, reflecting distinct morphological profiles according to sport type.

These findings underscore the value of somatotyping as a strategic tool for talent identification, training planning, and the personalization of nutritional interventions, particularly in disciplines with specific physical demands. It is recommended that somatotype assessment be incorporated into regular monitoring protocols in national sports centers and that these morphological profiles be considered in physical preparation programs to optimize performance and reduce injury risk. Likewise, an individualized approach is advised, considering sex, sports discipline, and competitive level. This study provides novel evidence of somatotypes in Mexican athletes, contributing to a field that has historically lacked comprehensive research in this population. Future research is encouraged to validate personalized interventions based on somatotype, explore its relationship with injury risk and training adaptation, and include comparisons across different levels of elite competition. The development of longitudinal studies is also recommended to assess the evolution of somatotype profiles in Mexican athletes and their long-term impact on athletic performance.

Supplementary Materials: The following supporting information can be downloaded at: https://www.mdpi.com/article/10.3390/jfmk10030329/s1, Figure S1: Somatocharts from different sports disciplines in male athletes; Figure S2: Somatocharts from different sports disciplines in male athletes; Table S1: Descriptive characteristics across Mexican male athletes; Table S2: Descriptive characteristics of Mexican female athletes.

Author Contributions: Conceptualization, X.M.-M., E.R. and R.L.-G.; methodology, X.M.-M., J.O.L.-C., V.C.-C., E.R. and R.L.-G.; software, X.M.-M. and E.R.; validation, X.M.-M., J.O.L.-C., V.C.-C., E.R., M.E.V. and R.S.-G.; formal analysis, X.M.-M., J.O.L.-C., V.C.-C., E.R., M.E.V., R.L.-G. and R.S.-G.; investigation, X.M.-M., J.O.L.-C., V.C.-C., X.O.-S., M.E.V., R.L.-G., S.G., C.B. and J.A.T.; resources, X.M.-M., J.O.L.-C., V.C.-C., X.O.-S., E.R. and R.L.-G.; data curation, X.M.-M., V.C.-C., X.O.-S., R.L.-G. and E.R.; writing—original draft preparation, X.M.-M., J.O.L.-C., E.R., R.L.-G. and J.A.T.; writing—review and editing, X.M.-M., J.O.L.-C., V.C.-C., E.R., R.L.-G., R.S.-G., S.G., C.B. and J.A.T.; visualization, X.M.-M., J.O.L.-C., V.C.-C. and R.S.-G.; supervision, E.R., R.S.-G.; project administration, X.M.-M. and E.R.; funding acquisition, S.G., C.B. and J.A.T. All authors have read and agreed to the published version of the manuscript.

Funding: S.G., C.B. and J.A.T. were funded by the Instituto de Salud Carlos III through the Fondo de Investigación para la Salud (CIBEROBN CB12/03/30038), which are co-funded by the European Regional Development Fund. Red EXERNET-Red de Ejercicio Físico y Salud (RED2022-134800-T) Agencia Estatal de Investigación (Ministerio de Ciencias e Innovación, Spain). IDISBA Grants (FOLIUM, PRIMUS, SYNERGIA, and LIBERI). The funding sponsors had no role in the design of the study, in the collection, analyses, or interpretation of the data; in the writing of the manuscript, or in the decision to publish the results.

Institutional Review Board Statement: The study was conducted in accordance with the Declaration of Helsinki and approved by the Ethics Committee of Facultad de Salud Pública y Nutrición from Universidad Autónoma de Nuevo León with the number 24-FaSPyN-SA-04; 11 June 2024.

Informed Consent Statement: Written informed consent has been obtained from the athletes to publish this paper.

Data Availability Statement: The original contributions and data created in this study are included in the article/supplementary materials. Further inquiries can be directed to the corresponding author (pep.tur@uib.es).

Acknowledgments: We extend our sincere gratitude to José Alberto Pérez García, Director of the Dirección General de Deportes UANL; sports nutritionist Mayra Cañamar; and the interns and students from the Department of Nutrition of the Dirección General de Deportes. We also thank sports nutritionist Roberto Benítez de la Rosa, Armando Salazar, and Nhilse Valdez of the Liga Mexicana de Powerlifting Nuevo León and Lifter's Sabinas, DHARMA Nutrition Center, as well as the team from the Body Composition Laboratory from Facultad de Salud Pública y Nutrición: Lucia Sandoval, Paulina Navarro, Brenda Veloz, Natania Lara, Neiry Dairyn Zaleta, Alana Ruíz, Jessica Herrera, Fernanda Borges, Ana Esquivel, Gabriela Domínguez, and Linda Burciaga, for their invaluable support.

Conflicts of Interest: The authors declare no conflicts of interest.

Abbreviations

The following abbreviations are used in this manuscript:

STROBE	Strengthening the Reporting of Observational Studies in Epidemiology
UANL	Universidad Autónoma de Nuevo León
CONADE	Comisión Nacional de Cultura Física y Deporte
CONADEIP	Comisión Nacional Deportiva Estudiantil de Instituciones Privadas
ONEFA	Organización Nacional Estudiantil de Futbol Americano
TEM	Technical Error of Measurement
ISAK	International Society for the Advancement of Kinanthropometry
BMI	Body Mass Index
%BF	Body Fat Percentage

Appendix A

Appendix A.1

The endomorphy component was obtained using the following equation:

$$Endomorphy = -0.7182 + 0.1451 \times \sum SS - 0.00068 \times \left(\sum SS\right)^2 + 0.0000014 \times \left(\sum SS\right)^3$$

where:

$$\sum SS = (skinfolds : triceps + subscapular + supraspinale) \times (170.18/height\ (cm))$$

The mesomorphy component was calculated using the following equation:

$$\begin{aligned} Mesomorphy = \ & (0.858 \times humerus\ breadth) + (0.601 \times femur\ breadth) \\ & +(0.188 \times corrected\ flexed\ arm\ girth) \\ & +(0.161 \times corrected\ calf\ girth) - (0.131 \times height(cm)) + 4.5 \end{aligned}$$

The ectomorphy component was calculated based on the height–weight ratio (HWR), defined as height divided by the cube root of body mass. The classification of ectomorphy was established according to the following criteria:

$$If\ HWR \geq 40.75,\ Ectomorphy = (0.732 \times HWR) - 28.58$$

$$If\ HWR > 38.25\ and < 40.75, \quad Ectomorphy = (0.463 \times HWR) - 17.63$$

$$If\ HWR \leq 38.25,\ Ectomorphy = 0.1$$

where:

$$HWR = \frac{height\ (cm)}{\sqrt[3]{weight\ (kg)}}$$

Appendix A.2

The X and Y axes are used for the somatochart, calculated using the following equations:

$$X = ectomorphy\ component - endomorphy\ component$$

$$Y = 2 \times mesomorphy\ component - (endomorphy\ component + ectomorphy\ component)$$

References

1. Boileau, R.A.; Lohman, T.G. The measurement of human physique and its effect on physical performance. *Orthop. Clin. N. Am.* **1977**, *8*, 563–581. [CrossRef]
2. Sheldon, W.H.; Stevens, S.S.; Tucker, W.B. *The Varieties of Human Physiquei*; Harper: New York, NY, USA, 1940.
3. Carter, J.E.L. *The Heath-Carter Anthropometric Somatotype: Instruction Manual*; San Diego State University, Department of Exercise and Nutritional Sciences: San Diego, CA, USA, 2002.
4. Drywień, M.; Górnicki, K.; Górnicka, M. Application of Artificial Neural Network to Somatotype Determination. *Appl. Sci.* **2021**, *11*, 1365. [CrossRef]
5. Peeters, M.W.; Thomis, M.A.; Loos, R.J.F.; Derom, C.A.; Fagard, R.; Claessens, A.L.; Vlietinck, R.F.; Beunen, G.P. Heritability of somatotype components: A multivariate analysis. *Int. J. Obes.* **2007**, *31*, 1295–1301. [CrossRef] [PubMed]
6. Baranauskas, M.; Kupčiūnaitė, I.; Lieponienė, J.; Stukas, R. Dominant Somatotype Development in Relation to Body Composition and Dietary Macronutrient Intake among High-Performance Athletes in Water, Cycling and Combat Sports. *Nutrients* **2024**, *16*, 1493. [CrossRef]
7. Terzi, E.; Kalkavan, A. To what extent do somatotype structures affect athletic performance in professional athletes? *J. Sports Med. Phys. Fit.* **2024**, *64*, 650–660. [CrossRef]
8. Vernillo, G.; Schena, F.; Berardelli, C.; Rosa, G.; Galvani, C.; Maggioni, M.; Agnello, L.; La Torre, A. Anthropometric characteristics of top-class Kenyan marathon runners. *J. Sports Med. Phys. Fit.* **2013**, *53*, 403–408.
9. Stewar, A.D.; Benson, P.J.; Michanikou, E.G.; Tsiota, D.G.; Narli, M.K. Body image perception, satisfaction and somatotype in male and female athletes and non-athletes: Results using a novel morphing technique. *J. Sports Sci.* **2003**, *21*, 815–823. [CrossRef]
10. Slankamenac, J.; Bjelica, D.; Jaksic, D.; Trivic, T.; Drapsin, M.; Vujkov, S.; Modric, T.; Milosevic, Z.; Drid, P. Somatotype Profiles of Montenegrin Karatekas: An Observational Study. *Int. J. Environ. Res. Public Health* **2021**, *18*, 12914. [CrossRef] [PubMed]
11. Lukaski, H.; Raymond-Pope, C.J. New Frontiers of Body Composition in Sport. *Int. J. Sports Med.* **2021**, *42*, 588–601. [CrossRef]
12. Salokun, S.O. Minimizing injury rates in soccer through preselection of players by somatotypes. *J. Sports Med. Phys. Fit.* **1994**, *34*, 64–69.
13. Hopper, D.M. Somatotype in high performance female netball players may influence player position and the incidence of lower limb and back injuries. *Br. J. Sports Med.* **1997**, *31*, 197–199. [CrossRef]
14. Xiong, J.; Qian, Y.; Yu, S.; Ji, H.; Teliewubai, J.; Chi, C.; Lu, Y.; Zhou, Y.; Fan, X.; Li, J.; et al. Somatotype and Its Impact on Asymptomatic Target Organ Damage in the Elderly Chinese: The Northern Shanghai Study. *Clin. Interv. Aging* **2021**, *16*, 887–895. [CrossRef]
15. Hohenauer, E.; Rucker, A.M.; Clarys, P.; Küng, U.M.; Stoop, R.; Clijsen, R. Anthropometric and performance characteristics of the German rugby union 7s team. *J. Sports Med. Phys. Fit.* **2017**, *57*, 1633–1641. [CrossRef]
16. Carter, J.E.L. The somatotypes of athletes—A review. *Hum. Biol.* **1970**, *42*, 535–569.
17. Ryan-Stewart, H.; Faulkner, J.; Jobson, S. The influence of somatotype on anaerobic performance. *PLoS ONE* **2018**, *13*, e0197761. [CrossRef] [PubMed]

18. Jakovljević, V.; Bošnjak, G.; Pašić, G.; Tešanović, G. Roll of somatotype in sports selection. *Acta Kinesiol.* **2022**, *16*, 84–92. [CrossRef]
19. Ochs, J.; Hohmann, A.; Pion, J. Harnessing coaches' expertise: Creating 11 sport-specific profiles for talent orientation. *Front. Sports Act. Living* **2025**, *7*, 1432903. [CrossRef] [PubMed]
20. Martínez-Mireles, X.; Nava-González, E.J.; López-Cabanillas Lomelí, M.; Puente-Hernández, D.S.; Gutiérrez-López, M.; Lagunes-Carrasco, J.O.; López-García, R.; Ramírez, E. The Shape of Success: A Scoping Review of Somatotype in Modern Elite Athletes Across Various Sports. *Sports* **2025**, *13*, 38. [CrossRef]
21. Peeters, M.W.; Thomis, M.A.; Claessens, A.L.; Loos, R.J.; Maes, H.H.; Lysens, R.; Vanden Eynde, B.; Vlietinck, R.; Beunen, G. Heritability of somatotype components from early adolescence into young adulthood: A multivariate analysis on a longitudinal twin study. *Ann. Hum. Biol.* **2003**, *30*, 402–418. [CrossRef]
22. Pereira, S.; Katzmarzyk, P.T.; Gomes, T.N.; Souza, M.; Chaves, R.N.; Santos, F.K.D.; Santos, D.; Hedeker, D.; Maia, J.A.R. Multilevel modelling of somatotype components: The Portuguese sibling study on growth, fitness, lifestyle and health. *Ann. Hum. Biol.* **2016**, *44*, 316–324. [CrossRef]
23. Kandel, M.; Baeyens, J.P.; Clarys, P. Somatotype, training and performance in Ironman athletes. *Eur. J. Sport Sci.* **2014**, *14*, 301–308. [CrossRef]
24. Castillo, M.; Sospedra, I.; González-Rodríguez, E.; Hurtado-Sánchez, J.A.; Lozano-Casanova, M.; Jiménez-Alfageme, R.; Martínez-Sanz, J.M. Body Composition and Determination of Somatotype of the Spanish Elite Female Futsal Players. *Appl. Sci.* **2022**, *12*, 5708. [CrossRef]
25. Queiroga, M.R.; Fernandes, D.; Ferreira, S.A.; Reis, M.; Fernandes, D.Z.; Cavazzotto, T.G.; Portela, B.S.; Tartaruga, M.P.; Nascimento, M.A.; Vieira, E.R. Characterization of Reproductive and Morphological Variables in Female Elite Futsal Players. *Front. Psychol.* **2021**, *12*, 625354. [CrossRef]
26. Caballero-Ruíz, A.; Carrasco-Legleu, C.E.; De León, L.G.; Candia-Luján, R.; Ortiz-Rodríguez, B. Somatotype of university female soccer players by playing position on the field. *Retos* **2019**, *36*, 228–230. [CrossRef]
27. Conde-Pipo, J.; Latorre, J.A.; Gimenez-Blasi, N.; Olea-Serrano, F.; Requena, B.; Mariscal-Arcas, M. Comparative Analysis of Body Composition Profiles among Latin American Elite Football Players Competing in Europe. *Appl. Sci.* **2023**, *1*, 6778. [CrossRef]
28. Gomes, C. Outside the classroom: The language of English and its impact on international student mental wellbeing in Australia. *J. Int. Stud.* **2020**, *10*, 934–953. [CrossRef]
29. Cornejo, M. Assessing the sociology of sport: On the Latin American Association of Sociocultural Studies in Sport and research development in Latin America. *Int. Rev. Sociol. Sport.* **2015**, *50*, 407–412. [CrossRef]
30. Casas, Y.G.; Schiller, B.C.; DeSouza, C.A.; Seals, D.R. Total and regional body composition across age in healthy Hispanic and white women of similar socioeconomic status. *Am. J. Clin. Nutr.* **2001**, *7*, 13–18. [CrossRef]
31. Mueller, W.H.; Shoup, R.F.; Malina, R.M. Fat patterning in athletes in relation to ethnic origin and sport. *Ann. Hum. Biol.* **1982**, *9*, 371–376. [CrossRef]
32. Ochoa Martínez, P.Y.; Hall López, J.A.; Alarcón Meza, E.I.; Arráyales Millán, E.M.; Sánchez León, R. Somatotype profile and body composition of players from the Mexican professional basketball league. *Int. J. Morphol.* **2014**, *32*, 1032–1035. [CrossRef]
33. Barajas-Pineda, L.T.; Del-Río-Valdivia, J.E.; Flores-Moreno, P.J.; Gómez-Figueroa, J.A.; Gómez-Gómez, E.; Salazar, C.M. Anthropometric Profile and Body Composition of the Mexican Men's Volleyball Team. *Int. J. Morphol.* **2021**, *39*, 90–94. [CrossRef]
34. Zuñiga, U.; Osorio, A.; Toledo, I.d.J.; Herrera, R. Somatotype of Mexican soccer players from different competition level. *Retos* **2018**, *34*, 100–102. [CrossRef]
35. Barajas-Pineda, L.; Salazar-C, C.M.; Flores-Moreno, P.J.; Gómez-Figueroa, J.A.; Andrade-Sánchez, A.I.; Del-Río-Valdivía, J.E.; Gómez-Gómez, E. Anthropometric Profile and Body Composition of the Mexican Olympic Beach Volleyball Team. *Int. J. Morphol.* **2023**, *41*, 225–230. [CrossRef]
36. Novoa-Vignau, M.F.; Salas-Fraire, O.; Salas-Longoria, K.; Hernández-Suárez, G.; Menchaca-Pérez, M. A comparison of anthropometric characteristics and somatotypes in a group of elite climbers, recreational climbers and non-climbers. *Med. Univ.* **2017**, *19*, 69–73. [CrossRef]
37. Manjarrez-Bastidas, I.A.; Hernández-Murúa, J.A.; Salazar, C.M.; Avendaño-Soto, M.; Salazar-Landeros, M.M.; Soto-Peña, J.L.; Barajas Pineda, L.T. Anthropometric Profile, Somatotype and Nutritional Status of Mexican Taekwondoin. *Int. J. Morphol.* **2022**, *40*, 959–965. [CrossRef]
38. Peralta-Sánchez, M.d.l.A. Anthropometry and Body Composition of Junior and Elite Mexican Triathletes. *Rev. Mex. Investig. Cult. Fís. Deporte* **2012**, *4*, 25–40.
39. von Elm, E.; Altman, D.G.; Egger, M.; Pocock, S.J.; Gøtzsche, P.C.; Vandenbroucke, J.P. The Strengthening the Reporting of Observational Studies in Epidemiology (STROBE) statement: Guidelines for reporting observational studies. *J. Clin. Epidemiol.* **2007**, *61*, 344–349. [CrossRef]
40. Secretaría de Salud. Decreto por el que se reforman, adicionan y derogan diversas disposiciones de los Títulos Tercero Bis y Décimo Octavo de la Ley General de Salud. In *Diario Oficial de la Federación*; Secretaría de Salud: Mexico City, Mexico, 2014. Available online: https://www.dof.gob.mx/nota_detalle.php?codigo=5345705&fecha=04/06/2014 (accessed on 8 May 2025).

41. Secretaría de Salud. NOM-012-SSA3-2012, Que establece los criterios para la ejecución de proyectos de investigación para la salud en seres humanos. In *Diario Oficial de la Federación*; Secretaría de Salud: Mexico City, Mexico, 2013. Available online: https://dof.gob.mx/nota_detalle.php?codigo=5284148&fecha=04/01/2013 (accessed on 8 May 2025).
42. World Medical Association. *Declaration of Helsinki–Ethical Principles for Medical Research Involving Human Subjects*; World Medical Association: Ferney-Voltaire, France, 2008. Available online: https://www.wma.net/what-we-do/medical-ethics/declaration-of-helsinki/doh-oct2008/ (accessed on 4 March 2025).
43. Araújo, C.G.; Scharhag, J. Athlete: A working definition for medical and health sciences research. *Scand. J. Med. Sci. Sports* **2016**, *26*, 4–7. [CrossRef]
44. McAuley, A.B.T.; Baker, J.; Kelly, A.L. Defining "elite" status in sport: From chaos to clarity. *Ger. J. Exerc. Sport Res.* **2021**, *52*, 193–197. [CrossRef]
45. Lean, M.E.J.; Han, T.S.; Deurenberg, P. Predicting body composition by densitometry from simple anthropometric measurements. *Am. J. Clin. Nutr.* **1996**, *63*, 4–14. [CrossRef]
46. Ryan-Stewart, H.; Faulkner, J.; Jobson, S. The Impact of Technical Error of Measurement on Somatotype Categorization. *Appl. Sci.* **2022**, *12*, 3056. [CrossRef]
47. Stewart, A.; Marfell-Jones, M.; Olds, T.; Ridder, H. *International Standards for Anthropometric Assessment*; International Society for the Advancement of Kinanthropometry: Lower Hutt, New Zealand, 2011.
48. Ramírez, E. NutriSolver. 2018. Available online: https://nutrisolver.com/fitness-composicion-corporal.php (accessed on 25 March 2025).
49. Siders, W.; Rue, M. Reuleaux triangle somatocharts. *Comput. Biol. Med.* **1992**, *22*, 363–368. [CrossRef]
50. Tukey, J.W. *Exploratory Data Analysis*; Addison-Wesley: Reading, MA, USA, 1977.
51. Hintze, J. NCSS 8. 2012. Available online: https://www.ncss.com (accessed on 20 May 2025).
52. Igbokwe, N.U. Somatotypes of Nigerian power athletes. *J. Sports Med. Phys. Fit.* **1991**, *31*, 439–441.
53. Kamath, S.; Adhikari, R.; Bawari, B.; Easow, J.; Kale, U.; Wong, F.Y.; Pullinger, S.A. Investigating Anthropometric characteristics and Somatotypes in Elite Indian Track & Field Athletes. *Int. J. Kinanthropometry* **2024**, *4*, 33–43. [CrossRef]
54. Bayios, I.A.; Bergeles, N.K.; Apostolidis, N.G.; Noutsos, K.S.; Koskolou, M.D.; Professor, A.; Also, K.A. Anthropometric, Body Composition and Somatotype Differences of Greek Elite Female Basketball, Volleyball and Handball Players. *J. Sports Med. Phys. Fit.* **2006**, *46*, 271–280.
55. Gholami, M.; Rad, L.S. Anthropometric, body composition and somatotype differences of Iranian elite female basketball and handball players. *Br. J. Sports Med.* **2010**, *44*, i35. [CrossRef]
56. Massidda, M.; Toselli, S.; Brasili, P.; Calò, C.M. Somatotype of elite Italian gymnasts. *Coll. Antropol.* **2013**, *37*, 853–857.
57. Gutnik, B.; Zuoza, A.; Zuozienė, I.; Alekrinskis, A.; Nash, D.; Scherbina, S. Body physique and dominant somatotype in elite and low-profile athletes with different specializations. *Medicina* **2015**, *51*, 247–252. [CrossRef] [PubMed]
58. Campa, F.; Silva, A.M.; Talluri, J.; Matias, C.N.; Badicu, G.; Toselli, S. Somatotype and Bioimpedance Vector Analysis: A New Target Zone for Male Athletes. *Sustainability* **2020**, *12*, 4365. [CrossRef]
59. Cinarli, F.S.; Buyukcelebi, H.; Esen, O.; Barasinska, M.; Cepicka, L.; Gabrys, T.; Nalbant, U.; Karayigit, R. Does Dominant Somatotype Differentiate Performance of Jumping and Sprinting Variables in Young Healthy Adults? *Int. J. Environ. Res. Public Health* **2022**, *19*, 11873. [CrossRef]
60. Giannopoulos, N.; Vagenas, G.; Noutsos, K.; Barzouka, K.; Bergeles, N. Somatotype, Level of Competition, and Performance in Attack in Elite Male Volleyball. *J. Hum. Kinet.* **2017**, *58*, 131–140. [CrossRef]
61. Han, J.; Waddington, G.; Anson, J.; Adams, R. Level of competitive success achieved by elite athletes and multi-joint proprioceptive ability. *J. Sci. Med. Sport* **2013**, *18*, 77–81. [CrossRef]
62. Barley, O.R.; Harms, C.A. Profiling Combat Sports Athletes: Competitive History and Outcomes According to Sports Type and Current Level of Competition. *Sports Med. Open* **2021**, *7*, 63. [CrossRef]
63. Carter, J.L.; Heath, B.H. *Somatotyping: Development and Applications*; Cambridge University Press: Cambridge, UK, 1990; p. 201.
64. Carter, J.E.L.; Ackland, T.E. Somatotype in Sport. In *Applied Anatomy and Biomechanics in Sport*, 2nd ed.; Ackland, T., Elliott, B.C., Bloomfield, J., Eds.; Human Kinetics: Champaign, IL, USA, 2009; pp. 47–66.
65. Weyerstraß, J.; Stewart, K.; Wesselius, A.; Zeegers, M. Nine genetic polymorphisms associated with power athlete status—A meta-analysis. *J. Sci. Med. Sport* **2018**, *21*, 213–220. [CrossRef]
66. Wilson, B.R.; Olson, H.W.; Sprague, H.A.; Van Huss, W.D.; Montoye, H.J. Somatotype and longevity of former university athletes and nonathletes. *Res. Q. Exerc. Sport* **1990**, *61*, 1–6. [CrossRef] [PubMed]
67. Henriksen, K.; Stambulova, N.; Roessler, K.K. Holistic approach to athletic talent development environments: A successful sailing milieu. *Psychol. Sport Exerc.* **2009**, *11*, 212–222. [CrossRef]
68. Velez, A.E.; Garcia-Roca, J.A.; Vaquero-Cristobal, R. What Variables Differentiate between Selected and Not Selected High-Performance Adolescent Track and Field Athletes? *J. Hum. Kinet.* **2025**, *97*, 263–280. [CrossRef] [PubMed]

69. Jacob, Y.; Spiteri, T.; Hart, N.H.; Anderton, R.S. The Potential Role of Genetic Markers in Talent Identification and Athlete Assessment in Elite Sport. *Sports* **2018**, *6*, 88. [CrossRef]
70. Till, K.; Jones, B.L.; Cobley, S.; Morley, D.; O'Hara, J.; Chapman, C.; Cooke, C.; Beggs, C.B. Identifying Talent in Youth Sport: A Novel Methodology Using Higher-Dimensional Analysis. *PLoS ONE* **2016**, *11*, e0155047. [CrossRef]
71. Han, M.; Gómez-Ruano, M.A.; Calvo, A.L.; Calvo, J.L. Basketball Talent Identification: A Systematic Review and Meta-Analysis of the Anthropometric, Physiological and Physical Performance Factors. *Front. Sports Act. Living* **2023**, *5*, 1264872. [CrossRef] [PubMed]
72. Reilly, T.; Bangsbo, J.; Franks, A. Anthropometric and Physiological Predispositions for Elite Soccer. *J. Sports Sci.* **2000**, *18*, 669–683. [CrossRef] [PubMed]
73. Schmitz, T.L.; Fleddermann, M.T.; Zentgraf, K. Talent Selection in 3×3 Basketball: Role of Anthropometrics, Maturation, and Motor Performance. *Front. Sports Act. Living* **2024**, *6*, 1459103. [CrossRef]
74. Hoare, D.G. Predicting Success in Junior Elite Basketball Players—The Contribution of Anthropometric and Physiological Attributes. *J. Sci. Med. Sport* **2000**, *3*, 391–405. [CrossRef] [PubMed]
75. Włodarczyk, M.P.; Kusy, K.; Zieliński, J. Changes in Eccentric Utilization Ratio, Reactive Strength Index and Leg Stiffness in Highly Trained Sprinters between Training Phases. *Acta Kinesiol.* **2024**, *18*, 23–30. [CrossRef]

Journal of
Functional Morphology and Kinesiology

MDPI

Article

Age-Related Changes in Predictors of BMI in 6, 9 and 12-Year-Old Boys and Girls: The NW-CHILD Study

Barry Gerber * and Anita Elizabeth Pienaar

Physical Activity, Sport and Recreation (PhASRec), Faculty of Health Sciences, Potchefstroom Campus, North-West University, Potchefstroom 2520, South Africa; anita.pienaar@nwu.ac.za
* Correspondence: barry.gerber@nwu.ac.za; Tel.: +27-72-581-9629

Abstract: Background: Information on childhood body composition is critical to understanding children's growth, development, and long-term health outcomes. BMI metrics, however, have several limitations for assessing and understanding changes in BMI. Therefore, understanding the influence of various body composition factors (covariates) that are linked to, and influence, BMI over time in growing children is important. This study aims to determine sex differences in longitudinal changes in covariates of BMI from 6 to 13 years. **Methods**: Participants (N = 332, 160 boys 172 girls) from North West Province in South Africa were assessed longitudinally at the following three time-points during their primary years of schooling: Grade 1 (6–7 years); Grade 4 (9–10 years); and Grade 7 (12–13 years). Covariates included: stature (cm); body weight (kg); sub-scapular-, calf-, and triceps skinfolds (mm); body fat percentage (%), relaxed forearm, waist and mid-upper arm circumferences; percentage fat weight; and percentage muscle weight. Correlational analysis and multiple stepwise regression analysis in SPSS analyzed the significance of the contributions of the different covariates to changes in BMI from 6 to 12 years. **Results**: Different covariates influence BMI in boys and girls at different ages and the covariates also change over time in boys and girls. Weight had the strongest influence on the BMI of boys and girls, although the prediction value decreased over time. Weight and stature were consistently the strongest BMI predictors across all ages in boys. In girls, a broader range of variables influences BMI from a younger age, where slightly higher BMI correlations with fat-related variables emerged, and the percentage of fat weight distribution was a strong influential factor. These findings indicate a more in-depth analysis of BMI to determine sound intervention strategies.

Keywords: BMI; covariates; predictors; gender differences; longitudinal

1. Introduction

Obesity in children and adolescents has become a major global issue over the past decade [1], resulting in rising global health issues, such as noncommunicable diseases, especially in low- to middle-income countries (LMICs) [2]. Recently, the World Health Organization [1] reported that 37 million children under the age of 5 were overweight in 2022, with 390 million children and adolescents aged 5–19 years being overweight in 2022, including 160 million who were living with obesity. Various studies have globally indicated that being overweight and obese is associated with detrimental health effects, with carry-over effects from childhood to adulthood [3–5]. Therefore, providing effective and compassionate care tailored to the child and family, focusing on healthy body composition, is vital [6].

Body composition is defined as the proportional composition of a person's total body weight, which consists of muscle, bone, fat, and other tissue [7]. Fat weight refers to the extractable fat in adipose and other body tissues, while fat-free body weight can be defined as the remaining fat-free chemical substances and tissues in the body, which include muscle, bone, connective tissue, and internal organs [8]. A pressing issue nowadays is determining the best measuring tool or protocol with which to assess body composition in children and adults. In this regard, various high-technology methods for measuring body composition are available, including dual-energy X-ray absorptiometry, air displacement plethysmography, and hydrostatic weighing. However, these methods are not suited for field-based testing and are expensive. Although other methods, such as bioelectrical impedance analysis (a technique grounded in the basic principles of electrical conductivity, introducing a small, weak alternating electrical current into the body at one or more frequencies via electrodes) and the widely used body weight index (BMI) over decades, are more practical, inexpensive, easy to use and affordable, the accuracy of these methods is still in question [9–12]. In this regard, various studies globally attest to the reliability, precision, and accuracy of impedance technologies and have found correlations greater than 0.95 with dual energy X-ray absorptiometry (DXA) or dilution-measured TBW with regards to TBW or FFM, although the level of agreement may be large (±5–10%) [9,13,14]. Furthermore, Feng et al. [15] found high correlations between BIA and dual-energy X-ray absorptiometry (DXA) measurements for fat weight (FM) and fat-free weight (FFM). However, BIA tended to underestimate FM by approximately 1.84 kg and overestimate FFM by about 2.56 kg. Another study, episodically focusing on children with obesity, revealed that multi-frequency octopolar BIA devices closely matched DXA results, with minimal differences in body fat percentage and lean weight; however, single-frequency devices significantly underestimated body fat percentage [16]

BMI, which is an indirect method for assessing body composition using only stature and weight variables, fails to distinguish between fat, muscle, and bone mass; consequently, it is prone to misclassification, particularly among individuals with a muscular build [17]. Various other researchers have also indicated over the past decade that the BMI metric has several limitations for assessing reasons for changes in BMI [18–21]. In this regard, the relationship between BMI and BMI z-scores (BMIz) is curvilinear, approaching a maximum value that varies by sex and age [22]. Therefore, Freedman et al. [22] believe that it is crucial to understand how BMI fluctuates over time in growing children to determine the most suitable scale for measuring BMI changes. Furthermore, it is essential to understand the covariates of BMI that influence these fluctuations and changes in BMI over time [23].

Sattar et al. [24] indicated that although BMI is based on two variables (stature and weight), it can be influenced by various covariates, including body fat, muscle weight, sex, race, age, and waist circumference. In this regard, referring to changes in specific covariates of BMI, the WHO guidelines indicate that at 5 years of age, the differences between the 15th and 85th percentiles for sub-scapular and triceps skinfolds are 3 mm and 4–5 mm, respectively [25]. Furthermore, Wernech et al. [26] found that the sum of skinfolds (triceps and sub-scapular) of boys and girls between 7 and 10 years old changes from 23.5 mm to 33.5 mm (boys) and 25.7 mm to 33.9 mm (girls). Regarding fat percentage, Cossio-Bolaños et al. [27] report that although they found small differences in fat percentage of 2.1% (girls) and 9% between the ages of 7.5 and 15.4 years, significant standard deviations (SD) were found within each year group, of up to 6.9% (girls) and 8.7%(boys). Lalucci et al. [28] investigated the correlations between and covariates and found correlations between BMI and fat to lean weight ratio (general, $r = 0.69$; female, $r = 0.74$; male, $r = 0.69$); BMI and fat-free weight (general, $r = 0.49$; female, $r = 0.67$; male, $r = 0.44$); BMI and skeletal muscle weight (general, $r = 0.50$; female, $r = 0.68$; male, $r = 0.44$); and BMI and body fat percentage (general, r = 0.47; female, r = 0.54; male, r = 0.40; all with $p < 0.01$). These results underline the theory that there is more to just BMI. As a consequence, in order for researchers to understand changes in BMI better, further investigation is needed to verify if the covariates of weight (body fat percentage, skinfolds, circumferences) play a vital role in BMI changes, or if the ratio of the percentage of change from the covariates changes over time, especially during childhood, as children grow and mature. Furthermore, it is essential to investigate if the covariates influence BMI differently in boys and girls. In this regard, Mast and coworkers [29] had already found in 1998 that WHR, the sum of four skinfolds and fat percentage, differed German children between 5 and 7 years of age. However, we found no studies that specifically focus on possible changes and influences of various anthropometric and body composition covariates concerning BMI, and whether this difference might, in turn, affect the BMI of different genders differently. Considering the above studies in the literature, and based on various methods that are available to determine body composition, including BMI, the accuracy of the BIA method in determining BMI and fat and muscle percentages in children was first analyzed as a sub-objective to decide whether we could use the additional body composition measures available, including muscle mass and fat mass, in our analysis. As a primary objective, this study aims to determine whether changes in the covariates of BMI occur in 6-, 9-, and 12-year-old boys and girls, and whether such changes differ between boys and girls.

2. Materials and Methods

2.1. Research Design

This research is a sub-study of the NW-CHILD (Child-Health-Integrated-Learning and Development) longitudinal study conducted in Northwest Province (NWP) in South Africa between 2010 and 2016. This longitudinal study had three time-points (2010, 2013, and 2016) during the primary school years. The primary research included demographic questionnaires, physical fitness, physical activity, anthropometric, physiological, and blood pressure measurements.

The NW-CHILD study employed a random and stratified sampling method to select districts, schools, and participants, stratified by gender and school quintile. As part of the design, schools were randomly selected from four of the eight districts within NWP. Five schools from each district were selected, totaling 20 schools, for the testing. Participants from NWP were assessed at three time-points during their primary years of schooling: Grade 1 (6–7 years); Grade 4 (9–10 years); and Grade 7 (12–13 years). Schools from each district represented a range of socioeconomic statuses, with schools from quintiles 1–3 categorized as low SES and those from quintiles 4–5 considered high SES, according to South Africa's classification system [30].

The NW-CHILD study received ethical clearance from North-West University's Health Research Ethics Committee (HREC) (00070-09-A1) and approval from the North-West Department of Basic Education (DBE). The participating school principals gave consent for testing during school hours and parents of all students were provided with an informed consent form. Additionally, children whose parents gave consent to their participation were required to give assent if they were under 8 years old or consent if they were older.

2.2. Investigating Group and Procedures

The initially recruited 860 participants, who were enrolled from 20 primary schools in 2010 (Grade 1) at baseline, and were approximately six years (±0.39) of age. Although 829 parents consented for the children to participate, on the day of testing, only 816 (419 boys and 397 girls) were available due to absence on the day of testing or exclusion because of inaccurate ages. For the first follow-up measurement, 574 participants (282 boys and 292 girls; dropout rate 29.7%) aged approximately 9 years (±0.38) consented to participate in the study. At the final time-point measurement in 2016, 381 participants (181 boys and 200 girls) consented, indicating a further drop-out of 33.6%. Consequently, the study had a total dropout rate of 57.2% (n = 467) over the 6-year follow-up period. Furthermore, although some participants were present at all time-points, others had incomplete data due to various reasons. Consequently, another 49 participants were excluded, resulting in a final group of 332 (38.6% of the initial participants) participants (160 boys and 172 girls). This study focused on all participants who participated in all three time-point measurements in 2010, 2013, and 2016. A previous study by Pienaar (2015) [31] from the same research group investigated possible bias due to lost subjects during follow-up analysis using *t*-tests. Insignificant Cohen's d values provided no evidence of bias in baseline height ($p = 0.553$, d = 0.04), mass ($p = 0.03$, d = 0.16), BMI ($p = 0.008$, d = 0.19) and fat percentage ($p = 0.223$, d = 0.09). All children who presented with physical disabilities and whose parents did not consent were excluded. The layout and set-up of the stations were kept as similar as possible but varied between school settings according to the available space (Figure 1).

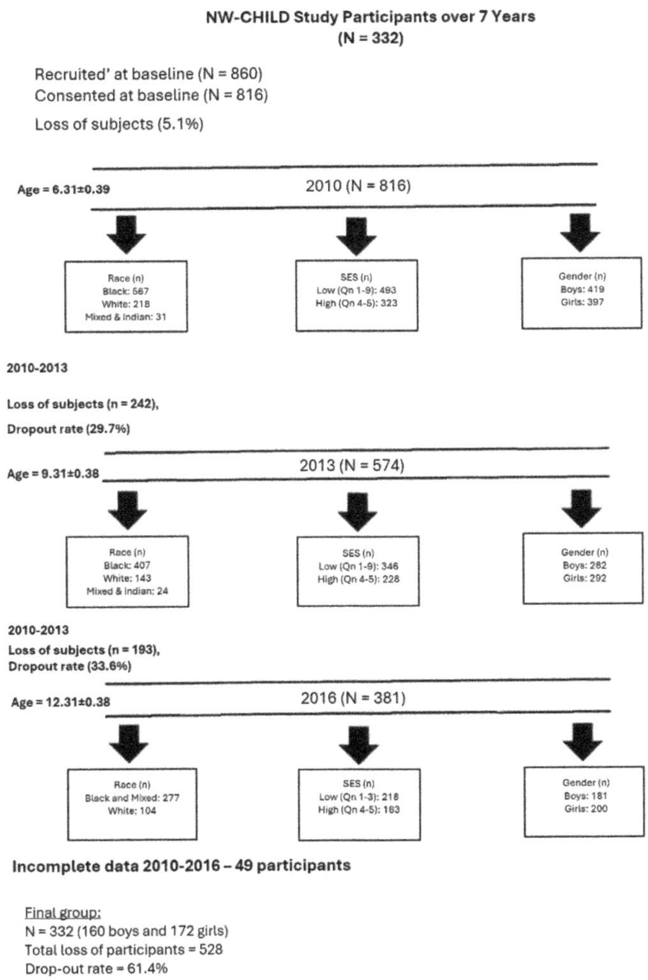

Figure 1. NW-CHILD investigation group over 7 years.

2.3. Measurement Instruments and Apparatus

Anthropometric measurements

Anthropometry, which included stature (cm), body weight (kg), measurement of the sub-scapular, calf, and triceps skinfolds (mm), relaxed forearm and waist circumferences (cm), was measured before any fitness testing commenced. The fat percentage was calculated by adding the triceps and sub-scapular skinfolds and assessed according to the gender-specific equation developed by Lohman (1992) [32].

All anthropometric tests were conducted by qualified level 2 Kinantropometrists who were postgraduate students specializing in Kinderkinetics following International Society for the Advancement of Kinanthropometry (ISAK) protocols [33].

A Harpenden portable stadiometer (Holstein Limited, London, UK) and two electronic scientific calibrated scales, including the SECA and Omron BF 511 [34], measured stature and body weight to the nearest 0.1 cm and 0.1 kg. BMI was calculated from these measurements (body weight in kg divided by length in m^2).

Body composition

The Omron BF 511 BIA body composition analyzer (OMRON, Milton Keynes, UK) was used to measure body composition characteristics, including fat-free weight (percentage muscle weight) and body fat (percentage body fat). The rationale for using the OMRON BF511 is that it is a non-invasive, quick, and easy-to-use clinically validated bioelectrical impedance analysis (BIA) device that offers a portable and cost-effective alternative to expensive methods, and can be operated with minimal training, making it ideal for use in community and school environments. Its application in pediatric populations, particularly in children aged 6 years and older, is supported by several methodological and practical advantages. No funding was received from the industry for using the equipment.

In this regard, participants wore minimal clothing, no jewelry that could influence the bioimpedance, and were barefoot. The participants stood on the Omron scale, positioned on the indicated footpads and held the hand devices with outstretched arms in front of their bodies while the measurements were taken. The mid-upper arm and waist circumferences (in cm) were measured using a metal measuring tape (Cescorf, Triteza, Brazil) and taken twice to ensure validity and reliability. The sub-scapular, calf, and triceps skinfolds (mm) were also taken twice with a Harpenden skinfold caliper to provide an average value. This takes into consideration technical errors in measurement.

If necessary, translators were made available when English or Afrikaans was not the participant's first language.

2.4. Statistical Analysis

The "Statistica for Windows" [35] and Statistical Package for the Social Sciences (SPSS) for Windows, version 27 [36], were used. A previous study by Pienaar [31] analyzed the same sample population and indicated that no bias was introduced by the loss of subjects over time. The data were assumed to be missing at random and did not significantly affect the results. Diagnostic checks were performed to ensure that the assumptions of linear regression were met, including the evaluation of residual plots for normality, linearity, and homoscedasticity. Standardized residuals were examined, and plots of residuals versus predicted values did not indicate patterns suggestive of heteroscedasticity.

Firstly, a collinearity diagnostics analysis was performed on the baseline measurements (T1) (see Supplementary Tables), with results revealing a condition index above 30 in the final dimension, suggesting potential multicollinearity. The variance proportion analysis identified that mid-upper arm circumference is strongly associated with the final dimension (variance = 0.98), indicating it may be collinear with the model intercept. A moderate multicollinearity between triceps and calf skinfolds was also found. To address this, a principal components analysis (PCA) of the body composition variables (excluding weight and stature) was performed to assess dimensionality and address multicollinearity (see Supplementary Tables). Results indicated a significant Kaiser-Meyer-Olkin (KMO) value of 0.818, and Bartlett's Test of Sphericity ($\chi^2 = 1412.94$, $p < 0.001$), indicating that the data were suitable for PCA. However, the 2 new loadings including component 1, adiposity (dominated by skinfolds and fat %), and component 2, muscle mass (loading almost exclusively on muscle %), were deemed not to be more interpretable predictors than the original variables, and it was therefore decided to continue with the original variables to investigate how these different body composition variables, although closely linked to each other, might contribute differently to changes in BMI over a longitudinal period.

Descriptive statistics were calculated to analyze means and standard deviations of the boys' and girls' age and anthropometric and body composition profiles. A correlation coefficient analysis was conducted to determine the possible correlations between BMI and body Weight and anthropometric and body composition covariates, including stature, Weight, sub-scapular skinfold, triceps skinfold, calf skinfold, waist-circumference, mid-upper arm circumference, percentage fat mass, and percentage muscle mass. Cut points used to indicate significance included $0.1 < r < 0.3$ (small correlation), $0.3 < r < 0.5$ (medium correlation), and $r > 0.5$ (strong correlation) [37]. A multiple stepwise regression analysis was conducted in SPSS to determine the significance of the contributions of the different covariates to changes in BMI. The following cut points indicate the percentage variance explained: R2 = 1% is interpreted as a small effect, R2 = 10% as a medium effect, while R2 $\geq$ 25% is considered a large effect. For statistical significance, p is set at ≤ 0.05.

3. Results

The study included 332 participants between the ages of six and thirteen, including 160 boys and 172 girls, with mean ages of 6.79 + 0.50 (Grade 1, T1), 9.89 + 0.38 (Grade 4, T2), and 12.90 + 0.38 (Grade 7, T3), at each of the three time-points, who had complete datasets. Measurements taken over the 7-year longitudinal period during the primary school years provided 6-year follow-up data for this study.

Before analyzing the contributions of different body composition characteristics to BMI at various ages by using regression analysis, a correlational analysis was performed to determine the association between direct (BMI equation and sum of skinfolds) and indirect estimates of BMI derived from bioelectrical impedance analysis, conducted with the Omron BF 511 body composition analyzer.

In this regard, the results from Table 1 indicate a high correlation at T1 ($r > 0.93$) between the direct and indirect methods to determine BMI. This association strengthened over time ($r = 0.97$, T2; $r = 0.99$, T3). Regarding the correlation between the fat percentage determined by skinfolds and BIA, a slightly lower, yet still strong, association was found between the two methods (T1: $r = 0.68$; T2: $r = 0.89$; T3: $r = 0.86$). As a high association was evident from both analyses, taking into consideration that both methods have limitations, BIA results obtained from the Omron BF511 were considered to demonstrate consistency in output between the two methods. Hence, for this study, BMI, fat percentage, and muscle percentage, as measured by the Omron BF511, were used in all further analyses.

Table 2 reports the changes and significance of changes ($p < 0.01$) in age and body composition measurements across the three time-points (T1: 6 years, T2: 9 years, T3: 12 years) for boys (N = 160) and girls (N = 172). Similar age distributions were observed at each time-point, with boys being slightly older at each time-point; however, this difference was insignificant ($p > 0.05$).

Table 1. Association between BMI and fat percentage scores obtained from the Omrom and mathematical equations.

	BMI (Equations)		
	T1	**T2**	**T3**
BMI(BIA) T1	0.93 ***	-	-
BMI(BIA) T2	-	0.97 ***	-
BMI(BIA) T3	-	-	0.99 ***
	Sum of Skinfolds		
	T1	**T2**	**T3**
Fat % (BIA) T1	0.68 ***	-	-
Fat% % (BIA) T2	-	0.89 ***	-
Fat% % (BIA) T3	-	-	0.86 ***

BMI = body weight index; BIA=Results obtained by bioelectrical impedance analysis; T1 = Baseline measurement (2010); T2 = first follow-up measurement (2013); T3 = second follow-up measurement (2016) 0; *** = r > 0.5 (strong correlation).

Table 2. Basic statistics and significance of changes in the body composition profiles of boys and girls over three time-points.

	Boys (N = 160)					
Variables	**T1**	**T2**	**T3**	**T1-T2**	**T2-T3**	**T1-T3**
Age (years)	6.93 ± 0.51	9.93 ± 0.36	12.93 ± 0.38	3 *	3 *	6 *
BMI	15.95 ± 2.25	17.76 ± 3.49	19.61 ± 4.34	1.81 *	1.85 *	3.66 *
Stature (cm)	120.85 ± 6.49	136.18 ± 6.89	152.60 ± 9.03	15.33 *	16.42 *	31.75 *
Weight (kg)	23.41 ± 5.14	33.25 ± 9.06	46.01 ± 13.87	9.84 *	12.76 *	22.6 *
Sub-scapular sf (mm)	6.29 ± 2.92	7.49 ± 4.92	9.15 ± 7.70	1.2 *	1.66 *	2.86 *
Triceps sf (mm)	8.28 ± 3.64	10.36 ± 5.54	11.05 ± 7.27	2.08 *	0.69	2.77 *
Calf sf (mm)	7.91 ± 3.74	11.81 ± 6.26	13.87 ± 8.53	3.9 *	2.06 *	5.96 *
Waist-circ (cm)	55.13 ± 6.01	61.19 ± 8.16	66.40 ± 10.03	6.06 *	5.21 *	11.27 *
Mid-upper arm circ (cm)	-	20.65 ± 3.70	23.01 ± 4.76	-	2.36 *	-
% Fat Weight	19.09 ± 6.89	20.95 ± 7.84	19.89 ± 8.36	1.86 *	−1.06	0.8
% Muscle Weight	25.23 ± 4.47	31.21 ± 3.03	35.83 ± 4.33	5.98 *	4.62 *	10.6 *
	Girls (N = 172)					
Age (years)	6.87 ± 0.48	9.87 ± 0.38	12.87 ± 0.37	3 *	3 *	6 *
BMI	15.69 ± 2.02	17.75 ± 3.41	20.28 ± 4.11	2.06 *	2.53 *	4.59 *
Stature (cm)	119.44 ± 5.93	136.35 ± 7.22	154.46 ± 7.08	16.19 *	18.11 *	35.02 *
Weight (kg)	22.51 ± 4.26	33.11 ± 8.16	48.60 ± 11.74	10.6 *	15.49 *	26.09 *
Sub-scapular sf (mm)	7.32 ± 3.14	8.79 ± 5.14	10.73 ± 5.73	1.47 *	1.94 *	3.41 *
Triceps sf (mm)	9.50 ± 3.43	12.48 ± 5.24	14.47 ± 6.92	2.98 *	1.99 *	4.97 *
Calf sf (mm)	9.59 ± 3.65	14.25 ± 5.97	18.67 ± 8.93	4.66 *	4.42 *	9.08 *
Waist-circ (cm)	54.03 ± 4.90	59.51 ± 7.25	66.21 ± 8.85	5.48 *	6.7 *	12.18 *
Mid-upper arm circ (cm)	-	20.81 ± 3.43	23.88 ± 4.70	-	3.07 *	-
% Fat mass	16.41 ± 6.93	21.80 ± 8.23	25.45 ± 7.99	5.39 *	3.65 *	9.04 *
% Muscle mass	26.62 ± 3.05	30.55 ± 2.60	32.71 ± 3.75	3.93 *	2.16 *	6.09 *

BMI = Body mass index; cm-centimeter; kg = kilogram; mm = millimeter; sf = skinfold; circ = circumference; SD = Standard deviation; N = Number of participants; T1 = Baseline measurement, 6 years; (2010); T2 = 9 years (2013); T3= 12 years (2016); * = Statistical significant ($p < 0.01$) Note: Values represent the absolute mean differences.

The boys' anthropometric and body composition profiles over the three time-points (Table 2) revealed statistically significant changes ($p < 0.01$) across most variables. Body mass index (BMI) increased from 15.95 ± 2.25 kg/m^2 at T1 to 19.61 ± 4.34 kg/m^2 at T3, reflecting a significant cumulative gain of 3.66 kg/m^2. Similarly, stature and body weight showed continuous and statistically significant increases over time, with stature increasing by 31.75 cm and body weight increasing by 22.6 kg from T1 to T3. Sub-scapular and calf skinfolds showed consistent and significant increases at all time-points, while triceps skinfold thickness increased significantly from T1 to T2 (+2.08 mm) and from T1 to T3 (+2.77 mm), but not between T2 and T3 (+0.69 mm, $p > 0.01$). Waist circumference increased from 55.13 ± 6.01 cm to 66.40 ± 10.03 cm across the three measurements, with significant changes ($p < 0.01$) observed at each interval. Although mid-upper arm circumference was only recorded from T2 onwards, it increased significantly (++2.36 cm), from 20.65 cm to 23.01 cm by T3 ($p < 0.01$).

Regarding body composition, boys demonstrated a significant increase in muscle mass percentage, rising from 25.23% at T1 to 35.83% at T3, indicating a total gain of 10.6% over the six years. However, examining the fat mass percentage revealed minor changes, with only the increase from T to T2 (+1.86%) being significant ($p < 0.01$). Furthermore, a non-significant decrease from T2 to T3 was found in fat mass percentage, resulting in a slight, insignificant overall increase of 0.8%.

Significant changes were also observed in girls across all anthropometric and body composition measures from 6 to 12 years of age. BMI changed significantly, with 4.59 kg/m^2 from 15.69 ± 2.02 kg/m^2 at T1 to 20.28 ± 4.11 kg/m^2 at T3 $p < 0.01$). Stature and weight showed even greater increases than those observed in boys, with stature rising by 35.02 cm and weight increasing by 26.09 kg over the study period. These changes were statistically significant across all time intervals. Subcutaneous fat levels measured by skinfolds increased significantly across all measurement sites for girls. Triceps, sub-scapular, and calf skinfolds all showed significant gains ($p < 0.01$) at each interval, with the calf skinfold demonstrating the highest overall increase (+9.08 mm). Waist circumference increased from 54.03 ± 4.90 cm at T1 to 66.21 ± 8.85 cm at T3, with statistically significant changes across all time-points.

Mid-upper arm circumference, measured from T2 onward, increased significantly ($p < 0.01$), by 3.07 cm, at T3. Girls also exhibited significant increases in body fat percentage, rising from 16.41% at T1 to 25.45% at T3, with each interval showing statistically significant changes ($p < 0.01$). Muscle mass percentage increased significantly, albeit to a lesser extent than in boys, rising from 26.62% to 32.71% over the six years (+6.09%, $p < 0.01$).

The increasing standard deviations observed across most variables from T1 to T3 in both boys and girls indicate growing variability in body composition with age. For both sexes, measurements, such as weight, skinfold thickness, and body fat percentage, showed wider distributions over time, reflecting individual differences in growth rates, maturation timing, and lifestyle factors. Notably, girls exhibited slightly higher variability in fat-related measures, while boys showed greater dispersion in muscle mass gains. These findings highlight the importance of considering inter-individual variability when interpreting developmental trends during late childhood and early adolescence.

Table 3 and Figures 2 and 3 display the results of a correlation analysis between the various body composition variables, weight, and BMI at the three age time-points by sex. The correlation patterns between body composition variables and weight and BMI show notable sex differences across the three measurement points. In boys, stature maintains a stronger relationship with weight than BMI, while in girls, this association is weaker overall, especially with BMI. Skinfold measures (sub-scapular, triceps, and calf) and waist circumference show consistently strong correlations with both weight and BMI for both genders. However, girls generally exhibit slightly higher BMI correlations with fat-related variables, notably with the percentage of fat mass. Muscle mass presents a contrasting trend; while it initially shows a moderate positive correlation with weight in boys and girls, this shifts to a negative correlation with BMI over time, which is more pronounced in girls.

Table 3. Changes in correlations between various body composition co-variates and BMI and weight of boys and girls over three time-points.

Variable	T1		T2		T3	
	Weight	BMI	Weight	BMI	Weight	BMI
			Boys			
Stature (cm)	0.78 ***	0.39 **	0.74 ***	0.49 **	0.70 ***	0.41 **
Weight (kg)	-	0.84 ***	-	0.93 ***	-	0.90 ***
Sub-scapular skinfold (mm)	0.70 ***	0.76 ***	0.88 ***	0.91 ***	0.72 ***	0.82 ***
Triceps skinfold (mm)	0.84 ***	0.79 ***	0.89 ***	0.91 ***	0.75 ***	0.87 ***
Calf skinfold (mm)	0.82 ***	0.77 ***	0.78 ***	0.81 ***	0.75 ***	0.85 ***
Waist-circumference (cm)	0.85 **	0.76 ***	0.94 ***	0.94 ***	0.89 ***	0.94 ***
Mid-upper-arm circumference (cm)	-	-	0.94 ***	0.94 ***	0.87 ***	0.91 ***
% fat mass	0.61 ***	0.77 ***	0.83 ***	0.94 **	0.65 ***	0.85 ***
% muscle mass	0.66 ***	0.34 **	0.20 *	0.00	−0.11 *	−0.29 *
			Girls			
Stature (cm)	0.73 ***	0.27 *	0.62 ***	0.29 **	0.55 ***	0.24 *
Weight (kg)	-	0.85 ***	-	0.88 ***	-	0.93 ***
Sub-scapular skinfold (mm)	0.62 ***	0.76 ***	0.84 ***	0.89 ****	0.78 ***	0.85 ***
Triceps skinfold (mm)	0.75 ***	0.77 ***	0.83 ***	0.83 ***	0.74 ***	0.82 ***
Calf skinfold (mm)	0.70 ***	0.72 ***	0.76 ***	0.79 ***	0.71 ***	0.77 ***
Waist-circumference (cm)	0.81 ***	0.81 ***	0.91 ***	0.90 ***	0.91 ***	0.94 ***
Mid-upper-arm circumference (cm)	-	-	0.91 ***	0.90 ***	0.88 ***	0.89 ***
% fat mass	0.71 ***	0.95 ***	0.82 ***	0.91 ***	0.78 ***	0.88 ***
% muscle mass	0.53 ***	0.11 *	−0.10 *	−0.35 **	−0.33 **	−0.46 **

BMI = Body mass index; sf = skinfold; circ = circumference; T1 = Baseline measurement (2010); T2 = first follow-up measurement (2013); T3 = second follow-up measurement (2016); * = $0.1 < r < 0.3$ (small correlation); ** = $0.3 < r < 0.5$ (medium correlation); *** = $r > 0.5$ (strong correlation). Note: Correlations are based on the absolute values of boys and girls.

Overall, while BMI and weight are both strongly correlated with several body composition variables, BMI appears to be a more sensitive indicator of fat accumulation and a less reliable measure of muscularity, especially in girls as they age.

BMI included the sub-scapular skinfolds, triceps skinfolds, calf skinfolds, and waist-circumference, which all showed strong positive correlations across the three time-points, with correlations often exceeding 0.75. Mid-upper-arm circumference also displayed strong associations, particularly at T2 and T3, which correlated highly with weight and BMI.

Body fat percentage demonstrated moderate to strong correlations with weight and BMI, with stronger associations observed at later (older) time-points (9 and 12 years). Conversely, the percentage of muscle mass exhibited weak or negative correlations, particularly at 12 years, suggesting that it has a minimal influence on either weight or BMI. Overall, body fat measurements (skinfolds and fat percentage) consistently showed stronger relationships with BMI than with muscle mass, emphasizing the greater role of fat in predicting both weight and BMI across time.

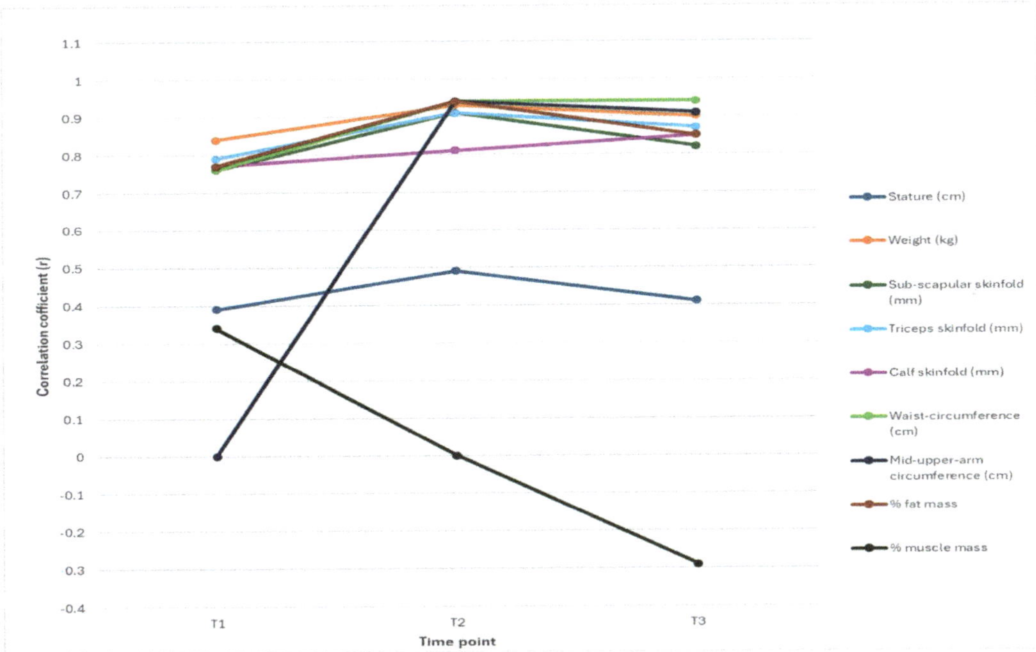

Figure 2. Temporal evolution of BMI correlations in boys.

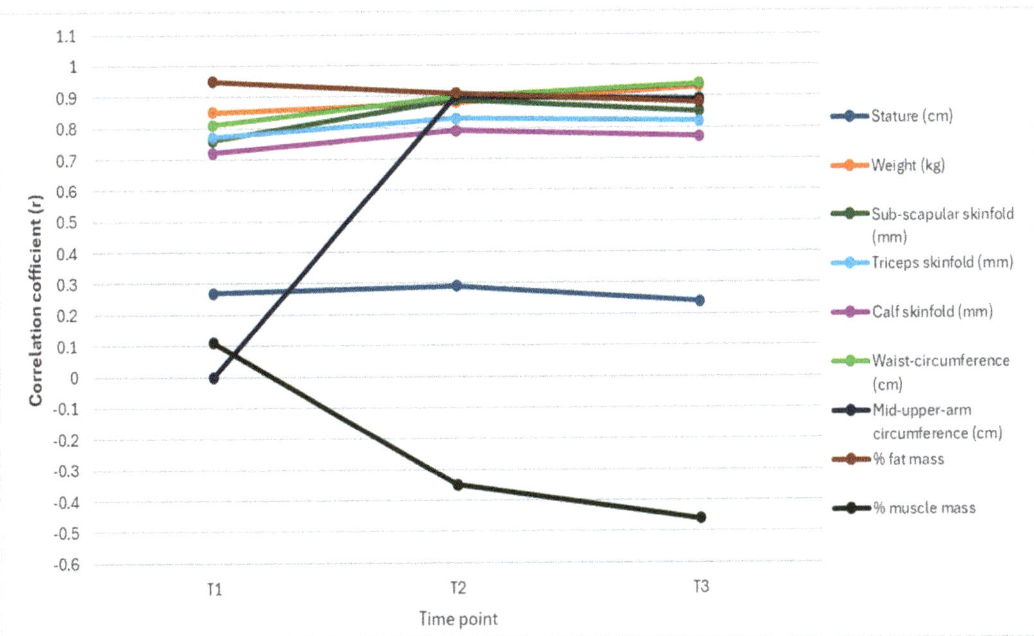

Figure 3. Temporal evolution of BMI correlations in girls.

In girls, the correlations reveal several key patterns. Weight and BMI exhibit a strong positive correlation at all time-points, with the strongest relationship observed at 12 years (r = 0.93). Body fat measurements, including sub-scapular, triceps, and calf skinfolds, consistently show moderate to strong correlations with both weight and BMI, particularly at later time-points (9 and 12 years), highlighting their significant role in predicting body composition. Waist circumference demonstrates the strongest association, with correlations exceeding 0.90 for both weight and BMI at 9 and 12 years. The percentage of fat mass also shows a strong positive correlation with BMI (ranging from 0.88 to 0.95) across all time-points, while also correlating moderately with weight. In contrast, muscle mass percentage exhibits weak or negative associations with BMI and weight correlations, particularly with both weight and BMI, especially at 9 and 12 years, indicating a diminishing relationship over time. Overall, fat-related variables showed stronger correlations with both weight and BMI, while muscle mass showed a limited influence, particularly with increasing age.

Figures 2 and 3 illustrate changes in correlations, revealing that BMI maintained consistently strong associations with subcutaneous fat measures (e.g., skinfolds, waist circumference) and fat mass percentage across all ages, indicating its continued relevance as a proxy for adiposity during childhood and early adolescence. However, the strength of association between BMI and muscle mass declined over time, becoming negative by age 12, particularly in girls, highlighting a shift in BMI's predictive utility. These findings underscore the importance of considering developmental stage and sex when interpreting BMI in relation to underlying body composition.

Tables 4 and 5 provide the results of a series of stepwise regression analyses conducted to examine the contribution of different body composition traits to BMI across different ages. Each regression model included the following independent variables: stature; weight; sub-scapular, triceps, and calf skinfolds; waist circumference; mid-upper arm circumference; fat mass percentage; and muscle mass percentage. In the multiple regression analysis, β (beta) coefficients were used to assess the independent contribution of each body composition variable to BMI while controlling for the effects of other predictors. Positive β values indicated that increases in a given variable were associated with a higher BMI. In contrast, negative β values reflected an inverse relationship, such as those observed in values for stature. The magnitude of the β coefficient denotes the strength of the association, with larger absolute values indicating stronger predictive power. These coefficients offer a clearer understanding of which variables most significantly influence BMI across different developmental stages.

The multiple regression analysis for boys (Table 4) across the three age time-points reveals shifting patterns in the predictors of BMI as boys progress through early to mid-childhood. Strong relationships between the dependent and independent variables were found across all three age models (6, 9, and 12 years). In this regard, a correlation coefficient between R = 0.9458 and R = 0.9876 was found, with a coefficient of determination between R^2 = 0.8946 and R^2 = 0.9755. This indicates a good-to-excellent fit, explaining between 89.46% and 97.55% of the variance in these dependent variables, with the highest fit at age 9. All three models are statistically significant ($p < 0.0000$), supported by high F-statistics between 160.32 and 644.95, reinforcing the robustness of the models. At age 6 (T1), only two variables—weight (b* = 1.398) and stature (b* = −0.773)—showed significant contributions ($p < 0.001$). By age 9 (T2), the model's explanatory power increases substantially (R^2 = 0.976) and more variables emerge as significant predictors. In addition to weight and stature, waist circumference, fat percentage, mid-upper arm circumference, and calf skinfold also become significant. At age 12, although the overall variance explained slightly decreases (R^2 = 0.964), the pattern becomes more refined, with waist circumference (b* = 0.324) becoming the strongest contributor after weight. At the same time, fat percentage and mid-upper arm circumference also remain significant, while the sub-scapular skinfold becomes a modest but significant predictor (b* = 0.067, p = 0.028). Notably, muscle mass percentage and most skinfold measures remain statistically nonsignificant across all ages.

Table 4. Multiple regression analysis of co-variants in predicting BMI of boys over three time-points.

	T1 (6 years) R = 0.945, R^2 = 0.894 Adjusted R^2 = 0.889, F(8,151) = 160.32 $p < 0.000$ * Std. error of estimate: 0.748		
Intercept	b = 32.554 Std. error of b = 3.158, p-value = 0.000 *		
Variables	b*	Std. error	p
Weight (kg)	1.398 *	0.110	<0.01 *
Stature (M)	−0.772 *	0.098	<0.01 *
Waist circumference (cm)	0.052	0.052	0.445
Muscle %	0.044	0.060	0.560
Calf sf (mm)	0.044	0.057	0.726
Fat %	−0.034	0.059	0.322
Sub-scap sf (mm)	−0.030	0.054	0.568
Triceps sf (mm)	−0.024	0.068	0.726
Mid-upper arm circumference (cm)	-	-	-
	T2 (9 years) R = 0.987, R^2 = 0.975 Adjusted R^2 = 0.974, F (9,150) = 664.95 $p < 0.000$ * Std. error of estimate: 0.562		
	b = 18.969 Std. error of b = 2.003, p-value = 0.000 *		
Weight (kg)	0.625 *	0.068	<0.01 *
Stature (cm)	−0.285 *	0.036	<0.01 *
Waist circumference (cm)	0.211 *	0.049	<0.01 *
Fat %	0.185 *	0.039	<0.01 *
Mid-upper arm circumference (cm)	0.105 *	0.053	<0.01 *
Calf sf (mm)	0.084	0.025	<0.01 *
Sub-scap sf (mm)	−0.031	0.039	0.42
Triceps sf (mm)	0.024	0.044	0.57
Muscle %	0.016	0.020	0.41
	T3 (12 years) R = 0.981, R^2 = 0.964 Adjusted R^2 = 0.961, F (9,150) = 446.74 $p < 0.000$ * Std. error of estimate: 0.848		
	B = 12.520 Std. error of b = 2.105, p-value = 0.000 *		
Weight (kg)	0.469 *	0.051	<0.01 *
Waist circumference (cm)	0.323 *	0.046	<0.01 *
Stature (cm)	−0.210 *	0.031	<0.01 *
Fat %	0.171 *	0.042	<0.01 *
Mid-upper arm circumference (cm)	0.154 *	0.040	<0.01 *
Sub-scap sf (mm)	0.066 *	0.030	<0.09 *
Triceps sf (mm)	−0.049	0.047	0.29
Calf sf (mm)	0.039	0.041	0.34
Muscle %	0.031	0.020	0.13

BMI = Body mass index; sf = skinfold; Circ = circumference; T1 = Baseline measurement (2010); T2 = first follow-up measurement (2013); T3 = second follow-up measurement (2016); std. error = standard error; p = statistically significant ($p < 0.05$); b* = beta coefficients; * = statistically significant ($p < 0.05$).

Table 5. Multiple regression analysis of co-variants in predicting BMI of girls.

	T1 (6 years) R = 0.996 R^2 = 0.992, Adjusted R^2 = 0.991, F(8,163) = 263.45 $p < 0.0000$ * Std. error of estimate: 0.18146		
Intercept	b = 25.995 Std. error of b = 0.737, *p*-value = 0.000 *		
Variables	b*	Std. error	*p*
Weight (kg)	1.061 *	0.030	<0.01 *
Stature (cm)	−0.585 *	0.024	<0.01 *
Fat %	0.219 *	0.020	<0.01 *
Muscle %	0.053 *	0.014	<0.01
Sub-scapular (mm)	0.044 *	0.014	<0.01 *
Calf sf (mm)	0.022	0.014	0.11
Triceps sf (mm)	−0.020	0.015	0.17
Waist circ (cm)	−0.000	0.015	0.97
Mid-upper arm circ (cm)	-	-	-
	T2 (9 years) R = 0.959 R^2 = 0.919 Adjusted R^2 = 0.915, F (9,162) = 206.81 $p < 0.000$ * Std. error of estimate: 0.993		
	B = 12.070 Std. error of b = 2.467, *p*-value = 0.000 *		
Weight (kg)	0.412 *	0.084	<0.01 *
Fat %	0.355 *	0.068	<0.01 *
Mid-upper arm circ (cm)	0.231 *	0.072	<0.01 *
Stature (cm)	−0.178 *	0.041	<0.01 *
Triceps sf (mm)	−0.173 *	0.061	<0.01 *
Waist circ (cm)	0.151 *	0.070	<0.01 *
Sub-scapular sf (mm)	0.092	0.062	0.14
Calf sf (mm)	0.003	0.045	0.94
Muscle %	0.000	0.036	0.98
	T3 (12 years) R = 0.988 R^2 = 0.976 Adjusted R^2 = 0.975, F (9,162) = 744.33 $p < 0.000$ * Std. error of estimate: 0.649		
	B = 20.688 Std. error of b = 2.049, *p*-value = 0.000 *		
Weight (kg)	0.656 *	0.054	<0.01 *
Stature (cm)	−0.243 *	0.021	<0.01 *
Waist circ (cm)	0.171 *	0.036	<0.01 *
Fat %	0.139 *	0.033	<0.01 *
Mid-upper arm circ (cm)	0.114 *	0.030	<0.01 *
Sub-scapular sf (mm)	0.037	0.026	0.16
Calf sf (mm)	0.014	0.027	0.59
Muscle %	0.009	0.015	0.51
Triceps sf (mm)	0.002	0.032	0.93

BMI = Body mass index; sf = skinfold; Circ = circumference; T1 = Baseline measurement (2010); T2 = first follow-up measurement (2013); T3 = second follow-up measurement (2016); std. error = standard error; *p* = statistically significant ($p < 0.05$); b* = beta coefficients; * = statistically significant ($p < 0.05$).

In girls (Table 5), the regression analyses reveal strong relationships between the independent variables and the dependent variable across the three age models. For Model T1 (age 6), the correlation coefficient (R = 0.9962) and coefficient of determination (R^2 = 0.9923) indicate an excellent fit, explaining 99.23% of the variance in the dependent variable. Similarly, Model T2 at age 9 (R = 0.9591, R^2 = 0.9199), and Model T3 at age 12 (R = 0.9881, R^2 = 0.9764), also demonstrate strong explanatory power, with R^2 values above 0.91. All three models are statistically significant ($p < 0.0000$), supported by high F-statistics (2634.5, 206.81, and 744.33, respectively), reinforcing the robustness of the models. The standard errors of the estimates are relatively low, with the smallest error (0.18146) at age 6, followed by ages 12 (0.64973) and 9 (0.99390), suggesting that the models' predictions are highly accurate. Notably, several independent variables consistently demonstrate statistically significant relationships across the models. Stature (b* = −0.585361, $p < 0.0001$ for T1) and fat percentage (b* = 0.053296, $p = 0.000385$ for T1) have strong and significant effects on the dependent variable, with negative and positive relationships, respectively.

The effect of weight is consistently significant and positive across all models (b* ranging from 0.412900 to 1.061365). However, other variables, such as the sub-scapular and triceps skinfolds, show less consistent significance.

At age 9, the triceps skinfold is significant (b* = −0.173048, $p = 0.005272$); however, at age 12, it becomes non-significant (b* = 0.002438, $p = 0.939485$). These findings suggest that while some variables, such as body fat percentage and stature, consistently influence the outcome, others may have context-dependent effects that vary across models. Overall, the results demonstrate the predictive power of the models, highlighting the importance of specific variables, such as stature, body fat percentage, and weight, in explaining the variations in the dependent variable.

When exploring sex differences in the predictive power of the independent variables and changes in the predictive power, as displayed in Figure 4 (boys) and Figure 5 (girls), the models demonstrate strong predictive power, with R^2 values consistently above 0.89 for boys and 0.91 for girls, indicating that the included variables explain a significant proportion of BMI variance. Weight (weight) emerges as the strongest predictor in both boys and girls, although its influence is initially higher in boys (β* = 1.40 at age 6) and stabilizes over time. In contrast, the impact of stature (height) is consistently negative, with a stronger effect on boys (β* = −0.77 at age 6) compared to girls (β* = −0.58 at T1).

Fat percentage significantly predicts BMI in both groups, but with a greater effect in girls, especially at age 9 (β* = 0.36, $p < 0.001$). Mid-upper arm and waist circumference measurements become increasingly relevant for boys over time, whereas they show earlier significance in girls. Overall, BMI prediction is slightly stronger in girls, particularly at age 6, suggesting sex-based differences in body composition dynamics. These findings highlight the importance of considering both adiposity and muscle distribution when modelling BMI trajectories in children.

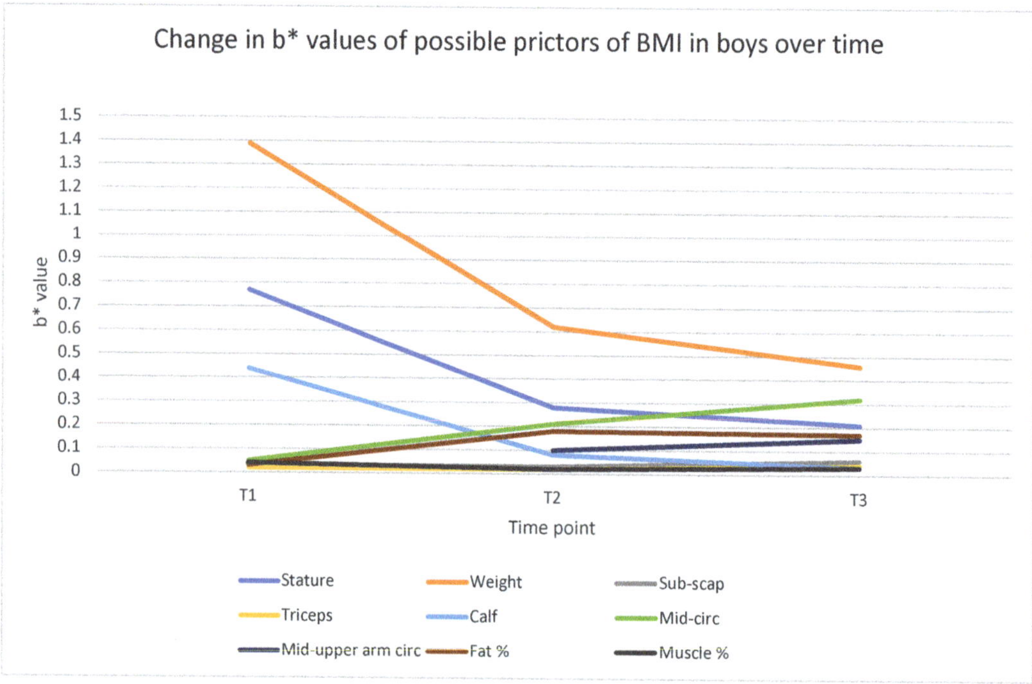

Figure 4. Changes in boys' predictive b* value over time.

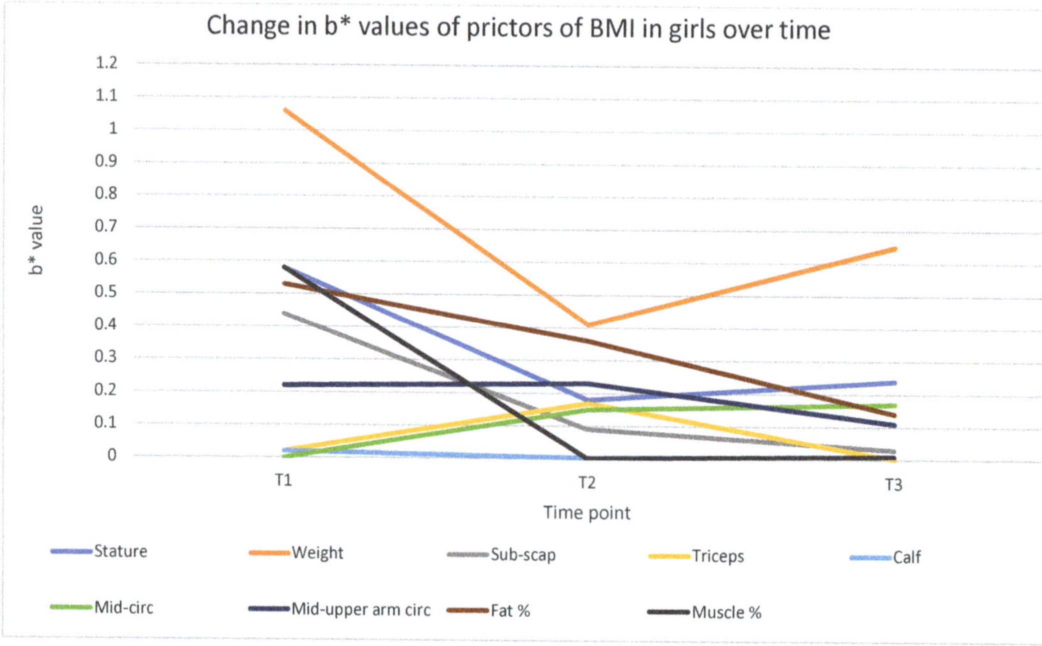

Figure 5. Changes in girls' predictive b* value over time.

4. Discussion

The primary objective of this study was to investigate potential changes in the associations between body composition covariates, specifically BMI, from 6 to 12 years in boys and girls. Taking into account the evidence of potential multicollinearity found as explained, the main finding indicated that weight and stature were consistently the strongest predictors of BMI in both sexes, particularly at younger ages. As children aged, additional variables, such as waist circumference, mid-upper arm circumference, and fat percentage, became more influential, with differences in the timing and magnitude of these effects between boys and girls. Girls showed a broader set of significant predictors earlier in development, while boys exhibited a more consistent pattern over time. These findings highlight sex-specific developmental trajectories in body composition and BMI determinants. In this regard, no studies were found to compare these results with, and no study specifically focuses on the changes and influences of various anthropometric and body composition covariates regarding BMI and sex differences. As a sub-objective, we first analyzed the validity of the OMRON BF 511 bioimpedance scale to determine BMI and fat percentage, as we were interested in body composition variables that are only available with this measuring tool. High correlations ($r > 0.93$) were found between the OMRON-generated BMI and BMI-calculated scores (Table 1). The validity of the Omron to determine other aspects of body composition, and more specifically the percentage of BF, from the Omron BF 511, further indicated significantly high correlations ($0.68 > r < 0.89$) with BF, as calculated with an equation. In agreement, another study conducted by Brtková [38], which focused on 52 older participants with an average age of 22.4 years, also found an even stronger correlation ($r = 0.93$) between skinfold measurements and the use of the Omron BF511 as parameters for body fat percentage. These results align with the parameters set out by Omron stipulating that the accuracy of Omron models as bioimpedance scales varies from model to model based on the 'Standard Error of Estimate' (SSE). The SSE stipulates that 68% of all measurements for different users are accurate to within 3.5–4.1%, relative to body fat percentage ($kg.m^2$) [34]. Our results agreed with this and verified that the OMROM B511 bioimpedance analyzer could be used in our study as an accurate measuring tool with which to determine the BMI and body composition of children relatively accurately.

Although the aim of this paper was not to investigate sex differences in anthropometric and body composition profiles (Table 2), boys and girls showed similar anthropometric profiles throughout the study, with boys being slightly taller at age 6 (+2.6 cm) and T2 (+0.01 cm), with girls surpassing them at age 12 (+0.67 cm). Similar results were found in weight, although girls surpassed boys at age 9. Consequently, these minor differences resulted in very similar BMI scores (0.04–0.06) in boys and girls throughout the study. These results align with an Australian study by Cochrane [39], which also found only slight differences between boys' and girls' anthropometric profiles between the ages of 6 and 13. It is, however, not expected that significant anthropometric differences will exist before the onset of puberty due to insignificant hormonal influences [40]. Regarding body composition, girls tend to show higher body composition scores regarding skinfolds and body fat percentage throughout the study, from age 6 to 12 years (Table 2). In this regard, Pelemis et al. [41] found that girls at an average age of 6.25 years already showed higher abdominal and skinfolds at the back than boys. Furthermore, earlier research already showed that girls tend to have higher body fat percentages and lower muscle mass percentages compared to boys before puberty [29], which in turn will be noticeable in their slightly higher circumferences and skinfolds.

The body composition characteristics of young developing children are critical to understanding growth, development, and long-term health outcomes. Differences in body fat percentage, lean weight, and fat distribution between boys and girls emerge during primary school and become more pronounced during puberty, according to various studies [42–44], which our findings also confirmed. Although both boys and girls exhibit steady increases in height and weight throughout childhood, body composition diverges as they approach puberty [45]. Boys develop greater lean mass and bone density, whereas girls tend to accumulate more fat mass, particularly in the gluteofemoral region [46]. These differences are primarily driven by hormonal changes, particularly by increases in *estrogen* in girls and *testosterone* in boys [43]. Results from the current study are well aligned with this; boys initially had a higher fat percentage and lower muscle mass percentage compared to girls at age 6; however, this changed at a later age (Table 2). At an average age of 9 years, boys already surpassed girls in terms of muscle mass percentage (*testosterone*-related), with the difference becoming even larger with increasing age. On the other hand, boys also initially had a lower fat percentage at age 6, with girls surpassing them from age 9, with sex differences increasing further at age 12. This might be due to girls entering the pubertal phase with increased *estrogenic* influences.

Our sex-specific analysis, which identified the body composition variables that influence BMI at 6, 9, and 12 years and whether these contributions varied by age, revealed notable differences between boys and girls (Tables 3–5 and Figures 2–5). For boys, weight and stature were consistently the strongest predictors across all time-points, with waist circumference and fat mass percentage gaining predictive strength with age. By age 12, the mid-upper arm circumference also became a significant indicator. Skinfolds and muscle mass percentage generally showed weak or non-significant contributions. In contrast, a broader range of significant predictors emerges in girls, particularly at younger ages. At the age of 6, fat mass percentage, muscle mass percentage, and sub-scapular skinfold thickness significantly influence BMI, alongside weight and stature. By the ages of 9 and 12 years, like boys, waist circumference, fat mass percentage, and mid-upper arm circumference become more influential, while skinfolds and muscle mass percentage lose significance. Overall, the BMI of girls is more strongly influenced by fat distribution and composition at younger ages, whereas the BMI of boys is more consistently driven by weight and stature. These patterns suggest developmental and physiological differences between boys and girls in terms of fat accumulation and body composition. The findings align with, and add to, existing studies in the literature on body composition, particularly regarding fat mass, lean muscle mass, and growth patterns [47,48].

The models consistently indicate that weight is the strongest and most significant predictor of body composition in both sexes throughout the study. However, the strength of this contribution declined by more than 60% in boys (Table 4) and by approximately 40% in girls (Table 5) between the ages of 6 and 12 years. This is consistent with studies showing that total body weight is a primary determinant of fat-free mass and fat mass in children [46,49]. Interestingly, stature (height) had a negative relationship with body composition in both genders across all three age models. This suggests that as boys and girls grow taller, they may not necessarily gain proportional fat mass or muscle mass increases in this age group. This finding is supported by research indicating that taller children tend to have lower body fat percentages relative to their weight [50,51]. However, sex differences exist, especially at a later age, as boys tend to develop more lean weight, while girls have a higher percentage of body fat [43]. Furthermore, stature was primarily found in the top three influences of BMI in boys and girls; however, in girls, stature contributed to only the 4th highest influence at age 9. This might be due to girls entering their pubertal phase at around the age of 9–10 years, with puberty influences, such as increased fat mass (as seen in Table 5), contributing more towards BMI.

In boys, waist circumference was one of the top three significant contributors in all three models (Table 4, Figure 4) compared to girls, where waist circumference was insignificant at age 6 (p = 0.97). Waist circumference was 6th in contributing to BMI at age 9 ($p < 0.01$) and only become a significant factor at age 12, where it was 3rd in the stepwise model (Table 5, Figure 5). This is well aligned with findings reported by Nazare et al. [52] that reveal a slightly stronger correlation between the waist circumference of boys compared to girls, with a BMI (r = 0.84) for females and (r = 0.87) for males. Neuhauser [53] reports that waist circumference in children and adolescents varies depending on age, gender, and ethnic group. In this regard, boys tend to accumulate more abdominal (central) fat than girls, especially in early childhood, where girls typically have more peripheral fat distribution (hips, thighs, buttocks), which is less directly reflected in waist circumference [54,55]. Since WC primarily measures central fat, it naturally correlates more strongly with BMI in boys. This suggests that waist circumference may reflect different types of fat in boys and girls, influencing its correlation with BMI differently across sexes. Furthermore, this also aligns with findings by Ronnecke et al. [56], showing similar trends of relatively constant increases in median percentiles from 3 to 16 years in both sexes, with boys showing higher values in all waist circumference percentile curves. Mid-upper arm circumference, although primarily located at mid-table in contribution, also emerged as a significant contributor to BMI in boys and girls from the ages of 9 to 12 years. These findings align with the findings reported by Frisancho [57] indicating that MUAC is an excellent indicator of total body fat and nutritional status in children. Since arm circumference reflects muscle and fat mass, its interrelationship with BMI is understandable, especially in girls, who tend to store subcutaneous fat in the arms [58]. The significance of MUAC, and its similar contribution compared to fat mass percentage at ages 9 and 12 years, is also confirmed by a study indicating that upper-arm measurements are more closely associated with total fat mass and lean mass distribution [51].

While fat percentage (Fat%) emerged as a significant contributor at ages 9 and 12 years in boys and girls, individual skinfold thickness measurements (Triceps sf, Calf sf, and Subscapular sf), although stronger in girls, show mixed results (Tables 4 and 5, Figures 4 and 5). This could be due to sex-specific and gene-related patterns of fat distribution, where girls tend to accumulate more subcutaneous fat, particularly in the gluteofemoral region. At the same time, boys have a more even distribution of fat [49]. Girls typically have higher fat mass than boys from early childhood, due to hormonal and metabolic differences, which explains why fat percentage strongly predicts BMI in girls [45].

Skinfolds are commonly used to assess adiposity. The lack of significance and/or the lower-ranked order of the individual skinfold measurements in the stepwise regression for both genders suggests that total fat percentage is a better predictor of body composition than individual skinfolds, which aligns with research by Moreno et al. [59]. The subscapular skinfold measurement in girls was a significant predictor at age 6 (β = 0.044, p = 0.0022), indicating that trunk fat plays a role in the BMI of girls. However, the triceps skinfold was negatively associated with BMI at age 9 ($\beta = -0.173$, p = 0.0052), which is an unusual finding, but may suggest that arm fat does not contribute significantly to BMI variations compared to trunk fat. Research by Moreno et al. [59] found, in this regard, that subcutaneous fat deposition patterns differ between boys and girls, with girls accumulating more peripheral fat and boys accumulating more central fat.

Muscle mass percentage was not a significant predictor in any model, for either boys and girls, which may indicate that muscle mass does not strongly influence the dependent variables being measured at this age. Since BMI is primarily driven by fat mass and weight during primary school years, lean mass may not significantly impact BMI calculations in boys and girls, as muscle development is not yet pronounced at this stage [45]. These findings are consistent with other studies showing that muscle mass differences between boys and girls become more significant only after puberty [49]. Boys have similar lean body mass levels to girls in early childhood but begin to develop more muscle mass as they approach puberty [46].

Because few studies were found that focused on providing an understanding of body composition co-variates that influence BMI in younger children and the changes in the contribution of these changes to BMI over a longitudinal period, our findings make a unique contribution to the scientific field to understanding changes in BMI in the age period between 6 and 13 years, and especially regarding difference in BMI changes in boys and girls in this developmental period. However, the study also has limitations that need to be considered. A significant limitation of the study that is acknowledged is the possible high degree of intercorrelation among several body composition variables, raising potential concerns regarding multicollinearity. The results from this statistical analysis would have ob-scured the ability to discern the unique contribution of each variable to the outcomes of interest. Future analyses could benefit from applying techniques such as Principal Components Analysis (PCA), to reduce redundancy and identify orthogonal compo-nents that better represent underlying constructs. Although a randomized and stratified research design was followed, the participants were recruited from only one province within South Africa, with a relatively small participant sample, which limits the generalizability of the findings to the broader South African population. Therefore, it is recommended that similar studies should be conducted in the other eight provinces for a stronger generalization. We acknowledge that not all factors that might influence BMI, such as eating habits, physical activity levels, pubertal status, and socioeconomic status, were included in our analyses. Therefore, it is recommended that future studies also investigate the influences of these covariates. Lastly, the loss of subjects over the study period was relatively large; we acknowledge that this limits the internal validity of the study findings.

5. Conclusions

This longitudinal study is the first to examine sex-specific changes and influences of BMI covariates in developing children aged 6 to 12 years. Weight and stature emerged as the strongest predictors of BMI across all ages, particularly at age 6. Over time, other body composition indicators—such as waist circumference, mid-upper arm circumference, and fat percentage—gained importance, with distinct patterns observed between boys and girls. Girls showed a broader range of early BMI influencers that shifted over time, while boys displayed a more consistent pattern, with increasing emphasis on fat-related variables. These findings enhance the understanding of childhood growth and body composition and have implications for early obesity screening. Furthermore, these findings highlight the complex interplay between anthropometric variables and BMI in children and the differences in this regard between boys and girls, which has implications for growth monitoring, nutritional assessment, and early interventions for childhood obesity. Furthermore, as waist circumference, mid-upper arm circumference, and fat mass percentage showed strong associations with BMI, future research should test the predictive value of these variables independently for identifying early obesity and metabolic risk, potentially replacing or supplementing BMI as a screening tool. Additionally, there is a need to translate these findings into user-friendly, cost-effective screening instruments for schools and healthcare services, particularly in low-resource settings. Future work could focus on developing and validating simplified screening algorithms that incorporate BMI covariates for the early identification of at-risk children.

Supplementary Materials: The following supporting information can be downloaded at: https://www.mdpi.com/article/10.3390/jfmk10030320/s1.

Author Contributions: Conceptualization, A.E.P.; methodology, A.E.P.; software, A.E.P. and B.G.; validation, A.E.P. and B.G.; formal analysis, A.E.P. and B.G.; investigation, A.E.P.; resources, A.E.P. ad B.G.; data curation, A.E.P. and B.G.; writing—original draft preparation, A.E.P. and B.G.; writing—review and editing, A.E.P. and B.G.; visualization, project administration, A.E.P.; funding acquisition, A.E.P. All authors have read and agreed to the published version of the manuscript.

Funding: This research project was funded by a self-initiated grant from the Medical Research Council of South Africa (MRC), the South African Sugar Association (SASA), and the National Research Foundation (NRF). Any opinions, findings, conclusions or recommendations emanating from the research are those of the authors; therefore, the MRC, NRF and SASA do not accept any responsibility or liability.

Institutional Review Board Statement: The study was conducted following the Declaration of Helsinki and approved by the Health Research Ethics Council of the North-West University (NWU-00070-09-A1; approval date: 30 April 2025).

Informed Consent Statement: Informed consent was obtained from all subjects involved in the study.

Data Availability Statement: The authors confirm that the data supporting these findings are not available online but are available from the authors upon reasonable request, in accordance with NWU policy guidelines.

Acknowledgments: The authors would like to express their sincere gratitude to all the senior researchers from the Kinderkinetics program and the 2010, 2013 and 2016 Kinderkinetics Honours students at North-West University for their assistance in collecting the data.

Conflicts of Interest: The authors declare no conflicts of interest.

Abbreviations

Abbreviation	Description
BIA	Bioelectrical impedance analysis
BMI	Body mass index
BMIz	Body mass index z-score
CIRC	Circumference
CM	Centimeter
DBE	Department of Basic Education
FM	Fat Weight
FFM	Fat-free Weight
HREC	Health research ethics committee
ISAK	International society for the advancement of Kinanthropometry
Kg	Kilogram
LMIC	Low- to middle-income country
NWP	North-West Province
MM	Millimeter
MUAC	Mid-upper arm circumference
NW-CHILD	North-West Child Study
SF	Skinfolds
SES	Socio-economic status
WHO	Worl health organization

References

1. WHO. Obesity and Overweight. 2024. Available online: https://www.who.int/news-room/fact-sheets/detail/obesity-and-overweight (accessed on 1 March 2024).
2. Lobstein, T.; Brinsden, H. *Atlas of Childhood Obesity*; World Obesity Federation: London, UK, 2019; Volume 211, p. 486.
3. Pulgarón, E.R. Childhood obesity: A review of increased risk for physical and psychological comorbidities. *Clin. Ther.* **2013**, *35*, A18–A32. [CrossRef] [PubMed]
4. Horesh, A.; Tsur, A.M.; Bardugo, A.; Twig, G. Adolescent and childhood obesity and excess morbidity and mortality in young adulthood—A systematic review. *Curr. Obes. Rep.* **2021**, *10*, 301–310. [CrossRef]
5. Jebeile, H.; Cardel, M.I.; Kyle, T.K.; Jastreboff, A.M. Addressing psychosocial health in the treatment and care of adolescents with obesity. *Obesity* **2021**, *29*, 1413–1422. [CrossRef]
6. Jebeile, H.; Kelly, A.S.; O'Malley, G.; Baur, L.A. Obesity in children and adolescents: Epidemiology, causes, assessment, and management. *Lancet Diabetes Endocrinol.* **2022**, *10*, 351–365. [CrossRef]
7. Whitney, E.; Rolfes, S.R. *Understanding Nutrition*, 13th ed.; Wadsworth (W+R): Belmont, CA, USA, 2012; Volume 13, p. 424.
8. McArdle, W.D.; Katch, F.I.; Katch, V.L. *Exercise Physiology, Nutrition, Energy, and Human Performance*, 7th ed.; Williams & Wilkins: Baltimore, MD, USA, 2010; p. 726.

9. Mulasi, U.; Kuchnia, A.J.; Col, A.J.; Earthman, C.P. Bioimpedance at the bedside: Current applications, limitations, and opportunities. *Nutr. Clin. Pract.* **2015**, *30*, 180–193. [CrossRef] [PubMed]
10. Adab, P.; Pallan, M.; Whincup, P.H. Is BMI the best measure of obesity. *Br. Med. J.* **2018**, *360*, k1274. [CrossRef]
11. Aubrey, A. Step Aside BMI, Body Composition Tests Are on the Rise. Here's What to Know. 2024. Available online: https://www.npr.org/sections/shots-health-news/2024/06/03/nx-s1-4973714/bmi-body-composition-scans-healthy-weight-muscle-fat (accessed on 3 June 2024).
12. Wang, Y. Epidemiology of childhood obesity—Methodological aspects and guidelines: What is new? *Int. J. Obes.* **2004**, *28*, S21–S28. [CrossRef] [PubMed]
13. Kyle, U.G.; Bosaeus, I.; De Lorenzo, A.D.; Deurenberg, P.; Elia, M.; Gómez, J.M.; Heitmann, B.L.; Kent-Smith, L.; Melchior, J.-C.; Pirlich, M.; et al. Bioelectrical impedance analysis-part II: Utilization in clinical practice. *Clin. Nutr.* **2004**, *23*, 1430–1453. [CrossRef]
14. Sergi, G.; De Rui, M.; Stubbs, B.; Verones, N.; Manzato, E. Measurement of lean body Weight using bioelectrical impedance analysis: A consideration of the pros and cons. *Aging Clin. Exp. Res.* **2017**, *29*, 591–597. [CrossRef]
15. Feng, Q.; Bešević, J.; Conroy, M.; Omiyale, W.; Lacey, B.; Allen, N. Comparison of body composition measures assessed by bioelectrical impedance analysis versus dual-energy X-ray absorptiometry in the United Kingdom Biobank. *Clin. Nutr. ESPEN* **2024**, *63*, 214–225. [CrossRef]
16. Khan, S.; Xanthakos, S.A.; Hornung, L.; Arce-Clachar, C.; Siegel, R.; Kalkwarf, H.J. Relative accuracy of bioelectrical impedance analysis for assessing body composition in children with severe obesity. *J. Pediatr. Gastroenterol. Nutr.* **2020**, *70*, e129–e135. [CrossRef]
17. Rothman, K.J. BMI-related errors in the measurement of obesity. *Int. J. Obes.* **2008**, *32*, S56–S59. [CrossRef]
18. Cole, T.J.; Faith, M.S.; Pietrobelli, A.; Heo, M. What is the best measure of adiposity change in growing children: BMI, BMI%, BMI z-score or BMI centile? *Eur. J. Clin. Nutr.* **2005**, *59*, 419–425. [CrossRef]
19. Berkey, C.S.; Colditz, G.A. Adiposity in adolescents: Change in actual BMI works better than change in BMI z score for longitudinal studies. *Ann. Epidemiol.* **2007**, *17*, 44–50. [CrossRef]
20. Paluch, R.A.; Epstein, L.H.; Roemmich, J.N. Comparison of methods to evaluate changes in relative body Weight index in pediatric weight control. *Am. J. Hum. Bio* **2007**, *19*, 487–494. [CrossRef] [PubMed]
21. Kolotourou, M.; Radley, D.; Chadwick, P.; Smith, L.; Orfanos, S.; Kapetanakis, V.; Singhal, A.; Cole, T.J.; Sacher, P.M. Is BMI alone a sufficient outcome to evaluate interventions for child obesity? *Child. Obes.* **2013**, *9*, 350–356. [CrossRef] [PubMed]
22. Freedman, D.S.; Goodwin Davies, A.J.; Phan, T.L.T.; Cole, F.S.; Pajor, N.; Rao, S.; Eneli, I.; Kompaniyets, L.; Lange, S.J.; Christakis, D.A.; et al. Measuring BMI change among children and adolescents. *Pediatr. Obes.* **2022**, *17*, e12889. [CrossRef] [PubMed]
23. Obeso, A.; Drouard, G.; Palviainen, T.; Wang, X.; Ollikainen, M.; Silventoinen, K.; Kaprio, J. Proteomic associations with fluctuation and long-term changes in BMI: A 40-year follow-up study. *Diabetes Obes. Metab.* **2025**, *27*, 4192–4202. [CrossRef]
24. Sattar, A.; Baig, S.; Rehman, N.; Bashir, B. Factors affecting BMI: Assessment of the effect of sociodemographic factors on BMI In the population of Ghulam Mohammad Abad Faisalabad. *Prof. Med. J.* **2013**, *20*, 956–964. [CrossRef]
25. WHO. Child Growth Standards. 2009. Available online: https://www.who.int/tools/child-growth-standards/standards/triceps-skinfold-for-age (accessed on 15 July 2025).
26. Werneck, A.O.; Silva, D.R.; Oyeyemi, A.L.; Fernandes, R.A.; Romanzini, M.; Cyrino, E.S.; de Arruda, M.; Ronque, E.R. Tracking of physical fitness in elementary school children: The role of changes in body fat. *Am. J. Hum. Biol.* **2019**, *31*, e23221. [CrossRef]
27. Cossio-Bolaños, M.; Gómez-Campos, R.; Campos, L.F.C.C.; Sulla-Torres, J.; Urra-Albornoz, C.; Lopes, V.P. Muscle strength and body fat percentage in children and adolescents from the Maule region, Chile. *Arch. Argent. Pediatr.* **2020**, *118*, 320–326.
28. Lalucci, M.P.D.P.S.; de Souza Marques, D.C.; Santos, I.C.; Caitano, J.Z.; Silva, B.F.; Valdés-Badilla, P.; Branco, B.H.M. Interrelationship among body Weight index, body composition, and biochemical profiles of overweight adolescents in south of Brazil: A cross-sectional study. *Nutr. Clínica Y Dietética Hosp.* **2023**, *43*, 31–38.
29. Mast, M.; Körtzinger, I.; König, E.; Müller, M.J. Gender differences in fat Weight of 5–7-year-old children. *Int. J. Obes.* **1998**, *22*, 878–884. [CrossRef]
30. Haywood, X.; Pienaar, A.E. Long-Term Influences of Stunting, Being Underweight, and Thinness on the Academic Performance of Primary School Girls: The NW-CHILD Study. *Int. J. Environ. Res. Public Health* **2001**, *18*, 8973. [CrossRef]
31. Pienaar, A.E. Prevalence of overweight and obesity among primary school children in a developing country: NW-CHILD longitudinal data of 6–9-yr-old children in South Africa. *BMC Obes.* **2015**, *2*, 2. [CrossRef]
32. Lohman, T.G. *Advances in Body Composition Assessment*; Human Kinetics: Champaign, IL, USA, 1992.
33. Steward, A.D.; Marfell-Jones, M.; Olds, T.; de Ridder, H. *International Standards for Anthropometric Assessment*; International Society for Advancement of Kinanthropometry: Glasgow, Scotland; National Library of Australia: Canberra, Australia, 2011; 137p.
34. Omron. Available online: https://www.omron-healthcare.com/en/products/weightmanagement (accessed on 28 May 2025).
35. Statsoft. *Statistica for Windows: General Convention and Statistics*; Statsoft: Tilsa, OK, USA, 2024.
36. IBM Corp. *IBM SPSS Statistics for Windows*; IBM Co.: Armonk, NY, USA, 2022.

37. Cohen, J. *Statistical Power Analysis for the Behavioral Sciences*, 2nd ed.; Lawrence Erlbaum Associates: Hillsdale, NJ, USA, 1988.
38. Brtková, M.; Bakalár, P.; Matúš, I.; Hančová, M.; Rimárová, K. Body composition of undergraduates—comparison of four different measurement methods. *Phys. Act. Rev.* **2014**, 2, 42–52.
39. Cochrane, T.; Davey, R.C.; de Castella, F.R. Anthropometric standards for Australian primary school children: Towards a system for monitoring and supporting children's development. *J. Sci. Med. Sport* **2017**, *20*, 284–289. [CrossRef]
40. Pretorius, S.S.; Neophytou, N.; Watson, E.D. Anthropometric profiles of 8–11-year-old children from a low-income setting in South Africa. *BMC Public Health* **2019**, *19*, 1–7. [CrossRef] [PubMed]
41. Pelemiš, V.; Macura, M.; Branković, P. Sex differences in morphological characteristics of preschool children. *Anthropol.* **2017**, *30*, 153–160.
42. Loomba-Albrecht, L.A.; Styne, D.M. Effect of puberty on body composition. *Curr. Opin. Endocrinol. Diabetes Obes.* **2009**, *16*, 10–15. [CrossRef] [PubMed]
43. Wells, J.C.K. Body composition and susceptibility to type 2 diabetes: An evolutionary perspective. *Nat. Rev. Endocrinol.* **2014**, *10*, 637–643. [CrossRef]
44. Rusek, W.; Baran, J.; Leszczak, J.; Adamczyk, M.; Baran, R.; Weres, A.; Inglot, G.; Czenczek-Lewandowska, E.; Pop, T. Changes in children's body composition and posture during puberty growth. *Children* **2021**, *8*, 288. [CrossRef]
45. Malina, R.M.; Bouchard, C.; Bar-Or, O. *Growth, Maturation, and Physical Activity*, 2nd ed.; Human Kinetics: Champaign, IL, USA, 2004.
46. Taylor, R.W.; Grant, A.M.; Williams, S.M.; Goulding, A. Sex differences in regional body fat distribution from pre- to post-puberty. *Obesity* **2010**, *18*, 1410–1416. [CrossRef]
47. Alves Junior, C.A.; Mocellin, M.C.; Gonçalves, E.C.A.; Silva, D.A.; Trindade, E.B. Anthropometric Indicators as Body Fat Discriminators in Children and Adolescents: A Systematic Review and Meta-Analysis. *Adv. Nutr.* **2017**, *8*, 718–727. [CrossRef] [PubMed] [PubMed Central]
48. Kobylińska, M.; Antosik, K.; Decyk, A.; Kurowska, K.; Skiba, D. Body Composition and Anthropometric Indicators in Children and Adolescents 6–15 Years Old. *Int. J. Environ. Res. Public Health* **2022**, *19*, 11591. [CrossRef] [PubMed]
49. Wells, J.C.K. The intricacy of childhood body composition: Implications for obesity risk. *Proc. Nutr. Soc.* **2017**, *76*, 253–264.
50. Freedman, D.S.; Wang, J.; Maynard, L.M.; Thornton, J.C.; Mei, Z.; Pierson, R.N., Jr.; Dietz, W.H.; Horlick, M. Relation of BMI to fat and fat-free Weight among children and adolescents. *Int. J. Obes.* **2005**, *29*, 1–8. [CrossRef]
51. Boeke, C.E.; Kleinman, K.P.; Gillman, M.W.; Rifas-Shiman, S.L.; Taveras, E.M. Predicting childhood obesity using maternal and early life risk factors: Results from a prospective cohort study. *BMC Pediatr.* **2013**, *13*, 133.
52. Nazare, J.A.; Smith, J.; Borel, A.L.; Aschner, P.; Barter, P.; Van Gaal, L.; Tan, C.E.; Wittchen, H.U.; Matsuzawa, Y.; Kadowaki, T.; et al. Usefulness of measuring both body Weight index and waist circumference for the estimation of visceral adiposity and related cardiometabolic risk profile (from the INSPIRE ME IAA study). *Am. J. Cardiol.* **2015**, *115*, 307–315. [CrossRef] [PubMed]
53. Neuhauser, H.; Schienkiewitz, A.; Rosario, A.S.; Dortschy, R.; Kurth, B.M. *Reference Percentiles for Anthropometric Measures and Blood Pressure from the German Health Interview and Examination Survey for Children and Adolescents (KiGGS)*; Sarganas, G., Neuhauser, H., Eds.; Robert Koch Institute Nordufer 20 13353 Berlin, Germany, 2013; ISBN 978-3-89606-231-4 . [CrossRef]
54. Barreira, T.V.; Broyles, S.T.; Gupta, A.K.; Katzmarzyk, P.T. Relationship of anthropometric indices to abdominal and total body fat in youth: Sex and race differences. *Obesity* **2014**, *22*, 1345–1350. [CrossRef]
55. Bredella, M.A. Sex differences in body composition. In *Sex and Gender Factors Affecting Metabolic Homeostasis, Diabetes and Obesity*; Springer: New York, NY, USA, 2017; pp. 9–27.
56. Rönnecke, E.; Vogel, M.; Bussler, S.; Grafe, N.; Jurkutat, A.; Schlingmann, M.; Koerner, A.; Kiess, W. Age-and sex-related percentiles of skinfold thickness, waist and hip circumference, waist-to-hip ratio and waist-to-height ratio: Results from a population-based pediatric cohort in Germany (LIFE Child). *Obes. Facts* **2019**, *12*, 25–39. [CrossRef]
57. Frisancho, A.R. Anthropometric standards: An interactive nutritional reference of body size and body composition for children and adults. *Am. J. Hum. Biol.* **2008**, *20*, 456–457.
58. Maynard, L.M.; Wisemandle, W.; Roche, A.F.; Chumlea, W.C.; Guo, S.S.; Siervogel, R.M. Childhood body composition in relation to body Weight index. *Pediatrics* **2001**, *107*, 344–350. [CrossRef] [PubMed]
59. Moreno, L.A.; Rodríguez, G.; Fleta, J.; Bueno-Lozano, M.; Lázaro, A.; Bueno, G. Trends of dietary habits in adolescents. *Crit. Rev. Food Sci. Nutr.* **2013**, *50*, 106–112. [CrossRef] [PubMed]

Journal of
Functional Morphology and Kinesiology

MDPI

Article

From Overweight to Severe Obesity: Physical Activity and Behavioural Profiles in a Large Clinical Cohort

Francesca Campoli [1,2], Elvira Padua [1], Lucio Caprioli [2], Saeid Edriss [2], Giuseppe Annino [3], Vincenzo Bonaiuto [2] and Mauro Lombardo [1,*]

1 Department of Human Sciences and Promotion of the Quality of Life, San Raffaele Open University, 00166 Rome, Italy; francesca.campoli@uniroma5.it (F.C.)
2 Sports Engineering Laboratory, Department of Industrial Engineering, University of Rome Tor Vergata, 00133 Rome, Italy; lucio.caprioli@uniroma2.it (L.C.); saeid.edriss@alumni.uniroma2.eu (S.E.)
3 Human Performance Laboratory, Centre of Space Bio-Medicine, Department of Medicine Systems, University of Rome Tor Vergata, 00133 Rome, Italy; g_annino@hotmail.com
* Correspondence: mauro.lombardo@uniroma5.it

Abstract: Background: Behavioural heterogeneity in obesity is increasingly recognised, but how specific dietary patterns, food preferences and physical activity vary between obesity classes remains poorly characterised. **Methods:** We analysed behavioural, dietary, and lifestyle data from 1366 adults attending a tertiary obesity clinic in Italy. Participants were stratified into five obesity classes defined by BMI. Age-adjusted regression models and chi-square tests with Bonferroni correction were used to examine associations between obesity severity and key behavioural outcomes, including food preferences, eating behaviours, physical activity, and self-reported sleep quality. **Results:** The prevalence of uncontrolled eating, skipping meals, and fast eating significantly increased with obesity severity after adjusting for age (all $p < 0.05$). Preference for yoghurt and legumes declined with increasing BMI, whereas preferences for meat and dairy remained stable. Age-adjusted sport participation decreased progressively, with significantly lower odds in Obesity I, II, and IIIA compared to the Overweight group. Sleep quality was highest among overweight participants and declined with obesity severity; night-time awakenings were most frequent in Obesity IIIB. **Conclusions:** Distinct behavioural and lifestyle traits, including lower sport participation, reduced preference for fibre-rich foods, and greater frequency of uncontrolled, fast, and irregular eating, showed overall trends across obesity classes. While these findings suggest the presence of behavioural phenotypes, their interpretation is limited by the cross-sectional design and the use of self-reported, non-validated measures. Future studies should incorporate objective assessments to inform targeted obesity interventions.

Keywords: obesity; behavioural phenotype; eating behaviour; food preferences; physical activity; sleep quality

1. Introduction

The global increase in the prevalence of obesity has underlined the urgent need to understand not only its biological determinants, but also the behavioural and lifestyle patterns that underpin it. As of 2022, 1 in 8 people worldwide were living with obesity, with over 890 million adults affected. Since 1990, global adult obesity has more than doubled, and adolescent obesity has quadrupled [1]. In addition to established metabolic and genetic contributions, it is increasingly recognised that discrete behavioural phenotypes may underlie distinct trajectories of weight gain and resistance to weight loss [2,3]. In particular, the extent to which maladaptive eating patterns, food preferences, and lifestyle behaviours cluster and evolve through increasing degrees of obesity remains poorly characterised [4].

A large body of literature has linked individual behaviours, such as starvation eating, frequent snacking and distracted eating, with altered homeostatic regulation and increased energy intake [5]. Similarly, poor preference for fibre-rich and plant-based foods has been associated with poor diet quality and elevated cardiometabolic risk [6,7]. Although these behaviours are common in people with obesity, the degree of systematic variation according to obesity class is not well defined. The concept of behavioural stratification within obesity, although often invoked, has rarely been studied with the granularity necessary to inform clinical interventions [8]. At the same time, lifestyle factors such as physical activity and sleep quality represent further areas in which behavioural heterogeneity may exist [9].

Physical activity is not only a determining factor in energy balance, but also a modifiable behaviour with important implications for weight management, cardiometabolic risk reduction, musculoskeletal function, and psychological resilience, especially in obese individuals [10]. Despite being a pillar of obesity treatment, structured physical activity is often underestimated, and its decline across different classes of obesity remains poorly quantified.

Obesity is also associated with altered sleep architecture, but it is unclear whether these disturbances follow a linear pattern with increasing adiposity or emerge at specific thresholds [11]. Recent research has focused on the identification of behavioural phenotypes in obesity—defined as recurring patterns of lifestyle traits such as food preferences, eating behaviour, physical activity, and sleep—that may both contribute to and result from excess body weight [12]. A clearer understanding of these trajectories could provide useful information for the development of phenotype-based strategies that more effectively integrate physical activity and other behavioural components as targeted interventions [13].

In this cross-sectional study we examined a spectrum of behavioural and lifestyle characteristics—including eating behaviour, food and taste preferences, self-reported sleep quality, and physical activity—across five BMI-defined obesity classes. Our aim was to delineate the behavioural architecture of obesity across its severity gradient and to assess whether higher degrees of adiposity are associated with consistent changes in preferences, behaviours, and perceived well-being. Such models, if present, may offer a conceptual framework for more individualised and behaviourally anchored approaches in the treatment of obesity.

2. Materials and Methods

2.1. Subjects

This cross-sectional study was conducted between January 2024 and March 2025 in a tertiary centre for the treatment of obesity in Rome, Italy. All participants were enrolled during scheduled clinical visits as part of an institutional programme for the assessment and management of excess body weight. Recruitment was limited to individuals attending the centre for an initial consultation, ensuring standardised data collection under clinical supervision. Eligible participants were adults aged 18–75 years who provided written informed consent and completed a structured behavioural and lifestyle questionnaire prior to medical assessment. Patients were excluded if they reported a history of cardiovascular or metabolic disease (n = 52), if they were pregnant or breastfeeding (n = 27), if they had recently participated in a supervised weight loss intervention (n = 25), or if they could not provide consistent responses (n = 185). No digital or community-based recruitment strategies were used and all data were collected on site. Figure 1 illustrates the flow of participant inclusion and exclusion.

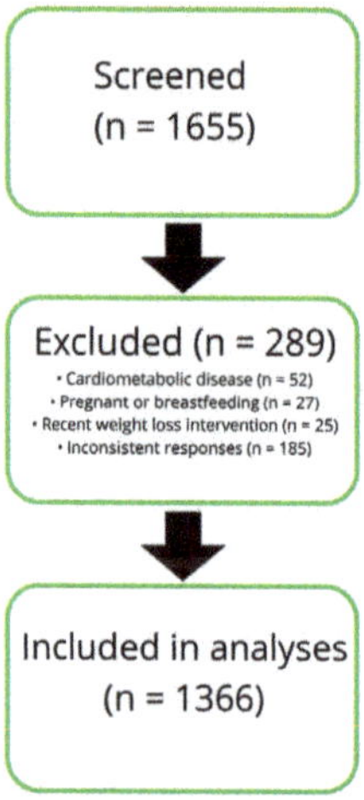

Figure 1. Flow diagram showing the number of participants screened, excluded, and included in the final analyses.

A total of 1366 participants met the inclusion criteria and provided complete data on anthropometric and behavioural variables. Food intake was assessed through structured food diaries, which were only available for participants who returned for a follow-up visit after receiving dietary recommendations at baseline. Therefore, analyses involving food frequency data were limited to this subgroup, while all other outcomes were examined in the entire cohort. Missing data were not imputed. Participants with incomplete responses or no follow-up diary were excluded from specific analyses, as appropriate.

Sample size considerations were based on power estimates performed using G*Power 3.1. Assuming a small to moderate effect size ($\eta^2 = 0.02$), a power of 0.80, an alpha level of 0.05, and five BMI-defined groups, the estimated required sample size was approximately 1600 participants. After applying exclusion criteria, the final sample included 1366 participants. A post hoc power analysis confirmed that this sample size retained a power above 0.78.

The study protocol was approved by the Lazio Area 5 Territorial Ethics Committee (Approval No. 57/SR/23; 7 November 2023) and adhered to the ethical standards outlined in the Declaration of Helsinki.

2.2. *Questionnaire and Behavioural Assessment*

Prior to the clinical assessment, participants completed a structured, self-administered questionnaire designed to capture behavioural, dietary, and lifestyle characteristics relevant to obesity risk and phenotype. The instrument was administered in person at the study site using digital tablets to ensure standardisation and data integrity. Completion time averaged 25–30 min. Informed consent was obtained during the initial clinical meeting and all responses were anonymised at the time of administration to protect the confidentiality of participants. The questionnaire was developed based on previous instruments used in nutritional epidemiology and behavioural research, including those described by Schulz et al. (2021), Hooson et al. (2020), and Carbonneau et al. (2017) [14–16]. Although not formally validated, its structure closely followed established instruments and was subjected to internal consistency checks during data cleaning. The section on eating behaviour probed the presence of key patterns involved in the dysregulation of energy intake. Participants were asked about the habit of skipping meals, the speed of eating, eating while distracted or away from the table, and the frequency of episodes characterised by loss of control in the absence of hunger. Additional questions concerned the tendency to eat at night. A separate section dealt with food and taste preferences. The food list was adapted to reflect both the principles of the Mediterranean diet and commonly consumed Western foods, following prior tools developed for Southern European cohorts [17,18]. Finally, participants provided information on structured physical activity, including type, frequency, duration, and preferred time of day. Sports participation was recorded as a dichotomous variable, and weekly activity levels were classified into ordinal bands.

2.3. Body Composition

Anthropometric and body composition assessments were conducted on-site using standardised procedures to ensure accuracy and reproducibility. All participants presented themselves fasting (minimum 8 h) and were measured wearing only light undergarments. Body weight was recorded to the nearest 0.1 kg using a calibrated bioelectrical impedance analyser (TANITA BC-420 MA; Tanita Corporation, Tokyo, Japan). Standing height was measured with a wall stadiometer, with the participant positioned according to the Frankfurt horizontal plane. Body mass index (BMI) was calculated as weight in kilograms divided by height in metres squared. Participants were stratified into five BMI-defined categories: Overweight (25.0–29.9 kg/m^2), Obesity class I (30.0–34.9 kg/m^2), class II (35.0–39.9 kg/m^2), class IIIA (40.0–44.9 kg/m^2), and class IIIB ($\geq$45.0 kg/m^2). Waist circumference was measured in triplicate at the midpoint between the lowest rib and the iliac crest, with participants standing upright and breathing normally. Body composition parameters, including fat mass (FM), fat-free mass (FFM) and basal metabolic rate (BMR), were obtained using the same bioimpedance device under controlled conditions. To minimise fluctuations related to hydration status or physical exertion, all measurements were taken at least three hours after waking up and food intake and a minimum of 12 h after any strenuous physical activity. Female participants were instructed to avoid measurements during menstruation. Each parameter was measured twice and the average values were used for analysis. Although bioelectrical impedance is widely used in clinical practice due to its ease of use and non-invasive nature, its accuracy in subjects with severe obesity may be limited due to alterations in body water distribution and increased truncal adiposity [19]. For this reason, body composition data were included in the descriptive analyses but were not used as primary outcomes in the inferential models.

2.4. Statistical Analysis

Descriptive statistics were calculated for all variables. Continuous data were summarised as means with standard deviations and compared using one-way ANOVA; categorical variables were compared using chi-square tests. Participants were stratified into five BMI categories. For each outcome, pairwise comparisons between BMI classes were performed using post hoc chi-square tests. To control for multiple testing, p-values from all pairwise comparisons were adjusted using the Bonferroni method. Age-adjusted logistic regression models were used to evaluate associations between obesity class and binary outcomes, including food preferences, eating behaviours, sleep quality (dichotomized), and sports participation. Ordinal logistic regression was applied to assess sleep quality on a six-point scale. Linear regression was used to analyse sweet–salty taste preference. All models included age as a covariate, and marginal estimates with 95% confidence intervals were reported. No covariates besides age were included in the regression models, as sex, income, and smoking status did not differ significantly between obesity classes and were therefore not considered confounding factors. No subgroup or sensitivity analyses were conducted. Statistical analyses were performed using Python (v3.11).

3. Results

The cohort comprised 1366 participants (Table 1), with a predominance of women (56.7%). Age, BMI, waist circumference, fat mass, and BMR increased progressively with obesity severity (all $p < 0.0001$).

Table 1. Clinical and socio-demographic characteristics by class of obesity.

	Total (n = 1366)	Overweight	Obesity I	Obesity II	Obesity IIIA	Obesity IIIB	p-Value
Male (n, %)	591 (43.3%)	324 (41.3%)	174 (45.7%)	72 (48.0%)	13 (37.1%)	8 (50.0%)	0.3672
Female (n, %)	775 (56.7%)	460 (58.7%)	207 (54.3%)	78 (52.0%)	22 (62.9%)	8 (50.0%)	
Age	42.8 ± 13.2	41.4 ± 13.0	43.2 ± 13.1	46.9 ± 13.3	50.0 ± 12.0	48.6 ± 13.1	<0.0001
Weight (kg)	86.4 ± 16.5	77.6 ± 10.0	92.1 ± 11.3	105.4 ± 13.2	115.5 ± 14.4	143.8 ± 18.2	<0.0001
BMI	30.4 ± 4.7	27.3 ± 1.5	32.2 ± 1.4	37.1 ± 1.4	42.1 ± 1.4	50.9 ± 5.2	<0.0001
FM (kg)	29.4 ± 10.1	23.5 ± 5.2	33.2 ± 6.0	41.4 ± 6.0	51.3 ± 6.4	70.5 ± 14.4	<0.0001
FM (%)	33.8 ± 8.1	30.7 ± 7.1	36.4 ± 7.0	39.7 ± 6.4	44.8 ± 6.1	49.7 ± 7.3	<0.0001
AC (cm)	102.7 ± 11.8	95.9 ± 7.4	107.8 ± 7.1	117.4 ± 8.1	124.8 ± 10.9	143.2 ± 10.3	<0.0001
FFM (kg)	54.0 ± 11.4	51.3 ± 10.2	55.8 ± 11.2	60.7 ± 12.4	61.1 ± 13.6	67.4 ± 12.3	<0.0001
BMR (kcal)	1721.5 ± 345.6	1623.2 ± 295.5	1789.2 ± 329.5	1955.2 ± 381.9	1978.8 ± 397.3	2270.6 ± 396.1	<0.0001

Clinical, anthropometric, and sociodemographic characteristics of participants by BMI class. Continuous values are reported as mean ± SD. Categorical values are given as number and percentage. p-values were calculated using one-way ANOVA (continuous variables) and chi-square tests (categorical variables). Abbreviations: BMI, body mass index; BMR, basal metabolic rate; FFM, fat-free mass; FM, fat mass; AC, abdominal circumference.

Among the 1366 participants, the prevalence of regular consumption ("Yes" response) for low-fat white yoghurt and legumes varied significantly across BMI classes. For low-fat white yoghurt, the proportion of "Yes" responses ranged from 39.4% in Obesity I to 50.0% in Obesity IIIB (p = 0.001, chi-square test). For legumes, "Yes" responses ranged from 62.5% in Obesity IIIB to 85.7% in Obesity IIIA (p = 0.046, chi-square test). The 95% confidence intervals for each proportion are shown in Figure 2.

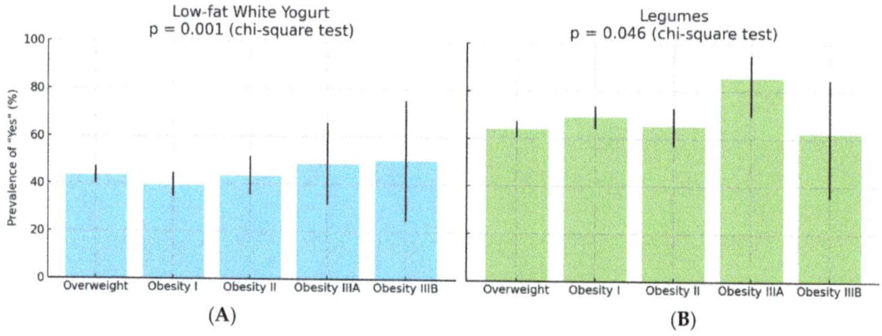

Figure 2. Prevalence of "Yes" responses for (**A**) low-fat white yoghurt and (**B**) legumes by BMI class. Bars indicate the observed percentage of participants in each BMI group reporting regular consumption. Error bars represent 95% binomial confidence intervals. Global comparison by chi-square test: p = 0.001 for yoghurt, p = 0.046 for legumes.

No significant differences were observed in sweet, salty, or indifferent taste preference across BMI classes (chi-square p = 0.206). This finding remained unchanged after adjustment for age using multinomial logistic regression and after Bonferroni correction for multiple comparisons.

The age-adjusted prevalence of maladaptive eating behaviours across BMI classes is shown in Figure 3 and Supplementary Table S1. Uncontrolled eating was highly prevalent and increased with adiposity, from 76.9% in the Overweight group to 100% in class IIIB Obesity ($p = 0.032$). Fast eating showed a non-linear trend, with prevalence ranging from 62.5% to 78.7% ($p = 0.039$). Skipping meals was reported by 31.2% to 43.8% of participants, with significant differences between BMI categories ($p = 0.034$). Other behaviours, including snacking between meals and distracted eating, were frequent across all BMI classes (range: 55.3–81.2% and 62.7–68.8%, respectively) but showed no significant association with adiposity after age adjustment ($p = 0.105$ and $p = 0.393$, respectively). Night eating was less common (14.5–25.7%) and showed a borderline association with BMI class ($p = 0.062$).

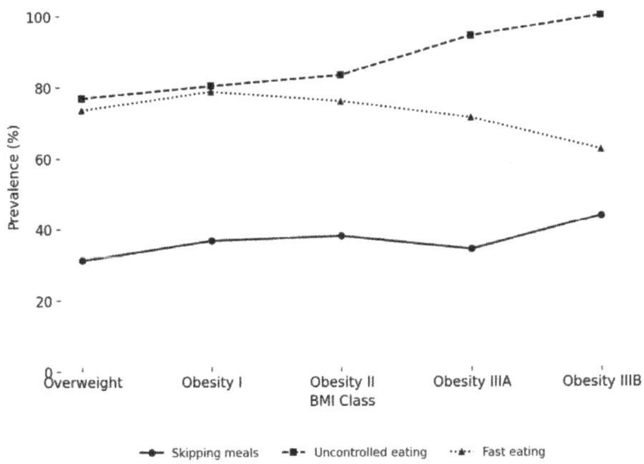

Figure 3. Prevalence of significant eating behaviours by BMI class.

Figure 3 reports the age-adjusted percentage of participants who reported skipping meals, uncontrolled eating despite not being hungry, and fast eating, stratified by BMI class. Both uncontrolled eating and fast eating showed significant differences across BMI classes after adjusting for age ($p = 0.032$ and $p = 0.039$, respectively), while skipping meals was also significant ($p = 0.034$).

Sleep quality varied significantly between BMI classes after adjustment for age (Figure 4). The percentage of participants reporting 'good' sleep was highest in the Overweight group (52.2%), followed by Obesity II (45.6%) and Obesity I (41.3%). In contrast, Obesity IIIA and IIIB groups reported substantially lower percentages (32.3% and 28.6%, respectively). The frequency of night-time awakenings increased from 37.0% in the Overweight group to 64.3% in Obesity class IIIB, while the prevalence of difficulty falling asleep ranged from 7.1% to 15.2% across BMI classes. The association between BMI class and sleep quality remained significant after adjustment for age ($p = 0.0038$, chi-square test). After Bonferroni correction for multiple comparisons, significant pairwise differences were observed between Obesity I and Overweight (adjusted $p = 0.045$) and between Overweight and Obesity IIIB (adjusted $p = 0.049$).

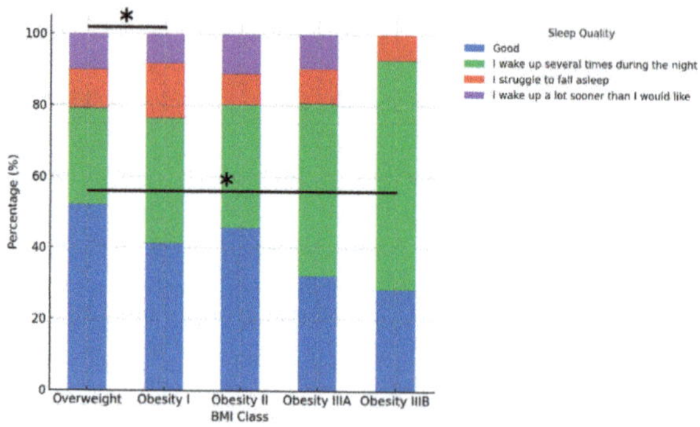

Figure 4. Sleep quality by BMI class. Asterisks indicate statistically significant pairwise differences after Bonferroni correction for multiple comparisons: Overweight vs. Obesity I ($p = 0.045$) and Overweight vs. Obesity IIIB ($p = 0.049$). Global association: chi-square $p = 0.0038$.

Sport participation decreased significantly with increasing BMI class (Figure 5). After adjustment for age, the probability of practising sport was highest in the Overweight group (48.8%), followed by Obesity I (37.8%), Obesity II (34.6%), and Obesity IIIA (15.9%). The global association between BMI class and sport participation was highly significant (chi-square $p < 0.0001$). Pairwise post hoc tests with Bonferroni correction confirmed that the probability of practising sport was significantly higher in the Overweight group than in Obesity I (adjusted $p = 0.0019$), Obesity II (adjusted $p = 0.0025$), and Obesity IIIA (adjusted $p = 0.0009$).

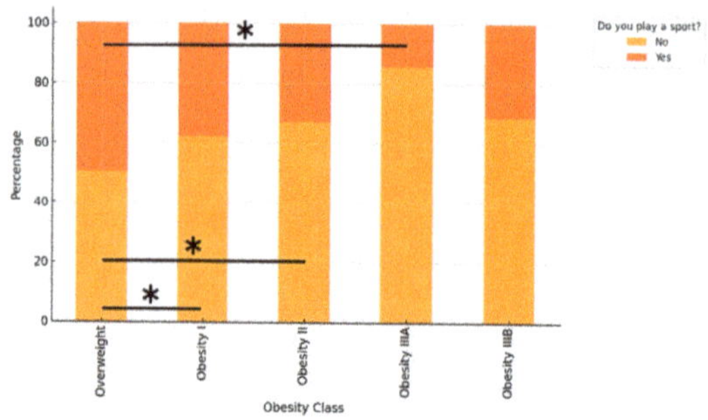

Figure 5. Bars indicate the age-adjusted probability of practising sport ("Yes") in each BMI class, with 95% confidence intervals. Asterisks indicate statistically significant pairwise differences (Bonferroni-adjusted $p < 0.01$) between the Overweight group and the corresponding Obesity class (I, II, or IIIA). The global association between BMI class and sport participation was highly significant (chi-square $p < 0.0001$). Asterisks indicate significant pairwise differences between Overweight and Obesity I ($p = 0.0019$), Overweight and Obesity II ($p = 0.0025$), and Overweight and Obesity IIIA ($p = 0.0009$), Bonferroni-adjusted.

4. Discussion

This study investigated how behavioural and lifestyle characteristics vary across different classes of obesity, aiming to identify systematic patterns that may reflect distinct behavioural phenotypes. Our findings highlight progressive changes in food preferences, eating behaviours, sleep quality, and physical activity levels with increasing BMI. These results support the hypothesis that behavioural traits cluster differently along the obesity spectrum, with potential implications for personalised intervention strategies.

4.1. Food Taste

yoghurt and legumes were the only food items to exhibit significant differences in preference across BMI categories, with markedly lower preference observed among individuals with more severe obesity. These findings contrast with prior epidemiological evidence, which has consistently demonstrated inverse associations between the consumption of these foods and measures of adiposity. Specifically, Eales et al. reported that yoghurt intake was associated with lower body mass index, body weight, and adiposity in both cross-sectional and cohort studies, although randomised trials showed mixed results regarding causality [20]. Similarly, a systematic review by Sayon-Orea and colleagues concluded that yoghurt consumption was linked to favourable weight trajectories and a reduced risk of metabolic syndrome [21]. Prospective analyses from the SUN cohort further supported an inverse relationship between yoghurt consumption and the risk of developing overweight or obesity, particularly when combined with higher fruit intake [22]. For legumes, Tucker found that higher intake was associated with reduced weight gain and lower body fat indices over a 10-year period, although these associations attenuated after adjustment for fibre intake [23]. Heshmatipour et al. reported similar findings in overweight and obese adolescents, with inverse associations between legume consumption and markers of metabolic dysfunction, though statistical significance diminished after full adjustment for confounders [24]. A likely explanation for this discrepancy is that our data reflect declared preferences rather than measured intake. Unlike population-based cohorts that assess actual food consumption, our assessment was conducted before any nutritional counselling, capturing spontaneous patient-reported preferences. These may reflect underlying behavioural traits—such as reduced interest in or aversion to certain healthy foods—that become more pronounced with increasing obesity severity.

4.2. Eating Behaviours

Despite extensive evidence linking obesity to taste alterations, our findings did not reveal significant differences in sweet, salty, or indifferent taste preferences across BMI categories. This is somewhat unexpected given that several studies have suggested that individuals with obesity may exhibit altered taste perception or hedonic response to specific flavours, which could influence food choices and energy intake. A recent systematic review concluded that approximately 40% of studies identified significant taste alterations in individuals with obesity, although methodological heterogeneity and limited sensitivity of taste assessments were noted as critical limitations in the current literature [25]. In contrast, maladaptive eating behaviours—including uncontrolled eating, fast eating, and meal skipping—showed significant variation by obesity class. Uncontrolled eating was reported by all patients with class IIIB obesity and was significantly more prevalent than in overweight individuals. These findings align with prior data showing a strong independent association between lack of food control and obesity risk in primary care settings. Rohrer et al. reported that individuals with trouble controlling their eating had a more than sixfold increased odds of being obese compared to those without such difficulties (OR = 6.67; 95% CI, 3.91–11.4) [26]. Fast eating was also more frequent among individuals with obesity, in line with a meta-analysis showing that individuals who eat quickly have significantly higher BMI and greater odds of obesity (OR = 2.15; 95% CI, 1.84–2.51) [27]. Rapid eating may disrupt satiety signalling, leading to increased caloric intake before the onset of fullness. Meal skipping was reported in 31–44% of participants and varied across BMI classes. Recent cohort evidence from over 26,000 Japanese university students suggests that skipping dinner—but not breakfast or lunch—was significantly associated with $\geq$10% weight gain and the onset of overweight/obesity, with adjusted incidence rate ratios between 1.42 and 1.74 [28]. Together, these findings support the relevance of maladaptive eating behaviours in the progression of obesity and highlight their potential utility in behavioural phenotyping.

4.3. Sleep Quality

Sleep quality showed a significant inverse association with obesity class in our cohort. Overweight participants reported sleeping better than individuals in all obesity categories, with those in Obesity IIIA and IIIB showing the highest prevalence of nocturnal awakenings. These findings are in line with previous studies showing that poor sleep duration and quality are associated with an increased risk of obesity [29,30]. Experimental research has shown that sleep restriction increases energy intake by more than 250 kcal per day, largely through alterations in reward processing rather than appetite hormones [31]. Furthermore, higher BMI has been linked to greater sleep fragmentation and lower sleep efficiency, even after adjusting for behavioural and emotional confounders [32]. Although causality cannot be established due to the cross-sectional design, these results suggest that sleep quality may be a relevant component in the behavioural phenotyping of obesity.

4.4. Physical Activity

Participants with higher degrees of obesity were less likely to engage in structured physical activity. The most marked drop was observed between Obesity class I and class IIIA, possibly indicating a threshold beyond which participation declines more sharply, although this interpretation remains speculative. These results are consistent with prior evidence suggesting that perceived physical limitations, social stigma, and motivational barriers may contribute to reduced physical activity in individuals with severe obesity [33]. Across all BMI classes, lower activity levels were also associated with older age, underlining the combined effects of obesity and ageing on sedentarism.

4.5. Strengths and Limitations

This study has several strengths, including its large sample size, standardised data collection in a clinical setting, and comprehensive assessment of behavioural and lifestyle dimensions. However, some limitations deserve consideration. Nonetheless, several limitations must be acknowledged. First, all behavioural data—including dietary intake, sleep quality, and physical activity—were self-reported, introducing the possibility of recall and social desirability bias. Future studies should employ validated and objective tools such as accelerometers or wearable devices for physical activity and sleep, and dietary apps or supervised 24 h recalls to enhance data accuracy and methodological rigour. Second, body composition was assessed using bioelectrical impedance analysis, which may be less accurate in individuals with severe obesity; therefore, these data were reported descriptively but excluded from inferential models. Third, the cross-sectional design limits causal inference and prevents the identification of temporal relationships. The observed associations may be bidirectional: certain behaviours may contribute to weight gain, but higher adiposity could also reinforce these behavioural patterns. Fourth, although regression models were adjusted for age and post hoc comparisons were corrected for multiple testing, other potential confounders such as sex and income were not included due to incomplete data and limited statistical power for subgroup analyses. As such, residual confounding cannot be ruled out. Fifth, while the behavioural questionnaire was based on previously used instruments and underwent internal consistency checks, it was not formally validated. This represents a methodological limitation; future studies should include psychometric validation and comparisons with gold-standard tools such as the TFEQ, PSQI, or IPAQ to ensure reliability and generalisability. Finally, participants were recruited from a tertiary-level obesity clinic, which may introduce selection bias and limit the generalisability of our findings to broader populations, including those in community or primary-care settings. Future studies should also investigate behavioural differences across sex and age subgroups to identify potentially important modifiers of obesity-related behaviours.

5. Conclusions

This study shows that specific behavioural and lifestyle traits cluster with increasing obesity severity (Table 2). Individuals with higher BMI reported reduced sports participation, poorer sleep quality, and lower consumption of foods such as yoghurt and legumes. Fast eating, uncontrolled eating, and meal skipping significantly increased across BMI classes. Recognising these profiles may help tailor clinical interventions to individual behavioural characteristics, enabling more personalised and effective obesity management. Future research should validate these behavioural phenotypes in independent cohorts and assess their utility in guiding personalised obesity treatment strategies.

Table 2. Summary of key findings by obesity class.

	Key Findings
Food Preferences	Lower preference for yoghurt and legumes with increasing obesity class. No relevant differences for meat or other dairy products.
Taste Preference	No association between BMI class and sweet–salty taste preference.
Eating behaviours	Fast eating, uncontrolled eating, and skipping meals increased with BMI class.
Sleep Quality	Overweight participants reported better sleep quality; the proportion reporting "Good" sleep declined and night-time awakenings increased with higher obesity class.
Sports Participation	The Overweight class was more likely to practice sports than all obesity classes.

Supplementary Materials: The following supporting information can be downloaded at: https://www.mdpi.com/article/10.3390/jfmk10030283/s1. Table S1. Age-Adjusted Prevalence of Self-Reported Eating Behaviours by BMI Class.

Author Contributions: Conceptualization, F.C., E.P., and M.L.; methodology, F.C., L.C., and S.E. data curation, F.C.; writing—original draft preparation, F.C. and M.L.; writing—review and editing, E.P., L.C., S.E., G.A., V.B., and M.L.; visualisation, F.C. and L.C.; supervision, E.P. and M.L. All authors have read and agreed to the published version of the manuscript.

Funding: This research received no external funding.

Institutional Review Board Statement: The study was conducted in accordance with local regulations and ethical standards. Ethical approval was obtained from the Lazio Area 5 Territorial Ethics Committee (Approval Code: N.57/SR/23; Approval Date: 7 November 2023).

Informed Consent Statement: All participants provided written informed consent prior to inclusion in the study.

Data Availability Statement: This study was registered at ClinicalTrials.gov (ID: NCT06654674). The dataset is publicly available at Mendeley Data: https://data.mendeley.com/datasets/d5sd9zx74d/1 (accessed on 1 July 2025).

Conflicts of Interest: The authors declare no conflict of interest.

References

1. World Health Organization. *Obesity and Overweight [Internet]*; Geneva: World Health Organization: Geneva, Switzerland, 2025. Available online: https://www.who.int/news-room/fact-sheets/detail/obesity-and-overweight (accessed on 18 July 2025).
2. Zandvakili, I.; Pulaski, M.; Pickett-Blakely, O. A phenotypic approach to obesity treatment. *Nutr. Clin. Pract.* **2023**, *38*, 959–975. [CrossRef] [PubMed]
3. Fitzpatrick, S.L.; Coughlin, J.W.; Appel, L.J.; Tyson, C.; Stevens, V.J.; Jerome, G.J.; Dalcin, A.; Brantley, P.J.; Hill-Briggs, F. Application of latent class analysis to identify behavioral patterns of response to behavioral lifestyle interventions in overweight and obese adults. *Int. J. Behav. Med.* **2015**, *22*, 471–480. [CrossRef] [PubMed]
4. Liberali, R.; Del Castanhel, F.; Kupek, E.; Assis, M.D. Latent class analysis of lifestyle risk factors and association with overweight and/or obesity in children and adolescents: Systematic review. *Child. Obes.* **2020**, *17*, 2–15. [CrossRef] [PubMed]
5. Fuente González, C.E.; Chávez-Servín, J.L.; de la Torre-Carbot, K.; Ronquillo González, D.; Aguilera Barreiro, M.L.Á.; Ojeda Navarro, L.R. Relationship between Emotional Eating, Consumption of Hyperpalatable Energy-Dense Foods, and Indicators of Nutritional Status: A Systematic Review. *J. Obes.* **2022**, *2022*, 4243868. [CrossRef] [PubMed]
6. Zhang, Y.; Wei, Y.; Tang, D.; Lu, J.; Zhang, N.; Hu, Y.; He, R.; Guan, H.; Xu, J.; Wang, S.; et al. Association of major dietary patterns and different obesity phenotypes in Southwest China: The China Multi-Ethnic Cohort (CMEC) Study. *Eur. J. Nutr.* **2022**, *62*, 465–476. [CrossRef] [PubMed]
7. Hendryx, M.; Dinh, P.; Chow, A.; Kroenke, C.H.; Hingle, M.; Shadyab, A.H.; Garcia, L.; Howard, B.V.; Luo, J. Lifestyle and psychosocial patterns and diabetes incidence among women with and without obesity: A prospective latent class analysis. *Prev. Sci.* **2020**, *21*, 850–860. [CrossRef] [PubMed]
8. St-Onge, M.P.; Grandner, M.A.; Brown, D.; Conroy, M.B.; Jean-Louis, G.; Coons, M.; Bhatt, D.L.; American Heart Association Obesity, Behavior Change, Diabetes, and Nutrition Committees of the Council on Lifestyle and Cardiometabolic Health; Council on Cardiovascular Disease in the Young; Council on Clinical Cardiology; et al. Sleep Duration and Quality: Impact on Lifestyle Behaviors and Cardiometabolic Health: A Scientific Statement From the American Heart Association. *Circulation* **2016**, *134*, e367–e386. [CrossRef] [PubMed] [PubMed Central]
9. Raiman, L.; Amarnani, R.; Abdur-Rahman, M.; Marshall, A.; Mani-Babu, S. The role of physical activity in obesity: Let's actively manage obesity. *Clin. Med.* **2023**, *23*, 311–317. [CrossRef] [PubMed] [PubMed Central]
10. Jacob, J.; Isaac, R. Behavioral therapy for management of obesity. *Indian J. Endocrinol. Metab.* **2012**, *16*, 28–32. [CrossRef] [PubMed]
11. Unick, J.; Beavers, D.; Jakicic, J.; Kitabchi, A.; Knowler, W.; Wadden, T.; Wing, R. Effectiveness of lifestyle interventions for individuals with severe obesity and type 2 diabetes. *Diabetes Care* **2011**, *34*, 2152–2157. [CrossRef] [PubMed]
12. Camacho-Barcia, L.; Lucas, I.; Miranda-Olivos, R.; Jiménez-Murcia, S.; Fernández-Aranda, F. Applying psycho-behavioural phenotyping in obesity characterization. *Rev. Endocr. Metab. Disord.* **2023**, *24*, 871–883. [CrossRef] [PubMed] [PubMed Central]
13. Cifuentes, L.; Ghusn, W.; Feris, F.; Campos, A.; Sacoto, D.; De la Rosa, A.; McRae, A.; Rieck, T.; Mansfield, S.; Ewoldt, J.; et al. Phenotype tailored lifestyle intervention on weight loss and cardiometabolic risk factors in adults with obesity: A single-centre, non-randomised, proof-of-concept study. *EClinicalMedicine* **2023**, *58*, 101923. [CrossRef] [PubMed]
14. Schulz, C.A.; Oluwagbemigun, K.; Nöthlings, U. Advances in dietary pattern analysis in nutritional epidemiology. *Eur. J. Nutr.* **2021**, *60*, 4115–4130. [CrossRef] [PubMed] [PubMed Central]
15. Hooson, J.; Hutchinson, J.; Warthon-Medina, M.; Hancock, N.; Greathead, K.; Knowles, B.; Vargas-Garcia, E.; Gibson, L.E.; Bush, L.A.; Margetts, B.; et al. A systematic review of reviews identifying UK validated dietary assessment tools for inclusion on an interactive guided website for researchers: www.nutritools.org. *Crit. Rev. Food. Sci. Nutr.* **2020**, *60*, 1265–1289. [CrossRef] [PubMed]
16. Carbonneau, E.; Bradette-Laplante, M.; Lamarche, B.; Provencher, V.; Bégin, C.; Robitaille, J.; Desroches, S.; Vohl, M.C.; Corneau, L.; Lemieux, S. Development and validation of the food liking questionnaire in a French-Canadian population. *Nutrients* **2017**, *9*, 1337. [CrossRef] [PubMed]
17. Kushner, R.F.; Hammond, M.M. Using psycho-behavioral phenotyping for overweight and obesity: Confirmation of the 6 factor questionnaire. *Obes. Sci. Pract.* **2021**, *8*, 185–189. [CrossRef] [PubMed]
18. Gnagnarella, P.; Dragà, D.; Misotti, A.M.; Sieri, S.; Spaggiari, L.; Cassano, E.; Baldini, F.; Soldati, L.; Maisonneuve, P. Validation of a short questionnaire to record adherence to the Mediterranean diet: An Italian experience. *Nutr. Metab. Cardiovasc. Dis.* **2018**, *28*, 1140–1147. [CrossRef] [PubMed]
19. Coppini, L.; Waitzberg, D.; Campos, A. Limitations and validation of bioelectrical impedance analysis in morbidly obese patients. *Curr. Opin. Clin. Nutr. Metab. Care* **2005**, *8*, 329–332. [CrossRef] [PubMed]
20. Eales, J.; Lenoir-Wijnkoop, I.; King, S.; Wood, H.; Kok, F.J.; Shamir, R.; Prentice, A.; Edwards, M.; Glanville, J.; Atkinson, R.L. Is Consuming Yoghurt Associated with Weight Management Outcomes? Results from a Systematic Review. *Int. J. Obes.* **2016**, *40*, 731–746. [CrossRef] [PubMed]

21. Sayon-Orea, C.; Martínez-González, M.A.; Ruiz-Canela, M.; Bes-Rastrollo, M. Associations between Yogurt Consumption and Weight Gain and Risk of Obesity and Metabolic Syndrome: A Systematic Review. *Adv. Nutr.* **2017**, *8*, 146S–154S. [CrossRef] [PubMed]
22. Martínez-González, M.A.; Sayon-Orea, C.; Ruiz-Canela, M.; de la Fuente, C.; Gea, A.; Bes-Rastrollo, M. Yogurt Consumption, Weight Change and Risk of Overweight/Obesity: The SUN Cohort Study. *Nutr. Metab. Cardiovasc. Dis.* **2014**, *24*, 1189–1196. [CrossRef] [PubMed]
23. Tucker, L.A. Legume Intake, Body Weight, and Abdominal Adiposity: 10-Year Weight Change and Cross-Sectional Results in 15,185 U.S. Adults. *Nutrients* **2023**, *15*, 460. [CrossRef] [PubMed]
24. Heshmatipour, H.; Hajhashemy, Z.; Mirzaei, S.; Asadi, A.; Akhlaghi, M.; Saneei, P. Association of Legumes and Nuts Consumption with Metabolic Health Status in Iranian Overweight and Obese Adolescents. *Sci. Rep.* **2023**, *13*, 5784. [CrossRef] [PubMed]
25. Peinado, B.R.R.; Frazão, D.R.; Bittencourt, L.O.; de Souza-Rodrigues, R.D.; Vidigal, M.T.C.; da Silva, D.T.; Paranhos, L.R.; Magno, M.B.; Fagundes, N.C.F.; Maia, L.C.; et al. Is Obesity Associated with Taste Alterations? A Systematic Review. *Front. Endocrinol.* **2023**, *14*, 1167119. [CrossRef] [PubMed]
26. Rohrer, J.E.; Vickers-Douglas, K.S.; Stroebel, R.J. Uncontrolled Eating and Obesity in Adult Primary Care Patients. *Obes. Res. Clin. Pract.* **2009**, *3*, 161–167. [CrossRef] [PubMed]
27. Ohkuma, T.; Hirakawa, Y.; Nakamura, U.; Kiyohara, Y.; Kitazono, T.; Ninomiya, T. Association Between Eating Rate and Obesity: A Systematic Review and Meta-Analysis. *Int. J. Obes.* **2015**, *39*, 1589–1596. [CrossRef] [PubMed]
28. Yamamoto, R.; Tomi, R.; Shinzawa, M.; Yoshimura, R.; Ozaki, S.; Nakanishi, K.; Ide, S.; Nagatomo, I.; Nishida, M.; Yamauchi-Takihara, K.; et al. Associations of Skipping Breakfast, Lunch, and Dinner with Weight Gain and Overweight/Obesity in University Students: A Retrospective Cohort Study. *Nutrients* **2021**, *13*, 271. [CrossRef] [PubMed]
29. Keramat, S.A.; Alam, K.; Basri, R.; Jansen, M.; Shaikh, M.; Renzaho, A.M.N. Sleep duration, sleep quality and the risk of being obese: Evidence from the Australian panel survey. *Sleep Med.* **2023**, *109*, 56–64. [CrossRef] [PubMed]
30. Huang, H.; Yu, T.; Liu, C.; Yang, J.; Yu, J. Poor sleep quality and overweight/obesity in healthcare professionals: A cross-sectional study. *Front. Public Health* **2024**, *12*, 1390643. [CrossRef] [PubMed]
31. Chaput, J.P.; McHill, A.W.; Cox, R.C.; Broussard, J.L.; Dutil, C.; St-Onge, M.P. The role of insufficient sleep and circadian misalignment in obesity. *Nat. Rev. Endocrinol.* **2023**, *19*, 82–97. [CrossRef] [PubMed]
32. Eid, S.W.; Brown, R.F.; Maloney, S.K.; Phillips, J.G.; Touyz, S. Can the relationship between overweight/obesity and sleep quality be explained by affect and behaviour? *Eat. Weight. Disord.* **2022**, *27*, 2821–2834. [CrossRef] [PubMed]
33. Swift, D.L.; McGee, J.E.; Earnest, C.P.; Carlisle, E.; Nygard, M.; Johannsen, N.M. The effects of exercise and physical activity on weight loss and maintenance. *Prog Cardiovasc. Dis.* **2014**, *56*, 441–447. [CrossRef] [PubMed]

Journal of
Functional Morphology and Kinesiology

MDPI

Article

Anthropometric Measurements for Predicting Low Appendicular Lean Mass Index for the Diagnosis of Sarcopenia: A Machine Learning Model

Ana M. González-Martin [1,*], Edgar Samid Limón-Villegas [1], Zyanya Reyes-Castillo [1], Francisco Esparza-Ros [2], Luis Alexis Hernández-Palma [1], Minerva Saraí Santillán-Rivera [3], Carlos Abraham Herrera-Amante [4,5], César Octavio Ramos-García [4,5] and Nicoletta Righini [1,*]

1 Instituto de Investigaciones en Comportamiento Alimentario y Nutrición (IICAN), Universidad de Guadalajara, Ciudad Guzmán 49000, Jalisco, Mexico; samid.limon@cusur.udg.mx (E.S.L.-V.); zyanya.reyes@cusur.udg.mx (Z.R.-C.); alexis.hernandez@cusur.udg.mx (L.A.H.-P.)
2 Injury Prevention in Sport Research Group, Universidad Católica San Antonio de Murcia (UCAM), 30107 Murcia, Spain; fesparza@ucam.edu
3 Departamento de Ciencias Clínicas, Universidad de Guadalajara, Ciudad Guzmán 49000, Jalisco, Mexico; minerva.santillan@cusur.udg.mx
4 Nutritional Assessment and Nutritional Care Laboratory (LECEN), Division of Health Sciences, Centro Universitario de Tonalá, Universidad de Guadalajara, Tonalá 45425, Jalisco, Mexico; carlos.amante@academicos.udg.mx (C.A.H.-A.); octavio.ramos@academicos.udg.mx (C.O.R.-G.)
5 Ibero-American Network of Researchers in Applied Anthropometry (RIBA2), 04120 Almería, Spain
* Correspondence: ana.gonzalez@cusur.udg.mx (A.M.G.-M.); nicoletta.righini@cusur.udg.mx (N.R.)

Abstract: Background: Sarcopenia is a progressive muscle disease that compromises mobility and quality of life in older adults. Although dual-energy X-ray absorptiometry (DXA) is the standard for assessing Appendicular Lean Mass Index (ALMI), it is costly and often inaccessible. This study aims to develop machine learning models using anthropometric measurements to predict low ALMI for the diagnosis of sarcopenia. **Methods**: A cross-sectional study was conducted on 183 Mexican adults (67.2% women and 32.8% men, ≥60 years old). ALMI was measured using DXA, and anthropometric data were collected following the International Society for the Advancement of Kinanthropometry (ISAK) protocols. Predictive models were developed using Logistic Regression (LR), Decision Trees (DTs), Random Forests (RFs), Artificial Neural Networks (ANNs), and LASSO regression. The dataset was split into training (70%) and testing (30%) sets. Model performance was evaluated using classification performance metrics and the area under the ROC curve (AUC). **Results**: ALMI indicated strong correlations with BMI, corrected calf girth, and arm relaxed girth. Among models, DT achieved the best performance in females (AUC = 0.84), and ANN indicated the highest AUC in males (0.92). Regarding the prediction of low ALMI, specificity values were highest in DT for females (100%), while RF performed best in males (92%). The key predictive variables varied depending on sex, with BMI and calf girth being the most relevant for females and arm girth for males. **Conclusions**: Anthropometry combined with machine learning provides an accurate, low-cost approach for identifying low ALMI in older adults. This method could facilitate sarcopenia screening in clinical settings with limited access to advanced diagnostic tools.

Keywords: anthropometry; aged; artificial intelligence; Mexico; muscles; sarcopenia

 https://doi.org/10.3390/jfmk10030276

1. Introduction

Currently, there has been a sustained increase in the global elderly population [1]. In some countries, for the first time, the number of adults over the age of 65 has surpassed that of children under five years old, and projections estimate that by 2050, the number of individuals in this age group will equal that of those under 14 years old [2]. This trend is significant because aging is associated with a higher prevalence of chronic diseases, which increases healthcare costs for both individuals and their families while also placing a greater financial burden on public health systems [3–5]. As a result, some have even proposed aging itself as a disease specific to this stage of life [3].

Sarcopenia is a muscular disease characterized and diagnosed by low muscle mass, decreased muscle strength, and poor physical performance, all of which negatively affect mobility and quality of life in older adults [6,7]. Proper identification and interpretation of the condition are essential for its prevention and treatment [8]. Several international consensus groups have established guidelines and methods for measuring these diagnostic criteria [6,9,10]; however, some of these methods—particularly those used to assess Appendicular Lean Mass Index (ALMI)—are not accessible in all clinical settings. For instance, dual-energy X-ray absorptiometry (DXA), while considered a standard reference, requires expensive and often unavailable equipment, making sarcopenia screening difficult in many regions [6,9,11]. Moreover, in "The Position Statements of the Sarcopenia Definition and Outcomes Consortium," published in 2020, the clinical utility of DXA for muscle mass estimation was questioned. Although muscle mass remains a central diagnostic criterion, panelists expressed concerns about relying on DXA-derived lean mass, citing evidence that it is not consistently associated with adverse health outcomes in community-dwelling older adults, even when adjusted for body size. They suggested that more accurate alternatives, such as D3-creatine dilution, may show stronger associations with clinical outcomes and could be prioritized in future practice [12].

Another challenge in diagnosing sarcopenia is the use of population-specific cut-off points established by various consensus groups, such as the Asian Working Group for Sarcopenia (AWGS) and the European Working Group on Sarcopenia in Older People (EWGSOP), the latter being the most widely used [6,9,13]. This is important because studies have shown variations in cut-off values across different populations [14,15]. Accordingly, several countries are working to develop diagnostic tools and validated cut-off points tailored to their populations [14–20].

Additionally, sarcopenia prevalence estimates can vary widely—between 4.6% and 41%—depending on the diagnostic criteria used. This may be due to significant differences in body composition across ethnicities [21]. The existence of multiple diagnostic methods and cut-off values leads to low concordance and may result in underdiagnosis or overdiagnosis [12,22,23]. Therefore, it is recommended to estimate prevalence using ethnicity-specific cut-off points [22].

Accurately measuring lean mass—both in terms of quantity and quality—remains a challenge, as the most reliable assessment tools are costly and often inaccessible [11]. This has led to increasing efforts to explore alternative methods for estimating sarcopenia-related variables, including the use of anthropometry to predict lean mass through formulas or girth measurements, particularly of the leg, which is the only anthropometric measure currently validated for case detection [6,11,24,25]. Quadriceps girth has also emerged as a promising technique to estimate both the quantity and quality of muscle mass [11]. Anthropometry offers key information on health status and is a cost-effective, non-invasive, and universally applicable method for measuring body mass, proportions, and composition [24]. However, further research is needed to evaluate and compare lean mass estimation methods, particularly those adapted to specific ethnic groups. Differences in body composition between ethnic groups may explain the need for such adaptations. For example, Alemán-Mateo et al. [21] found that Mexican individuals had higher fat mass and lower lean mass compared to African American and Caucasian individuals in the United States. Moreover, body composition variability has been documented even among different regions within Mexico, further complicating the establishment of standardized cut-off points [26]. These variations likely contribute to inconsistencies in diagnostic outcomes, emphasizing the importance of conducting validation procedures specific to each demographic group [11]. In this context, research focused on older Mexican people remains limited, particularly regarding body composition estimation methods that are both validated for this demographic and practical for use in clinical settings. This limitation is primarily due to restricted access to advanced validation tools, such as DXA, as well as the heterogeneity of existing studies and a predominant focus on younger and general adult populations. These factors present significant challenges for standardization and meaningful comparisons across studies. Therefore, additional studies are necessary to evaluate and compare various muscle mass assessment methods, tailored to the specific characteristics of this population.

The use of artificial intelligence (AI) tools—particularly machine learning (ML)—has become increasingly prevalent in health-related predictive modeling [27]. One of ML's key advantages in healthcare is the ability to automate data processing, thereby improving the efficiency of assessment, diagnosis, treatment, and monitoring [27–29]. AI algorithms such as Support Vector Machines (SVMs), Decision Trees (DTs), and Random Forests (RFs), among others, have proven useful for diagnosing sarcopenia and osteoporosis, offering predictive variables adapted to specific populations [14,29,30]. These methods have also been applied in models for diagnosing cardiovascular diseases, overweight and obesity, atrial fibrillation, endometrial lesions, and more [14,29–36]. A notable advantage of predictive risk models in clinical practice is their ability to provide more individualized risk assessments, thus enhancing the efficacy of interventions [35].

Considering the above, there is a growing need to develop accessible, accurate, and adaptable computational models capable of reliably estimating ALMI in resource-limited clinical settings. This study proposes the design of a predictive model based on anthropometric variables to detect low ALMI in Mexican adults. Such a tool could facilitate early sarcopenia detection and improve clinical decision-making, particularly in populations with limited access to advanced technologies. Furthermore, it could contribute to improving intervention methods and reducing the healthcare costs associated with functional decline in old age.

2. Materials and Methods

2.1. Study Design and Participants

Participants were selected using a non-probabilistic purposive sampling method, and participation was entirely voluntary. The minimum size was calculated to be 96 older adults from the state of Jalisco, including both males and females. The sample size was estimated using the following standard formula for descriptive studies, assuming a 95% confidence level:

$$n = \frac{(Z_{\alpha/2})^2 (p)(q)}{d^2}$$

where $Z_{\alpha/2} = 1.96$, $q = 1 - p$, estimated prevalence (10%): $p = 0.10$, and standard deviation of the prevalence (6%): $d = 0.06$.

This study adhered to the Strengthening Reporting of Observational Studies in Epidemiology (STROBE) guidelines [37]. A cross-sectional design was implemented with a sample of older Mexican adults aged 60 years and above, of both sexes. The recruitment strategy involved visits to institutions that gather older adults, such as health centers and recreational facilities. This study was named "Sarcopenia in Older Adults from Jalisco (SAMJ)". Individuals with limb loss, inability to move independently, or the presence of edema were excluded.

2.2. Instruments

2.2.1. Body Composition Assessment Using DXA

A Lunar iDXA densitometer (General Electric) was used. This equipment emits X-rays at two different energy levels, which are attenuated as they pass through tissues depending on their density. Using these attenuation coefficients, an R value is obtained, enabling the calculation of total and regional lean mass, fat mass, bone mass, and fat-free mass. From these data, Appendicular Lean Mass (ALM) was calculated for sarcopenia diagnosis.

The DXA evaluation process included the following three phases:

1. Participant preparation: participants were asked to avoid intense physical activity.
2. Positioning: participants were set in a supine position on the scanning table, free of external metal objects. The limbs were positioned alongside the body, palms facing downward, feet in a neutral or slightly inward position, and face facing upward in a neutral position.
3. Post-processing: scans were performed by trained personnel following the manufacturer's instructions [38].

2.2.2. Anthropometry

All measurements were taken on the participants' right side by a Level 3 ISAK-certified anthropometrist (A.M.G.-M.), following the standardized protocol of the International Society for the Advancement of Kinanthropometry (ISAK) [39]. Anthropometric measurements were obtained using a portable stable stadiometer (Seca 217, Seca GmbH & Co. KG, Hamburg, Germany), a wide platform digital weighing scale (model XL-700, Detecto, Webb City, Missouri, USA), a steel anthropometric tape (Lufkin W606PM, Lufkin, Missouri City, Texas, USA), and a skinfold caliper (Baty International, Harpenden, UK).

Before any measurement, the following anatomical landmarks were identified and marked with a dermatographic pencil:

1. Acromiale: upper border of the most lateral part of the acromion.
2. Radiale: proximal–lateral border of the radial head.
3. Mid acromiale–radiale: midpoint between the acromiale and radiale marks.
4. Triceps skinfold: posterior midline of the upper arm, aligned with the mid acromiale–radiale point.
5. Calf skinfold: medial aspect of the leg at its maximum girth.

Measurements taken:

1. Body mass: measured using a calibrated digital scale, with the participants in anthropometric position (standing, feet shoulder-width apart, arms relaxed at the sides).
2. Stretch stature: measured with a stadiometer, participants barefoot and upright, feet together, head in the Frankfurt plane.
3. Triceps skinfold (TSF): taken with a skinfold caliper, vertically at the mid acromiale–radiale point, parallel to the arm's longitudinal axis.
4. Calf skinfold (CSF): participant standing with the right foot on the anthropometric box and right knee flexed at 90°. The skinfold was measured vertically at the designated point on the medial leg.
5. Arm Relaxed Girth (ARG): measured at the mid acromiale–radiale point with a non-elastic tape, arm hanging naturally.

6. Arm Flexed and Tensed Girth (AFTG): arm flexed at 90° in front of the body. Participants were asked to contract the biceps maximally, and the girth was measured at the point of greatest muscle prominence.
7. Forearm Girth (FG): with the forearm slightly flexed and in a supine position, the maximum girth was measured.
8. Calf Girth (CG): with the participants standing on the anthropometric box, measured at the level of the leg skinfold point.

Corrected Girths (CoGs) were calculated using the following formula: CoG = Limb girth (cm) − (Skinfold (cm) × π) [40–42]. Each measurement was taken twice, and if the difference exceeded 5%, a third measurement was performed. The average of the closest two was recorded. The equipment was calibrated daily, and all measurements were taken by the same evaluator (A.M.G.-M.) to ensure reliability.

2.3. Procedures

This study was carried out in collaboration with the Laboratorio de Evaluación y Cuidado del Estado Nutricio (LECEN) at the Universidad de Guadalajara (CUTonalá) and the Centro de Atención Integral al Adulto Mayor (CAIAM) of the Sistema Nacional para el Desarrollo Integral de la Familia (DIF) Tonalá and Tlaquepaque. Initial recruitment occurred at the CAIAM in Guadalajara, where older adults attending activities were invited to participate. An agreement was made with the institution's director to arrange transportation to CUTonalá for anthropometric assessments by scheduled appointments.

Evaluations were conducted twice a week, with 10 to 17 participants assessed per session. Upon arrival at LECEN (10:00 am), participants were welcomed in a large classroom adjacent to the laboratory where measurements were taken. Participants were informed about this study and its implications, and informed consent was obtained. A patient rotation system was then organized. Participants first attended the anthropometry station, followed by the DXA scan, and then proceeded to other assessment stations (results not discussed in this article). Once the first participant had completed the full circuit, the next participant was evaluated, continuing in this manner until all individuals who attended that day had been assessed. While waiting, participants engaged in games and had access to snacks, coffee, tea, and water. Anthropometric and body composition evaluations were carried out following standardized protocols. For DXA scans, participants wore a medical gown and avoided underwear containing metal. For anthropometric measurements, they wore lightweight pants that could be rolled up to the knee and a sleeveless shirt. Minimal clothing was not required, prioritizing participant comfort.

2.4. Data Analysis

Microsoft Excel and RStudio (version 2024.09.1+394) were used for data analysis. Cut-off points for ALM were determined using the 20th percentile [43–46]. Normality of the data was assessed via histograms, Q-Q plots, and the Kolmogorov–Smirnov test. Pearson or Spearman correlations were used based on data distribution to assess relationships between DXA-derived ALM and anthropometric variables.

Subsequently, artificial intelligence tools were implemented using machine learning models, including Random Forests (RFs), Logistic Regression (LR), Decision Trees (DTs), Artificial Neural Networks (ANNs), and Least Absolute Shrinkage and Selection Operator (LASSO) regression. Models were trained on 70% of the dataset and tested on the remaining 30%. The target variable was binary ("normal" vs. "low" ALMI) and was predicted using anthropometric variables. Full details of the analysis can be reviewed in the R Markdown available in the Supplementary Materials.

2.5. Ethical and Biosafety Considerations

The SAMJ research project was approved by the Research and Graduate Studies Committee (registration number SAC/CIP/DOAN/027/2023) and by the Ethics Committee (registration CEI/77/2023, approved on 8 March 2025) of the Centro Universitario del Sur, Universidad de Guadalajara. This study followed the ethical principles outlined in the Declaration of Helsinki [47] and the Reglamento de la Ley General de Salud en Investigación en Salud (1987) of Mexico [48]. Informed consent and/or assent were obtained in accordance with Articles 20–22 of the aforementioned regulation.

Participation was entirely voluntary. Recruitment was conducted by health and research staff through direct invitations and informational posters at participating institutions. As a benefit, participants received a health report including body composition, diet, and physical performance assessment, valued at approximately 13,000 Mexican pesos. Results were delivered in person or via the participant's preferred method (phone call, WhatsApp, text, or email).

3. Results

3.1. General and Descriptive Result

The sample consisted of 183 older adults, 67.2% women and 32.8% men. Most participants were either married (41%) or widowed (35%), lived with their family (67.8%), did not smoke (90.2%), did not consume alcohol (84.7%), did not have insulin resistance (85.8%), and had no diagnosed depression (80.9%) (see Appendix A.1). Sex-based comparisons of the variables are shown in Tables 1 and 2. Significant sex differences were found in most variables, except for BMI and age.

Table 1. Sex differences in quantitative body composition variables with normal distribution in older adults from the SAMJ study.

	Male Mean (SD)	Female Mean (SD)	*p*-Value
	Body composition		
ALMI (kg/m^2)	8.13 (0.94)	6.66 (0.85)	<0.001
	Anthropometry		
BMI (kg/m^2)	28.6 (3.85)	28.99 (5.05)	0.542
ARG (cm)	30.94 (3.24)	30.84 (4.05)	<0.001
AFTG (cm)	31.62 (3.03)	30.04 (3.98)	<0.001
FG (cm)	27.03 (1.99)	24.08 (2.12)	<0.001
CG (cm)	36.27 (2.9)	34.59 (3.47)	<0.001
CoG arm (cm)	26.94 (2.35)	24.07 (2.64)	<0.001
CoG calf (cm)	36.27 (2.55)	28.06 (2.5)	<0.001

ALMI: Appendicular Lean Mass Index; BMI: Body Mass Index; ARG: Arm Relaxed Girth; AFTG: Arm Flexed and Tensed Girth; FG: Forearm Girth; CG: Calf Girth; CoG: Corrected Girth. *p*-values are from Student's *t*-tests.

Table 2. Sex differences in quantitative body composition variables without normal distribution in older adults from the SAMJ study.

Body Composition	Male Median (IQR)	Female Median (IQR)	*p*-Value
ALM (kg)	22.13 (3.7)	11.68 (3.37)	<0.001
ALM/BMI	0.71 (0.13)	0.53 (0.1)	<0.001

ALM: Appendicular Lean Mass; BMI: Body Mass Index; IQR: Interquartile Range. *p*-values are from the Mann–Whitney U test, with statistically significant results in bold.

Sarcopenia

Table 3 presents the cut-off points derived from the SAMJ study sample, which were used to identify and categorize ALMI levels as low or normal.

Table 3. Cut-off points for sarcopenia diagnosis in older adults from the SAMJ study.

Sex	ALMI (kg/m^2)	ALM/BMI
Male	<7.49	<0.69
Female	<5.93	<0.46

ALMI: Appendicular Lean Mass Index; ALM: Appendicular Lean Mass; BMI: Body Mass Index; Cut-off points were determined using the 20th percentile.

3.2. Body Composition Values Related to Anthropometry

Table 4 shows the correlations between anthropometric measurements and body composition variables assessed via DXA. The following variables indicated strong ($r > 0.7$; [49]) and statistically significant positive correlations: (a) body mass with ALM and lean mass in both arms and legs, including ALMI; (b) BMI, ARG, triceps, and leg skinfolds with fat mass values; (c) forearm girth with lean mass values, except for the ALM/BMI ratio; (d) calf girth with ALMI; and (e) corrected calf and arm girths with ALM and lean mass in the arms and legs (Figure 1).

Table 4. Correlation matrix between anthropometric and DXA variables of the older persons from the SAMJ study.

Anthropometry	DXA Fat Mass (kg)			DXA Lean Mass (kg)				
	Appendicular r	Arms r	Legs r	Appendicular r	Arms (kg) r	Legs (kg) r	ALMI (kg/m^2) r	ALM/BMI r
Body mass (kg)	0.496 **	0.474 **	0.479 **	0.802 **	0.744 **	0.809 **	0.798 **	0.317 **
BMI (kg/m^2)	0.731 **	0.702 **	0.704 **	0.731 **	0.287 **	0.352 **	0.589 **	−0.287 **
TSF (mm)	0.772 **	0.754 **	0.740 **	−0.235 **	−0.283 **	−0.209 *	−0.065	−0.607 **
CSF (mm)	0.721 **	0.613 **	0.718 **	−0.299 **	−0.373 **	−0.269 **	−0.159 *	−0.579 **
ARG (cm)	0.698 **	0.733 **	0.641 **	0.400 **	0.400 **	0.396 **	0.547	−0.137 **
AFTG (cm)	0.574 **	0.654 **	0.522 **	0.527 **	0.529 **	0.515 **	0.646 **	0.037
FG (cm)	0.303 **	0.359 **	0.271 **	0.825 **	0.824 **	0.818 **	0.809 **	0.461**
CG (cm)	0.549 **	0.444 **	0.554 **	0.630 **	0.540 **	0.655 **	0.701 **	0.202 *
Calf CoG (cm)	−0.562	−0.080	−0.049	0.831 **	0.805 **	0.825 **	0.786 **	0.650 **
Arm CoG (cm)	0.317 **	0.430 **	0.267 **	0.710 **	0.685 **	0.773 **	0.293 **	0.202 *

ALMI: Appendicular Lean Mass Index; ALM: Appendicular Lean Mass; BMI: Body Mass Index; TSF: Triceps Skinfold; CSF: Calf Skinfold; ARG: Arm Relaxed Girth; AFTG: Arm Flexed and Tensed Girth; FG: Forearm Girth; CG: Calf Girth; CoG: Corrected Girth; DXA: Dual-energy X-ray Absorptiometry; r = correlation coefficient. Pearson or Spearman correlation, depending on data normality, * $p < 0.05$, ** $p < 0.001$.

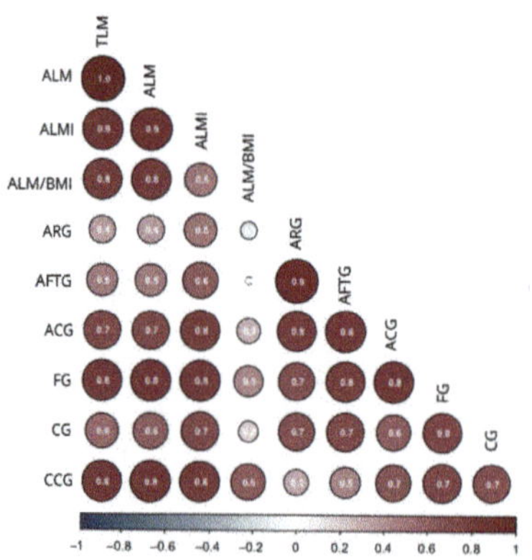

Figure 1. Correlation diagram between anthropometric and body composition variables of the older persons from the SAMJ study. ALM: Appendicular Lean Mass; ALMI: Appendicular Lean Mass Index; BMI: Body Mass Index; ARG: Arm Relaxed Girth; AFTG: Arm Flexed and Tensed Girth; ACG: Arm Corrected Girth; FG: Forearm Girth; CG: Calf Girth; CCG: Calf Corrected Girth. Pearson or Spearman correlation, depending on data normality. The numbers represent the correlation coefficient (r).

3.3. Predictive Models for Appendicular Lean Mass Index (ALMI) Using Anthropometry

ALM is a key diagnostic criterion for sarcopenia, yet one of the most difficult variables to assess in clinical practice. Therefore, estimating it using anthropometric variables is a priority and also a practical alternative. In this study, the aim was to predict the binary classification of ALMI as "normal" or "low." The following machine learning (ML) models were applied: Decision Trees (DTs), Logistic Regression Models (LR), Random Forests (RFs), Artificial Neural Networks (ANNs), and LASSO regression (LASSO). The models were evaluated using AUCs and specificity values.

3.3.1. Logistic Regression Models

The predictive values of the variables included in the model were obtained (Table 5). When classifying the cases in the test process, the model achieved a classification AUC value of 0.76 for females and 0.77 for males, which means they are able to identify 76% and 77% of the cases with low ALMI in the test sample (see Section 3.3.6).

Table 5. Logistic Regression model for predicting ALMI level using anthropometric variables stratified by sex among older persons from the SAMJ study.

	Estimate		Standard Error		Z Value		*p* Value	
	F	M	F	M	F	M	F	M
Intercept	7376.80	1802.22	706,862.64	973,842.13	0.010	0.002	0.992	0.999
BMI	−185.10	−3.13	15,623.58	14,513.51	−0.012	0.000	0.991	1.000
ARG	506.38	−56.27	40,247.17	99,777.97	0.013	−0.001	0.990	1.000
AFTG	−331.97	24.36	28,252.76	44,860.02	−0.012	0.001	0.991	1.000
Arm CoG	−222.99	23.28	19,574.73	45,842.11	−0.011	0.001	0.991	1.000
FG	−48.82	−21.73	6982.93	33,221.02	−0.007	−0.001	0.994	0.999
CG	−48.11	19.88	10,491.04	26,609.56	−0.005	0.001	0.996	0.999
Calf CoG	−14.11	−47.95	3338.89	29,310.10	−0.004	−0.002	0.997	0.999

BMI: Body Mass Index; ARG: Arm Relaxed Girth; AFTG: Arm Flexed and Tensed Girth; FG: Forearm Girth; CG: Calf Girth; CoG: Corrected Girth; F: Female; M: Male.

3.3.2. Decision Trees

In Figure 2 and Table 6, BMI and CCG were identified as the most relevant predictors of low ALMI for women, while ARG was the most relevant for men. Cut-off points for each variable were established to detect low ALMI. Model validation through case classification tests yielded an AUC of 0.84 for women and 0.80 for men, with slightly lower performance observed in males (Table 7).

Table 6. Decision Tree model for ALMI prediction by sex in older adults from the SAMJ study.

Female						Male					
Node	split	n	loss	yval	(yprob)	Node	split	n	loss	yval	(yprob)
(1)	root	87	18	1	(0.793–0.207)	(1)	root	42	8	1	(0.809–0.191)
(2)	BMI ≥ 25.83	61	1	1	(0.984–0.016) †	(2)	ARG ≥ 28.75	30	0	1	(0.000–0.000) †
(3)	BMI < 25.83	26	9	2	(0.346–0.333)	(3)	ARG < 28.75	12	4	2	(0.333–0.666) †
(4)	CCG ≤ 26.89	12	4	1	(0.666–0.333) †						
(5)	CCG < 26.89	14	1	2	(0.071–0.929) †						

† Denotes terminal node. split: splits; n: number of cases; yval: the mean response value of all observations in the training dataset; yprob: estimated probability of each class at each terminal node of the tree; BMI: Body Mass Index; CCG: Calf Corrected Girth; ARG: Arm Relaxed Girth.

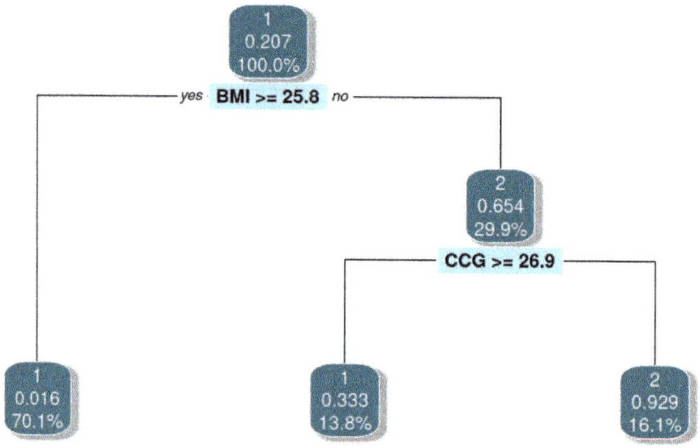

(a)

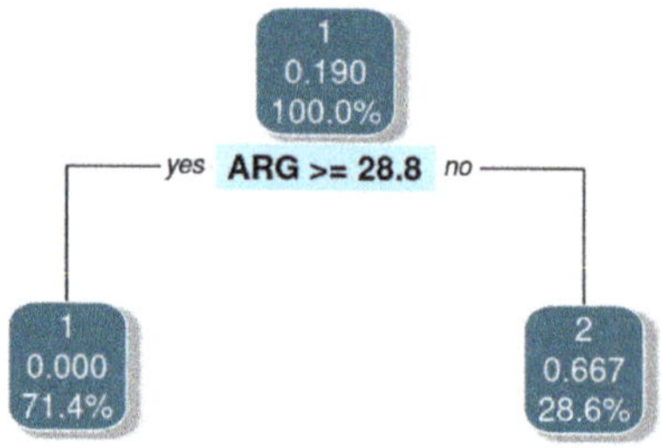

(b)

Figure 2. Decision Tree for Appendicular Lean Mass Index by sex in older adults from the SAMJ study. (**a**) Females. (**b**) Males. 1 = Normal ALMI; 2 = Low ALMI; Number of trees = 87; BMI: Body Mass Index; CCG: Calf Corrected Girth; ARG: Arm Relaxed Girth.

3.3.3. Random Forest

When the modeling was carried out, the variables with the greatest predictivity for ALMI level were identified as BMI and CG for females and CG and ARG for males (Figure 3). Table 7 shows the results of case classification during the test phase. The BAs indicated an AUC of 0.82 for females and 0.79 for males (Table 7).

3.3.4. Artificial Neural Network

Figure 4 displays the ANNs, each consisting of a single hidden layer with five neurons (H1 to H5) and one output neuron (O1) responsible for predicting the binary outcome: low vs. normal ALMI. Bias nodes B1 and B2 are included to adjust activation thresholds and enhance the learning process. The connecting lines represent the synaptic weights, where thicker lines indicate stronger connections. Line color denotes the direction of the weight: black for positive and burgundy for negative weights. The distribution of weights from input to hidden neurons appears more balanced in the network shown in Figure 4b. In the female model (Figure 4a), AFTG emerges as the most influential variable, followed by ARG and BMI, indicating a greater predictive contribution from thigh-related anthropometric measures. In the male model (Figure 4b), CCG appears to be the most important predictor, followed by CG and ARG. Regarding performance, the models achieved an AUC of 0.82 for ALMI classification in females and 0.92 for males on the test dataset (Table 7).

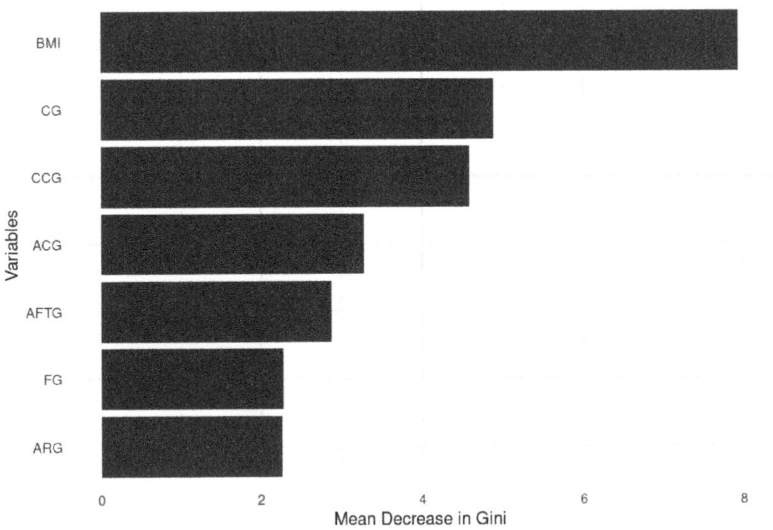

(**a**)

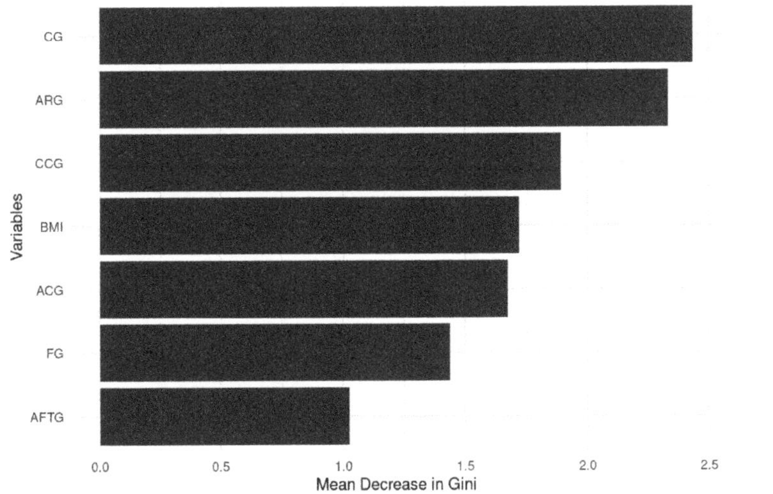

(**b**)

Figure 3. Variables with major significance for ALMI level in Random Forests. (**a**) Females, (**b**) Males. BMI: Body Mass Index; ARG: Arm Relaxed Girth; AFTG: Arm Flexed and Tensed Girth; FG: Forearm Girth; CG: Calf Girth; ACG: Arm Corrected Girth; CCG: Calf Corrected Girth. Mean Decrease in Gini: It is a measure of node purity used to construct Decision Trees, representing how mixed the classes are in a node. High values indicate that the variable contributes significantly to improving node purity, which implies that it is an important variable for predicting the response or classifying correctly.

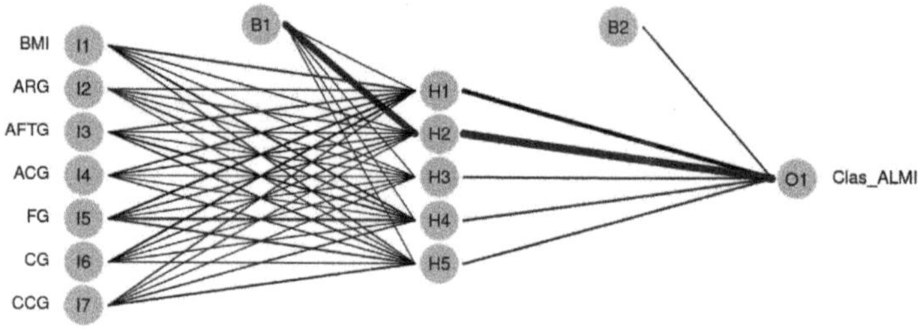

(**a**)

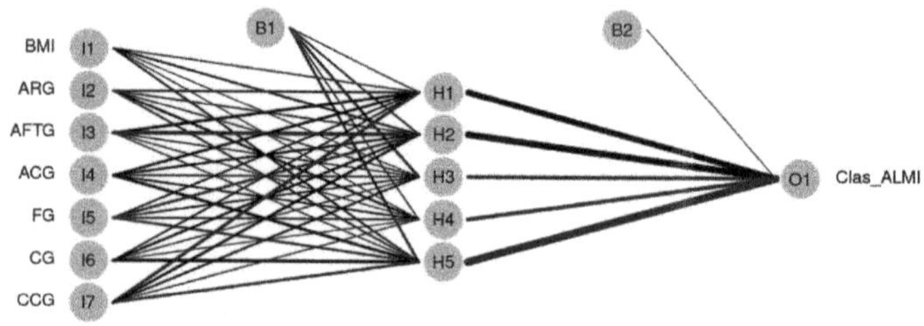

(**b**)

Figure 4. Artificial Neural Network for the prediction of ALMI level. (**a**) Females. (**b**) Males. BMI: Body Mass Index; ARG: Arm Relaxed Girth; AFTG: Arm Flexed and Tensed Girth; FG: Forearm Girth; CG: Calf Girth; ACG: Arm Corrected Girth; CCG: Calf Corrected Girth; 1: normal; 2: low. Line color denotes the direction of the weight: black for positive and burgundy for negative weights.

3.3.5. LASSO Regression

LASSO regression is a technique that combines coefficient shrinkage and variable selection to enhance model performance and interpretability in regression models [50]. Table 7 summarized the performance metrics of the model for ALMI classification, showing an AUC of 0.84 for females and 0.87 for males.

3.3.6. Performance Metrics and Model Comparison

Table 7 shows the performance metrics of all the evaluated models for ALMI classification. For females, both the DT and LASSO models demonstrated the best predictive performance, each achieving an AUC of 0.84. In the case of males, the ANN model outperformed others with an AUC of 0.92. These results are visually represented in Figure 5 through the ROC curves. Given that the primary goal in predicting sarcopenia is to accurately identify low ALMI cases, specificity is a particularly important metric. All models for females achieved high specificity scores above 0.93, with the DT model exhibiting perfect specificity (1.00), meaning it correctly identified all true negative cases (i.e., correctly identifying all the cases with low ALMI). For male participants, specificity values were slightly lower but still clinically relevant. LR had the lowest specificity at 0.77, whereas the RF model performed best, reaching a specificity of 0.92 in detecting negative cases.

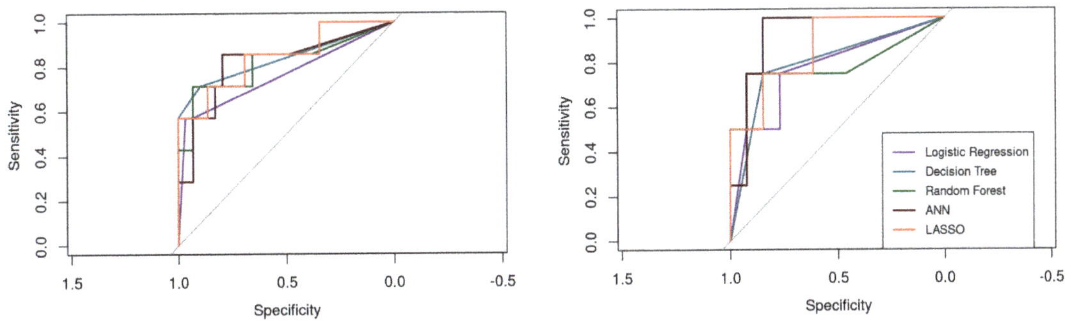

(a) (b)

Figure 5. ROC curves of the machine learning models to predict ALMI in older adults from the SAMJ study. (**a**) Females. (**b**) Males.

Table 7. Classification performance metrics to predict ALMI level with anthropometric variables in older adults from the SAMJ study.

	Female					Male				
	LR	**DT**	**RF**	**ANN**	**LASSO**	**LR**	**DT**	**RF**	**ANN**	**LASSO**
Accuracy	0.86	0.92	0.86	0.86	0.89	0.77	0.82	0.88	0.88	0.82
Sensitivity	0.57	0.57	0.57	0.57	0.57	0.75	0.75	0.75	0.75	0.75
Specificity	0.93	1.00	0.93	0.93	0.97	0.77	0.85	0.92	0.85	0.85
Precision	0.67	1.00	0.67	0.67	0.80	0.50	0.60	0.75	0.60	0.60
AUC	0.76	0.84	0.82	0.82	0.84	0.77	0.80	0.79	0.92	0.87

LR: Logistic Regression; DT: Decision Tree; RF: Random Forest; ANN: Artificial Neural Network; LASSO: LASSO regression.

4. Discussion

The results of ALMI level modeling are presented based on the performance of five machine learning approaches: DT, LR, RF, ANN, and LASSO regression. In DT models, the most important predictive variables were BMI and calf girth (CCG) for females and arm girth (ARG) for males. The DT model achieved the highest performance in females, with an AUC of 0.84. For males, the ANN model demonstrated the strongest predictive capability, achieving a test accuracy of 0.92 with low error rates. Overall, BMI emerged as a particularly influential predictor in models for women. In terms of specificity for detecting low ALMI, the DT model achieved perfect classification (100%) in females, showing high accuracy in identifying true negatives while maintaining good detection of low ALMI cases; in males, the RF model performed best, with a specificity of 93%.

The results of this modeling suggest that DT, RF, and ANN are promising tools for predicting low ALMI using anthropometry. Further training and validation could generate normative reference values useful in clinical practice. These ML models are commonly used for case classification, offering strong potential in clinical settings—particularly for distinguishing between the presence and absence of disease—by capturing complex, non-linear, and heterogeneous relationships among variables, even when the underlying physiological mechanisms are not fully understood due to biological complexity or pathological variability [51].

Among these, DTs are especially well-suited for clinical applications due to their intuitive structure, high classification accuracy, ease of validation by clinical experts, and user-friendly, human-readable format. They classify new instances based on patterns identified from previously labeled data, making them valuable for disease diagnosis and risk stratification. In our study, DTs enabled the identification of a small set of highly predictive features (e.g., BMI, CCG, and ARG) that could guide the design of simple, low-cost screening tools. However, DTs can be sensitive to noisy data and prone to overfitting, which may compromise generalizability—limitations that are less common in algorithms such as ANNs. While ANNs are more robust to noise, they often require longer training times and are less interpretable due to their "black box" nature [52]. RF, an ensemble method built from multiple DTs through bootstrap sampling, mitigates some of the weaknesses of single-tree models. It enhances predictive performance by reducing model variance and increasing stability [51,53]. Each tree in the forest independently generates a prediction, and the final output is determined by a majority vote (for classification) or by averaging predictions (for regression). As a non-parametric algorithm, RF is well-suited for both continuous and categorical variables and is relatively simple to tune [53]. Additional advantages include its robustness to overfitting, tolerance to outliers, and ability to compute ancillary metrics such as classification error and variable importance through permutation testing [51,53]. This capacity to quantify feature relevance, combined with its strong generalization capabilities, makes RF a valuable tool in biomedical research and clinical predictions [53]. Despite these strengths, the applicability of our models is currently limited by the absence of external validation. While internal cross-validation was applied, future research should involve independent datasets from diverse populations to evaluate reproducibility, generalizability, and key requirements for clinical implementation. In summary, both DT and RF models offer significant advantages for practical implementation in clinical contexts, given their interpretability and support for informed decision-making. However, the limited sample size and sex imbalance in our study may affect the stability and reliability of the models, particularly among male participants, underscoring the need for larger, more balanced datasets and external validation efforts.

Our results are consistent with those reported by Olshvang et al. [54], who applied RF, LR, LassoNet, and XGBoost models to predict lean mass based on anthropometric and sociodemographic data (age, ethnicity, body mass, stature, and waist girth) using data from the National Health and Nutrition Examination Survey (NHANES). In their study, body mass and male sex were among the most important predictors, and they concluded that RF, XGBoost, and LassoNet accurately predicted both total and appendicular lean masses. Similarly, Cichosz et al. [55] used NHANES data and ANN to predict ALMI and fat mass in adults, reporting strong correlations with DXA measurements. Buccheri et al. [56] also used NHANES data to develop a near-zero-cost screening tool using DT to emulate DXA performance for identifying low muscle mass. They found that anthropometric measurements of the lower limbs provided a simpler yet effective alternative, with an AUC of 0.88–0.90.

Marazzato et al. [57] demonstrated the feasibility of predicting ALMI via DXA using a recurrent neural network (RNN) in a diverse population consisting of 576 children, adolescents, and adults. The model included 10 demographic dimensions (sex, age, and seven ethnic groups) and 43 anthropometric dimensions obtained from a 3D optical scanner. The predicted and measured ALMI values were highly correlated, with small mean, absolute, and squared prediction errors, highlighting the potential of ANN in body composition prediction using large-scale digital anthropometric data. In India, Birk et al. [58] developed a machine learning model to estimate body composition using bioimpedance, skinfolds, body girths, and grip strength. Their model indicated lower prediction errors than traditional equations. This is consistent with the study by Guo et al. [59], who developed an online calculator using ML to predict low ALMI. Their model, based on XGBoost and using only stature, waist girth, age, and race, achieved an AUC above 0.85 in validation tests.

The tool was designed for community-level use in the U.S., facilitating early sarcopenia detection and intervention.

In a follow-up study, Buccheri et al. [60] developed computationally simple equations to estimate DXA-measured ALMI using 38 non-laboratory variables in older adults. Using only body mass, sex, and anthropometric measures (thigh and arm girth), they achieved an AUC-ROC of 0.89. Interestingly, they found that adding more than three variables did not improve model performance. Shi et al. [61] developed an anthropometric equation using LASSO regression to estimate ALMI in elderly women in India. The model included body mass, stature, BMI, sitting stature, waist-to-hip ratio (WHR), upper arm length, and other limb length summaries. The final equation—based on body mass, WHR, upper arm length, and sitting stature—demonstrated good agreement with DXA, with 95% limits of agreement. This is consistent with the findings of our study, in which BMI emerged as an important predictor of lean mass in women. Similarly, Kang et al. [62] also identified BMI as a key variable for predicting muscle mass in women and reported the highest accuracy using boosted algorithms. Multiple recent studies have focused on predicting sarcopenia, pre-sarcopenia, or similar conditions using accessible data sources, yielding promising and practical results, showing the importance of continuing the research in this field [63–67]. It is worth noting that most of these studies focused on predicting ALMI as a continuous variable (in kilograms) rather than its categorical classification (normal vs. low). In contrast, we employed a categorical prediction approach to avoid the limitations associated with defining population-specific cut-off points for sarcopenia diagnosis. As highlighted by Rangel-Peniche et al. [26], these cut-offs may vary across ethnicities and populations. Predicting ALMI status using anthropometric data may thus offer practical advantages in the clinical screening and diagnosis of sarcopenia.

As shown, the heterogeneity in model types and anthropometric variables across studies makes direct comparisons difficult. Nevertheless, ML tools are valuable alternatives when DXA is unavailable, potentially offering scalable solutions for sarcopenia screening via mobile applications or clinical software. Predicting ALMI status categorically is advantageous, as it directly informs the diagnostic component of sarcopenia and reduces variability and bias due to different cut-off definitions. The variability in feature importance across sex groups (BMI and CCG in women and ARG in men) further suggests that sex-specific models may be more appropriate than unisex models. These differences may be due to distinct patterns of fat and muscle distribution between men and women and merit further study in larger and more diverse samples [68,69].

One of the main limitations of this study was the sample size, particularly in the male group. This issue is not uncommon in the literature, as large-scale studies with comprehensive assessments are often constrained by high costs and limited resources [70,71]. Determining an appropriate sample size for predictive models developed using ML techniques is not straightforward [72]. Several studies have reported small sample sizes (with a median of 88 participants), and, in certain cases, higher accuracy has been observed in models trained on smaller datasets [71]. For example, Castillo-Olea et al. [73] used 166 participants in Tijuana, Mexico; Pineda-Zuluaga et al. [74] worked with 237 in Manizales, Colombia; and Abdalla et al. [75] studied 125 individuals in São Paulo, Brazil. Some datasets are vulnerable when the ratio of features to sample size is high, increasing the likelihood that the model fits noise rather than the underlying data patterns. Consequently, ML models may produce overly optimistic results when trained on small datasets, resulting in poor generalizability [71,76]. Although small samples are more prone to overfitting and less likely to detect true effects, high-quality data may compensate for this limitation [70]. Specifically, for neural networks, it has been recommended that the sample size be at least 50 times the number of model weights to ensure robustness and reduce bias [76].

Rajput et al. [70] found that predictive error decreased with sample sizes above 120, and optimal performance was achieved with more than 1000 participants. In our study, the female group met the minimum threshold for reliable modeling, while the smaller male sample likely limited model stability. Notably, once the minimum sample size is reached, increasing the number of participants does not significantly improve model performance, thus offering a favorable cost–benefit ratio [70].

Furthermore, external validation is essential to confirm the generalizability of ML models that demonstrate potential clinical value [56]. In this study, internal validation was performed by dividing the dataset into training and testing sets, which helps minimize optimistic bias. Nevertheless, even with reasonable sample sizes, this approach may be insufficient. The next step should involve validation using an independent sample, ideally from a different region or population subgroup. Cross-validation remains a widely used approach when available data are limited [70].

Another limitation is that the use of anthropometric measurements is prone to various sources of error, including lack of equipment calibration, data recording inaccuracies, and insufficient training or expertise of the anthropometrist. Even environmental conditions, such as room temperature, can influence measurement outcomes. In addition, skinfold and girth measurements can be affected by physiological factors (e.g., hydration status, fat redistribution in aging) and inter-observer variability, limiting precision. Despite these challenges, the use of standardized protocols, such as those established by ISAK, and proper certification can substantially reduce both intra- and inter-observer measurement error [39,77,78]. These considerations are particularly important in older adults, where physiological changes and greater variability in body composition increase the need for precise and reliable assessments [79]. In the context of clinical applicability, ensuring reproducibility and precision is essential, especially when working with diverse populations. This further underscores the importance of adopting validated, standardized methods tailored to specific populations, such as older Mexican adults, to support accurate diagnosis and appropriate clinical decision-making.

Despite these limitations, the models still achieved high accuracy, emphasizing the value of continuing to expand the dataset to improve predictive performance. Future research should consider adding new predictive variables or exploring data augmentation techniques such as SMOTE (Synthetic Minority Oversampling Technique) to handle imbalanced class distributions [80]. The clinical applicability of these models lies in their potential to be implemented in digital platforms or mobile applications for use by trained health professionals. Such tools could provide automated assessments of ALMI levels or flag individuals at risk of sarcopenia. For nationwide implementation, however, these tools must undergo rigorous external validation and standardization across population subgroups. This study represents a significant step toward developing anthropometry-based sarcopenia screening tools tailored to the Mexican older adult population.

Additionally, the lack of a global consensus on sarcopenia definitions and diagnostic criteria remains a challenge [81]. This uncertainty hinders comparability across studies and emphasizes the need for accessible, replicable tools based on interpretable and reliable predictors. Accurate evaluation of body composition using accessible methods is crucial for understanding both the health and disease of an older person, as variations in muscle and fat distribution are associated with conditions such as sarcopenia, obesity, and metabolic disorders. Conventional approaches, such as DXA and bioelectrical impedance analysis (BIA), have limited capacity to comprehensively and efficiently assess body composition in clinical settings. AI, particularly through ML, enhances the segmentation and analysis of these techniques, offering greater precision and accuracy. This enables more personalized healthcare by identifying specific patterns of muscle and fat distribution associated with disease risk [82].

5. Conclusions

There is an urgent need for non-invasive, cost-effective, and scalable methods for population-level screening while also providing practical support to clinicians. Anthropometry stands out as a valuable tool for identifying low ALMI in older adults with sarcopenia. Additionally, ML algorithms are gaining prominence, demonstrating strong predictive performance [83]. As highlighted in prior research, ML is becoming an increasingly powerful asset in healthcare, especially for the early detection of sarcopenia in older populations. These techniques excel at analyzing large and complex datasets, facilitating timely diagnosis and intervention.

Continued research is essential to refine and validate diagnostic tools for sarcopenia, including integrating emerging technologies such as wearable activity trackers and smartwatches. A recent comprehensive review highlights that these devices can offer valuable insights into sarcopenia progression, support monitoring, and indicate the need for early intervention [81]. AI and ML offer promising solutions to global healthcare challenges by driving innovation, improving efficiency, and expanding access to care. Moreover, they have the potential to revolutionize healthcare systems through enhanced disease prediction, earlier diagnosis, personalized treatment strategies, and improved equity. In this context, the clinical applicability of the models developed in this study will depend on the creation of a user-friendly platform—possibly in the form of a mobile or desktop application—for use by trained healthcare professionals. Such tools could estimate ALMI levels or predict sarcopenia diagnosis based on simple anthropometric inputs. Although further research and external validation are needed for national-level implementation, this study represents an important step toward improving sarcopenia detection in the older Mexican adult population. Moreover, this work adds to the growing evidence that AI-supported anthropometric assessments can help bridge current diagnostic gaps and improve health outcomes in aging populations. However, successful implementation depends on responsible and ethical governance, integration into healthcare infrastructure, secure data management, and robust user engagement [32,82]. Ethical reviews emphasize the need to address privacy, transparency, bias, and user autonomy when applying AI in clinical contexts [83]. Challenges remain, including ensuring input data accuracy and encouraging user adherence, which must be addressed to fully realize the benefits and potential of these technologies [81].

Supplementary Materials: The full R Markdown analysis, including all data processing, modeling steps, and results, is available at the following link: https://rpubs.com/anglez02/1317079 (accessed on 14 July 2025).

Author Contributions: Conceptualization, A.M.G.-M. and N.R.; methodology, A.M.G.-M., N.R., Z.R.-C., F.E.-R., and L.A.H.-P.; software, A.M.G.-M. and E.S.L.-V.; validation, A.M.G.-M. and E.S.L.-V.; formal analysis, A.M.G.-M. and E.S.L.-V.; investigation, A.M.G.-M., C.A.H.-A., and C.O.R.-G.; resources, A.M.G.-M., N.R., C.A.H.-A., and C.O.R.-G.; data curation, A.M.G.-M. and E.S.L.-V.; writing—original draft preparation, A.M.G.-M. and N.R.; writing—review and editing, A.M.G.-M., N.R., and M.S.S.-R.; visualization, A.M.G.-M. and N.R.; supervision, N.R.; project administration, N.R. and C.O.R.-G.; funding acquisition, A.M.G.-M., N.R., C.A.H.-A., and C.O.R.-G. All authors have read and agreed to the published version of the manuscript.

Funding: This research was supported by the Mexican Secretaría de Ciencia, Humanidades, Tecnología e Innovación (Secihti) through the doctoral scholarship No. 864093 to A.M.G.-M. Additional support was provided by the Laboratorio de Evaluación y Cuidado del Estado Nutricio (LECEN), which contributed equipment and personnel. Partial funding was also provided by the Instituto de Investigaciones en Comportamiento Alimentario y Nutrición (IICAN) and the Doctorado en Ciencia del Comportamiento con orientación en Alimentación y Nutrición.

Institutional Review Board Statement: This study was conducted in accordance with the Declaration of Helsinki and approved by the Research and Graduate Studies Committee (registration number SAC/CIP/DOAN/027/2023) and by the Ethics Committee (registration CEI/77/2023, approved on 8 March 2023) of the Centro Universitario del Sur, Universidad de Guadalajara.

Informed Consent Statement: Written informed consent was obtained from all subjects involved in this study.

Data Availability Statement: The data presented in this study are available on request from the corresponding authors due to agreements ensuring participant confidentiality. The full R Markdown analysis, including all data processing, modeling steps, and results, is publicly accessible at: https://rpubs.com/anglez02/1317079 (accessed on 14 July 2025).

Acknowledgments: The authors would like to thank the Universidad de Guadalajara, Centro Universitario del Sur, the Instituto de Investigaciones en Comportamiento Alimentario y Nutrición, and the Doctorado en Ciencia del Comportamiento con orientación en Alimentación y Nutrición for their institutional support and partial funding. A.M.G.-M. gratefully acknowledges the Secretaría de Ciencia, Humanidades, Tecnología e Innovación (Secihti) for the doctoral scholarship (No. 864093). Special thanks are extended to Heliodoro Alemán-Mateo for his invaluable mentorship and guidance during A.M.G.-M.'s research stay at Centro de Investigación en Alimentación y Desarrollo (CIAD). The authors are also grateful to Centro de Atención Integral al Adulto Mayor (CAIAM) for allowing access to their facilities and older adult population, which enabled the data collection process. Finally, sincere appreciation goes to Luis Navarro and Carmen Nuño for their kind support and assistance during data collection.

Conflicts of Interest: The authors declare no conflicts of interest.

Abbreviations

The following abbreviations are used in this manuscript:

ACG	Arm Corrected Girth
AFTG	Arm Flexed and Tensed Girth
AI	Artificial Intelligence
ALM	Appendicular Lean Mass
ALMI	Appendicular Lean Mass Index
ANN	Artificial Neural Network
ARG	Arm Relaxed Girth
AUC	ROC curve
AWGS	Asian Working Group for Sarcopenia
BAs	Boosted algorithms
BIA	Bioelectrical Impedance Analysis
BMI	Body Mass Index
BUN	Blood Urea Nitrogen
CAIAM	Centro de Atención Integral al Adulto Mayor
CCG	Calf Corrected Girth
CG	Calf Girth
CoGs	Corrected Girths
CoG	Corrected Girth
CSF	Calf Skinfold
CUTonalá	University of Guadalajara Tonalá Campus
DIF	Sistema Nacional para el Desarrollo Integral de la Familia

DT	Decision Trees
DXA	Dual-Energy X-ray Absorptiometry
EWGSOP	European Working Group on Sarcopenia in Older People
F	Female
FG	Forearm Girth
IQR	Interquartile Range
ISAK	International Society for the Advancement of Kinanthropometry
LASSO	Least Absolute Shrinkage and Selection Operator
LECEN	Laboratorio de Evaluación y Cuidado del Estado Nutricio
LR	Logistic Regression
M	Male
ML	Machine Learning
MS	Muscle Strength
n	number of cases
RBCs	Red Blood Cells
RFs	Random Forests
SAMJ	Sarcopenia in Older Adults from Jalisco
SPPB	Short Physical Performance Battery
SVMs	Support Vector Machines
TSF	Triceps Skinfold
WBCs	White Blood Cells
WHR	Waist-to-Hip Ratio

Appendix A

Appendix A.1

Table A1. Descriptive characteristics of the sample of older adults from the SAMJ study.

Variables		n (SD)	Female n (%)	Male n (%)
Sex		183	123 (67.2)	60 (32.8)
Age (years)		70 (6.37)	70.1 (6.28)	70.1 (6.41)
Monthly income (MNX)		7023.1 (7974.0)	6181.9 (5441.3)	7028.3 (8033.0)
Marital status	Single	23 (12.6)	20 (16.3)	3 (5.0)
	Married	75 (41.0)	33 (26.8)	42 (70.0)
	Domestic partnership	5 (2.7)	5 (4.1)	0 (0)
	Widow	64 (35.0)	54 (43.9)	10 (16.7)
	Separated	7 (3.8)	5 (4.1)	2 (3.3)
	Divorced	9 (4.9)	6 (4.9)	3 (5.0)
Living arrangement	Alone	37 (20.2)	27 (22.0)	10 (16.7)
	Family	90 (49.2)	76 (61.8)	14 (23.3)
	Partner and family	34 (18.6)	12 (9.8)	22 (36.7)
	Partner	21 (11.5)	7 (5.7)	14 (23.3)
	Friend	1 (0.5)	1 (0.8)	0 (0)
Smoker	Yes	18 (9.8)	10 (8.1)	8 (13.3)
	No	165 (90.2)	113 (91.9)	52 (86.7)
Alcohol consumption	Yes	28 (15.3)	14 (11.4)	14 (23.3)
	No	155 (84.7)	109 (88.6)	46 (76.7)
Insulin resistance	Yes	26 (14.2)	16 (13.0)	10 (16.7)
	No	157 (85.8)	107 (87)	50 (83.3)
Diagnosed depression	Yes	35 (19.1)	30 (24.4)	5 (8.3)
	No	148 (80.9)	93 (75.6)	55 (91.7)

n: sample size; SD: Standard Deviation.

References

1. Soto-Perez-de-Celis, E.; de Glas, N.A.; Hsu, T.; Kanesvaran, R.; Steer, C.; Navarrete-Reyes, A.P.; Battisti, N.M.L.; Chavarri-Guerra, Y.; O'Donovan, A.; Avila-Funes, J.A.; et al. Global geriatric oncology: Achievements and challenges. *J. Geriatr. Oncol.* **2017**, *8*, 374–386. [CrossRef] [PubMed]
2. World Health Organization. *Decade of Healthy Ageing: Plan of Action 2021-2030*; World Health Organization: Geneva, Switzerland, 2025.
3. De Cabo, R.; Le Couteur, D. Kasper, D., Fauci, A., Hauser, S., Longo, D., Jameson, J., Loscalzo, J., Eds.; The biology of aging. In *Harrison's Principles of Internal Medicine*; McGraw-Hill Education: New York, NY, USA, 2015; pp. 94e1–94e8.
4. Manrique-Espinoza, B.; Salinas-Rodríguez, A.; Moreno-Tamayo, K.M.; Acosta-Castillo, I.; Sosa-Ortiz, A.L.; Gutiérrez-Robledo, L.M.; Téllez-Rojo, M.M. Condiciones de salud y estado funcional de los adultos mayores en México. *Salud Public. Mex.* **2013**, *55*, S323–S331. [CrossRef]
5. Steffl, M.; Sima, J.; Shiells, K.; Holmerova, I. The increase in health care costs associated with muscle weakness in older people without long-term illnesses in the Czech Republic: Results from the Survey of Health, Ageing and Retirement in Europe (SHARE). *Clin. Int. Aging* **2017**, *12*, 2003–2007. [CrossRef] [PubMed]
6. Cruz-Jentoft, A.J.; Bahat, G.; Bauer, J.; Boirie, Y.; Bruyère, O.; Cederholm, T.; Cooper, C.; Landi, F.; Rolland, Y.; Sayer, A.A.; et al. Sarcopenia: Revised European consensus on definition and diagnosis. *Age Ageing* **2019**, *48*, 16–31. [CrossRef]
7. Musumeci, G. Sarcopenia and Exercise "The State of the Art". *J. Funct. Morphol. Kinesiol.* **2017**, *2*, 40. [CrossRef]
8. Cruz-Jentoft, A.J.; Baeyens, J.P.; Bauer, J.M.; Boirie, Y.; Cederholm, T.; Landi, F.; Martin, F.C.; Michel, J.P.; Rolland, Y.; Schneider, S.M.; et al. Sarcopenia: European consensus on definition and diagnosis: Report of the European Working Group on Sarcopenia in Older People. *Age Ageing* **2010**, *39*, 412–423. [CrossRef]
9. Chen, L.-K.; Woo, J.; Assantachai, P.; Auyeung, T.-W.; Chou, M.-Y.; Iijima, K.; Jang, H.C.; Kang, L.; Kim, M.; Kim, S.; et al. Asian Working Group for Sarcopenia: 2019 consensus update on sarcopenia diagnosis and treatment. *JAMDA* **2020**, *21*, 300–307.e2. [CrossRef]
10. Morley, J.E.; Vellas, B.; Abellan van Kan, G.; Anker, S.D.; Bauer, J.M.; Bernabei, R.; Cesari, M.; Chumlea, W.C.; Doehner, W.; Evans, J.; et al. Frailty consensus: A call to action. *JAMDA* **2013**, *14*, 392–397. [CrossRef]
11. Bauer, J.; Morley, J.E.; Schols, A.M.W.J.; Ferrucci, L.; Cruz-Jentoft, A.J.; Dent, E.; Baracos, V.E.; Crawford, J.A.; Doehner, W.; Heymsfield, S.B.; et al. Sarcopenia: A time for action. An SCWD position paper. *J. Cachexia Sarcopenia Muscle* **2019**, *10*, 956–961. [CrossRef]
12. Bhasin, S.; Travison, T.G.; Manini, T.M.; Patel, S.; Pencina, K.M.; Fielding, R.A.; Magaziner, J.M.; Newman, A.B.; Kiel, D.P.; Cooper, C.; et al. Sarcopenia definition: The position statements of the sarcopenia definition and outcomes consortium. *JAGS* **2020**, *68*, 1410–1418. [CrossRef]
13. Suzan, V.; Suzan, A.A. A bibliometric analysis of sarcopenia: Top 100 articles. *Eur. Geriatr. Med.* **2021**, *12*, 185–191. [CrossRef] [PubMed]
14. Kemmler, W.; von Stengel, S.; Kohl, M. Developing sarcopenia criteria and cutoffs for an older caucasian cohort—A strictly biometrical approach; a strictly biometrical approach. *Clin. Int. Aging* **2018**, *13*, 1365–1373. [CrossRef]
15. Woo, J.; Leung, J. Anthropometric cut points for definition of sarcopenia based on incident mobility and physical limitation in older chinese people. *J. Gerontol. A Biol. Sci. Med. Sci.* **2016**, *71*, 935–940. [CrossRef]
16. Calvani, R.; Picca, A.; Marini, F.; Biancolillo, A.; Cesari, M.; Pesce, V.; Lezza, A.M.S.; Bossola, M.; Leeuwenburgh, C.; Bernabei, R.; et al. The "BIOmarkers associated with sarcopenia and physical frailty in elderly persons" (BIOSPHERE) study: Rationale, design and methods. *Eur. J. Intern. Med.* **2018**, *56*, 19–25. [CrossRef] [PubMed]
17. Confortin, S.C.; Ono, L.M.; Barbosa, A.R.; d'Orsi, E. Sarcopenia e sua associação com mudanças nos fatores socioeconômicos, comportamentais e de saúde: Estudo EpiFloripa Idoso. *Cad. Saúde Pública* **2018**, *34*. [CrossRef]
18. Espinel-Bermúdez, M.C.; Sánchez-García, S.; García Peña, C.; Trujillo, X.; Huerta-Viera, M.; Granados-García, V.; Hernández-González, S.; Arias-Merino, E.D. Factores asociados a sarcopenia en adultos mayores mexicanos: Encuesta nacional de salud y nutrición 2012. *Rev. Med. Inst. Mex. Seguro Soc.* **2018**, *56*, 46–53.
19. Martín Sierra, J.A.; Calderón Loeza, G.Y.; Zapata Vázquez, R.E.; Novelo Tec, J.F. Sarcopenia y factores asociados en los adultos mayores de una unidad de medicina familiar en Yucatán, México. *Atención Fam.* **2021**, *28*, 191. [CrossRef]
20. Palus, S.; Springer, J.I.; Doehner, W.; von Haehling, S.; Anker, M.; Anker, S.D.; Springer, J. Models of sarcopenia: Short review. *Int. J. Cardiol.* **2017**, *238*, 19–21. [CrossRef]
21. Alemán-Mateo, H.; Lee, S.Y.; Javed, F.; Thornton, J.; Heymsfield, S.B.; Pierson, R.N.; Pi-Sunyer, F.X.; Wang, Z.M.; Wang, J.; Gallagher, D. Elderly Mexicans have less muscle and greater total and truncal fat compared to African-Americans and Caucasians with the same BMI. *J. Nutr. Health Aging* **2009**, *13*, 919–923. [CrossRef]
22. Alemán-Mateo, H.; López-Teros, M.T.; Ruiz-Valenzuela, R.E.; Ramírez-Torres, M.; Urquidez-Romero, R. Sarcopenia: Influence of regional skeletal muscle cutoff points and fat-free mass in older mexican people—A pilot study. *Curr. Gerontol. Geriatr.* **2020**, *8037503*, 1–9. [CrossRef]

23. Alley, D.E.; Shardell, M.D.; Peters, K.W.; McLean, R.R.; Dam, T.-T.L.; Kenny, A.M.; Fragala, M.S.; Harris, T.B.; Kiel, D.P.; Guralnik, J.M.; et al. Grip strength cutpoints for the identification of clinically relevant weakness. *J. Gerontol. A Biol. Sci. Med. Sci.* **2014**, *69*, 559–566. [CrossRef]
24. Pinheiro, P.A.; da Silva Coqueiro, R.; Carneiro, J.A.O.; Correia, T.M.L.; Pereira, R.; Fernandes, M.H. Anthropometric indicators as screening tools for sarcopenia in older adult women. *Enferm* **2020**, *30*, 269–274.
25. Rech, C.R.; Dellagrana, R.A.; Marucci, M.D.F.N.; Petroski, E.L. Validade de equações antropométricas para estimar a massa muscular em idosos. *Rev. Bras. Cineantropometria Desempenho Hum.* **2012**, *14*, 23–31.
26. Rangel Peniche, D.B.; Alemán Mateo, H.; de los Angeles Aguilera Barreiro, M.; Ruiz Valenzuela, R.E.; Ramírez-Torres, M.; Urquidez-Romero, R. Differences in body composition in older people from two regions of Mexico: Implications for diagnoses of sarcopenia and sarcopenic obesity. *Biomed. Res. Int.* **2018**, 1–9. [CrossRef] [PubMed]
27. Greener, J.G.; Kandathil, S.M.; Moffat, L.; Jones, D.T. A guide to machine learning for biologists. *Nat. Rev. Mol. Cell Biol.* **2022**, *23*, 40–55. [CrossRef]
28. Panch, T.; Szolovits, P.; Atun, R. Artificial intelligence, machine learning and health systems. *J. Glob. Health* **2018**, *8*, 020303. [CrossRef]
29. Zhang, Y.; Wang, Z.; Zhang, J.; Wang, C.; Wang, Y.; Chen, H.; Shan, L.; Huo, J.; Gu, J.; Ma, X. Deep learning model for classifying endometrial lesions. *J. Transl. Med.* **2021**, *19*, 10. [CrossRef]
30. Ou Yang, W.-Y.; Lai, C.-C.; Tsou, M.-T.; Hwang, L.-C. Development of machine learning models for prediction of osteoporosis from clinical health examination data. *Int. J. Environ. Res. Public. Health* **2021**, *18*, 7635. [CrossRef]
31. Albores-Mendez, E.M.; Aguilera Hernández, A.D.; Melo-González, A.; Vargas-Hernández, M.A.; Gutierrez de la Cruz, N.; Vazquez-Guzman, M.A.; Castro-Marín, M.; Romero-Morelos, P.; Winkler, R. A diagnostic model for overweight and obesity from untargeted urine metabolomics of soldiers. *PeerJ* **2022**, *10*, e13754. [CrossRef]
32. Butler, É.M.; Derraik, J.G.B.; Taylor, R.W.; Cutfield, W.S. Prediction models for early childhood obesity: Applicability and existing issues. *Horm. Res. Paediatr.* **2018**, *90*, 358–367. [CrossRef]
33. Chung, H.; Jo, Y.; Ryu, D.; Jeong, C.; Choe, S.; Lee, J. Artificial-intelligence-driven discovery of prognostic biomarker for sarcopenia. *J. Cachexia Sarcopenia Muscle* **2021**, *12*, 2220–2230. [CrossRef] [PubMed]
34. Cook, N.R. Statistical evaluation of prognostic versus diagnostic models: Beyond the ROC curve. *Clin. Chem.* **2008**, *54*, 17–23. [CrossRef]
35. Hendriksen, J.M.T.; Geersing, G.J.; Moons, K.G.M.; de Groot, J.A.H. Diagnostic and prognostic prediction models. *J. Thromb. Haemost.* **2013**, *11*, 129–141. [CrossRef] [PubMed]
36. Wolf, M.B.; DeLand, E.C. A comprehensive, computer-model-based approach for diagnosis and treatment of complex acid–base disorders in critically-ill patients. *J. Clin. Monit. Comput.* **2011**, *25*, 353–364. [CrossRef]
37. von Elm, E.; Altman, D.G.; Egger, M.; Pocock, S.J.; Gøtzsche, P.C.; Vandenbroucke, J.P. The Strengthening the Reporting of Observational Studies in Epidemiology (STROBE) statement: Guidelines for reporting observational studies. *Lancet* **2007**, *370*, 1453–1457. [CrossRef]
38. Messina, C.; Albano, D.; Gitto, S.; Tofanelli, L.; Bazzocchi, A.; Ulivieri, F.M.; Guglielmi, G.; Sconfienza, L.M. Body composition with dual energy X-ray absorptiometry: From basics to new tools. *Quant. Imaging Med. Surg.* **2020**, *10*, 1687–1698. [CrossRef]
39. Esparza-Ros, F.; Vaquero-Cristóbal, R.; Marfell-Jones, M. *International Standards for Anthropometric Assessment*; International Society for the Advancement of Kinanthropometry (ISAK): Glasgow, Scotland, 2019.
40. Barraza Gómez, F.; Hadler Garcés, A.; Jeria Fernández, J.; Riffo Esquivel, C. Cuantificación de la masa muscular de los miembros apendiculares, por medio ecuaciones antropométricas. *J. Mov. Health.* **2009**, *10*, 44–48. [CrossRef]
41. Herrera-Amante, C.A.; Ramos-García, C.O.; Alacid, F.; Quiroga-Morales, L.A.; Martínez-Rubio, A.J.; Bonilla, D.A. Development of alternatives to estimate resting metabolic rate from anthropometric variables in paralympic swimmers. *J. Sports Sci.* **2021**, *39*, 2133–2143. [CrossRef]
42. Bonilla, D.A.; De León, L.G.; Alexander-Cortez, P.; Odriozola-Martínez, A.; Herrera-Amante, C.A.; Vargas-Molina, S.; Petro, J.L. Simple anthropometry-based calculations to monitor body composition in athletes: Scoping review and reference values. *Nutr. Health* **2022**, *28*, 95–109. [CrossRef]
43. Ge, S.; Du, Q.; Feng, X.; Liu, Y.; Wang, H.; Hai, S.; Shi, X.; Sun, W.; Ma, A.; Lv, T.; et al. Optimal cutoffs for the diagnosis of sarcopenia in older Chinese adults. *Front. Nutr.* **2022**, *9*, 853323. [CrossRef] [PubMed]
44. Kim, Y.-P.; Joh, J.-Y.; Kim, S.; Hwang, H.-S.; Shin, I.-S. The application of different appendicular skeletal muscle cutoff points and research definitions associated with health-related quality of life in Korean older people: Data from KNHANES 2008–2011. *BMC Geriatr.* **2014**, *14*, 144. [CrossRef] [PubMed]
45. López-Teros, M.T.; Rosas-Carrasco, O.; Sánchez-García, S.; Castro-Porras, L.; Luna-López, A.; Agudelo-Botero, M. The association of osteosarcopenia with functional disability in community-dwelling Mexican adults 50 and older. *Front. Med.* **2021**, *8*, 674724. [CrossRef]

46. Viana, J.U.; Dias, J.M.D.; Pereira, L.S.M.; Silva, S.L.A.D.; Hoelzle, L.F.; Dias, R.C. Pontos de corte alternativos para massa muscular apendicular para verificação da sarcopenia em idosos brasileiros: Dados da Rede Fibra—Belo Horizonte/Brasil. *Fisioter. E Pesqui.* **2018**, *25*, 166–172. [CrossRef]
47. World Medical Association. World medical association declaration of Helsinki. Ethical principles for medical research involving human subjects. *JAMA* **2025**, *333*, 71–74. [CrossRef] [PubMed]
48. Mukaka, M.M. Statistics corner: A guide to appropriate use of correlation coefficient in medical research. *Malawi Med. J.* **2012**, *24*, 69–71.
49. Secretaría de Salud. In *Reglamento de la Ley General de Salud en Materia de Investigación para la Salud*; Diario Oficial de la Federación: Mexico City, Mexico, 6 January 1987. Available online: https://www.diputados.gob.mx/LeyesBiblio/regley/Reg_LGS_MIS.pdf (accessed on 1 June 2025).
50. Ranstam, J.; Cook, J.A. LASSO regression. *Br. J. Surg.* **2018**, *105*, 1348. [CrossRef]
51. Shaikhina, T.; Lowe, D.; Daga, S.; Briggs, D.; Higgings, R.; Khovanova, N. Decision tree and random forest models for outcome prediction in antibody incompatible kidney transplantation. *Biomed. Signal Process. Control.* **2019**, *52*, 456–462. [CrossRef]
52. Podgorelec, V.; Kokol, P.; Stiglic, B.; Rozman, I. Decision trees: An overview and their use in medicine. *J. Med. Syst.* **2002**, *26*, 445–463. [CrossRef]
53. Horning, N. Random Forests: An algorithm for image classification and generation of continuous fields data sets. In Proceedings of the International Conference on Geoinformatics for Spatial Infrastructure Development in Earth and Allied Sciences, Osaka, Japan, 9–11 December 2010; Volume 911, pp. 1–6.
54. Olshvang, D.; Harris, C.; Chellappa, R.; Santhanam, P. Predictive modeling of lean body mass, appendicular lean mass, and appendicular skeletal muscle mass using machine learning techniques: A comprehensive analysis utilizing NHANES data and the Look AHEAD study. *PLoS ONE* **2024**, *19*, e0309830. [CrossRef]
55. Cichosz, S.L.; Rasmussen, N.H.; Vestergaard, P.; Hejlesen, O. Precise prediction of total body lean and fat mass from anthropometric and demographic data: Development and validation of neural network models. *J. Diabetes Sci. Technol.* **2021**, *15*, 1337–1343. [CrossRef] [PubMed]
56. Buccheri, E.; Dell'Aquila, D.; Russo, M.; Chiaramonte, R.; Musumeci, G.; Vecchio, M. Can artificial intelligence simplify the screening of muscle mass loss? *Heliyon* **2023**, *9*, e16323. [CrossRef]
57. Marazzato, F.; McCarthy, C.; Field, R.H.; Nguyen, H.; Nguyen, T.; Shepherd, J.A.; Tinsley, G.M.; Heymsfield, S.B. Advances in digital anthropometric body composition assessment: Neural network algorithm prediction of appendicular lean mass. *Eur. J. Clin. Nutr.* **2024**, *78*, 452–454. [CrossRef] [PubMed]
58. Birk, N.; Kulkarni, B.; Bhogadi, S.; Aggarwal, A.; Walia, G.K.; Gupta, V.; Rani, U.; Mahajan, H.; Kinra, S.; Mallinson, P.A. Machine learning-based equations for improved body composition estimation in Indian adults. *medRxiv* **2024**. [CrossRef]
59. Guo, J.; He, Q.; She, C.; Liu, H.; Li, Y. A machine learning–based online web calculator to aid in the diagnosis of sarcopenia in the US community. *Digit. Health* **2024**, *10*, 1–14. [CrossRef]
60. Buccheri, E.; Dell'Aquila, D.; Russo, M.; Chiaramonte, R.; Vecchio, M. Appendicular skeletal muscle mass in older adults can be estimated with a simple equation using a few zero-cost variables. *J. Geriatr. Phys. Ther.* **2024**, *47*, E149–E158. [CrossRef] [PubMed]
61. Shi, J.; He, Q.; Pan, Y.; Zhang, X.; Li, M.; Chen, S. Estimation of appendicular skeletal muscle mass for women aged 60-70 years using a machine learning approach. *J. Am. Med. Dir. Assoc.* **2022**, *23*, 1985.e1–1985.e7. [CrossRef]
62. Kang, Y.-J.; Yoo, J.-I.; Ha, Y. Sarcopenia feature selection and risk prediction using machine learning: A cross-sectional study. *Medicine* **2019**, *98*, e17699. [CrossRef]
63. Du, J.; Tao, X.; Zhu, L.; Wang, H.; Qi, W.; Min, X.; Wei, S.; Zhang, X.; Liu, Q. Development of a visualized risk prediction system for sarcopenia in older adults using machine learning: A cohort study based on CHARLS. *Front. Public Health* **2025**, *13*, 1–8. [CrossRef]
64. Li, N.; Ou, J.; He, H.; He, J.; Zhang, L.; Peng, Z.; Zhong, J.; Jiang, N. Exploration of a machine learning approach for diagnosing sarcopenia among Chinese community-dwelling older adults using sEMG-based data. *J. Neuroeng. Rehabil.* **2024**, *21*, 69. [CrossRef]
65. Yang, S.; Yu, J.; Chen, Q.; Sun, X.; Hu, Y.; Su, T.; Li, J.; Jin, L. Development and external validation of a machine-learning based model to predict pre-sarcopenia in MASLD population: Results from NHANES 2017–2018. *Ann. Hepatol.* **2024**, *30*, 101585. [CrossRef] [PubMed]
66. Yu, P.; Zhang, X.; Sun, G.; Zeng, P.; Zheng, C.; Wang, K. Sarcopenia prediction model based on machine learning and SHAP values for community-based older adults with cardiovascular disease in China. *Front. Public Health* **2025**, *13*, 1527304. [CrossRef]
67. Johnson, T.; Su, J.; Andres, J.; Henning, A.; Ren, J. Sex differences in fat distribution and muscle fat infiltration in the lower extremity: A Retrospective Diverse-Ethnicity 7T MRI Study in a Research Institute Setting in the USA. *Diagnostics* **2024**, *14*, 2260. [CrossRef]
68. Fragala, M.S.; Clark, M.H.; Walsh, S.J.; Kleppinger, A.; Judge, J.O.; Kuchel, G.A.; Kenny, A.M. Gender differences in anthropometric predictors of physical performance in older adults. *Gend. Med.* **2012**, *9*, 445–456. [CrossRef]

69. Chawla, N.V.; Bowyer, K.W.; Hall, L.O.; Kegelmeyer, W.P. SMOTE: Synthetic minority over-sampling technique. *J. Artif. Intell. Res.* **2002**, *16*, 321–357. [CrossRef]
70. Rajput, D.; Wang, W.J.; Chen, C.C. Evaluation of a decided sample size in machine learning applications. *BMC Bioinform.* **2023**, *24*, 48. [CrossRef] [PubMed]
71. Vabalas, A.; Gowen, E.; Poliakoff, E.; Casson, A.J. Machine learning algorithm validation with a limited sample size. *PLoS ONE* **2019**, *14*, e0224365. [CrossRef] [PubMed]
72. Goldenholz, D.M.; Sun, H.; Ganglberger, W.; Westover, M.B. Sample Size Analysis for Machine Learning Clinical Validation Studies. *Biomedicines* **2023**, *11*, 685. [CrossRef]
73. Castillo-Olea, C.; Garcia-Zapirain Soto, B.; Zuñiga, C. Evaluation of prevalence of the sarcopenia level using machine learning techniques: Case study in Tijuana Baja California, Mexico. *Int. J. Environ. Res. Public Health* **2020**, *17*, 1917. [CrossRef]
74. Pineda-Zuluaga, M.C.; González-Correa, C.H.; Sepulveda-Gallego, L.E. Cut-off points for low skeletal muscle mass in older adults: Colombia versus other populations. *F1000Res* **2022**, *11*, 304. [CrossRef]
75. Abdalla, P.P.; Silva, A.M.; Venturini, A.C.R.; Santos, A.P.D.; Carvalho, A.D.S.; Siqueira, V.A.A.A.; Gomide, E.B.; Alves, N.M.D.C.; Ferriolli, E.; Machado, D.R.L. Cut-off points of appendicular lean soft tissue for identifying sarcopenia in older adults in Brazil: A cross-sectional study. *Nutr. Hosp.* **2020**, *37*, 306–312. [CrossRef]
76. Alwosheel, A.; van Cranenburgh, S.; Chorus, C.G. Is your dataset big enough? Sample size requirements when using artificial neural networks for discrete choice analysis. *J. Choice Model.* **2018**, 167–182. [CrossRef]
77. Adão, T.; Lameira, G.; dos Santos, J.; Palha, F. Technical error of measurement in anthropometry. *Rev. Bras. Med. Esporte.* **2005**, *11*, 81–85.
78. Esparza-Ros, F. *Manual de Cineantropometría*; Grupo Español De Cineantropometría: Zaragoza, España, 1993.
79. Padilla, C.J.; Ferreyro, F.A.; Arnold, W.D. Anthropometry as a readily accessible health assessment of older adults. *Exp. Gerontol.* **2021**, *153*, 111464. [CrossRef] [PubMed]
80. Turimov Mustapoevich, D.; Kim, W. Machine learning applications in sarcopenia detection and management: A comprehensive survey. *Healthcare* **2023**, *11*, 2483. [CrossRef] [PubMed]
81. Wang, B.; Torriani, M. Artificial intelligence in the evaluation of body composition. *Semin. Musculoskelet. Radiol.* **2020**, *24*, 030–037. [CrossRef] [PubMed]
82. Ozgur, S.; Altinok, Y.A.; Bozkurt, D.; Saraç, Z.F.; Akçiçek, S.F. Performance evaluation of machine learning algorithms for sarcopenia diagnosis in older adults. *Healthcare* **2023**, *11*, 2699. [CrossRef]
83. Hui, A.T.; Ahn, S.S.; Lye, C.T.; Deng, J. Ethical challenges of artificial intelligence in health care: A narrative review. *EBEM* **2021**, *12*, 55–71. [CrossRef]

Journal of
Functional Morphology and Kinesiology

MDPI

Article

Changes in Body Composition and Body Image Perception in Adolescent Soccer Players Examined with Repeated Measurements During Pre-Season and In-Season Training

Luciana Zaccagni [1,2], Mattia Reggiani [1], Stefania Toselli [3,*] and Emanuela Gualdi-Russo [1]

1 Department of Neuroscience and Rehabilitation, Faculty of Medicine, Pharmacy and Prevention, University of Ferrara, 44121 Ferrara, Italy; luciana.zaccagni@unife.it (L.Z.); mattia.reggiani@edu.unife.it (M.R.); emanuela.gualdi@unife.it (E.G.-R.)
2 Center for Exercise Science and Sports, University of Ferrara, 44123 Ferrara, Italy
3 Department for Life Quality Studies, University of Bologna, 47921 Rimini, Italy
* Correspondence: stefania.toselli@unibo.it

Abstract: Objectives: Adolescents' health is positively influenced by the performance of physical activity. Regarding soccer, a very popular sport, the aims of the study were to assess changes in body composition and body image of late adolescent players during pre- and in-season training periods, analyzing the relationships between dissatisfaction and body composition parameters. **Methods**: A sample of 16–19-year-old male soccer players was examined longitudinally by three surveys. The body composition was assessed by anthropometric method. Body image perception was evaluated using two different figural scales related to shape and muscularity. **Results**: During the period examined, fat-free mass increased, and changes in perceived and ideal figures occurred, showing a desire toward more robust and muscular figures. Body image dissatisfaction was generally low, with a decrease in muscle dissatisfaction during the period. Body composition parameters significantly predicted body image dissatisfaction. **Conclusions**: Our findings suggest anthropometric and body image perception changes during soccer training with improvements in body composition parameters and a decrease in musculature dissatisfaction. These results highlight the importance of incorporating physical and psychological monitoring into training programs to support the healthy development of athletes' body image and body composition.

Keywords: adolescence; football; body dissatisfaction; body fat; body fat-free mass

1. Introduction

Physical activity (PA) is deemed to positively affect physical and mental health [1]. This is especially important during adolescence, a critical period for developing a positive or negative body image perception and ideals, as it is characterized by puberty and rapid and diverse physical changes in body shape, size, composition, and identity development.

Body image (BI) refers to people's subjective view of their bodies, including their thoughts, perceptions, and feelings. When the person does not achieve their ideals, body image dissatisfaction (BID) emerges. In childhood, BID is associated with negative physical and mental health outcomes that include a rise in depressive symptoms, poor self-esteem, obesity and overweight, eating disorders, physical inactivity, and poor fitness [2]. During adolescence, achieving a positive self-concept and fostering mental wellbeing can be facilitated by engaging in PA, which improves body perceptions and satisfaction [3]. On the contrary, BID, teasing, and problems with gender identification may become determinants for continuing PA, more so than actual motor skills [4]. It is well known that a physically active lifestyle during developmental age has favorable repercussions in adulthood [5] and improves cardiovascular health, self-confidence, muscle strength and endurance, bone density, and body weight [6]. In general, increased participation in PA and sports promotes a positive BI: BID decreases as the number of hours spent in sports activities increases [7] and increases with increasing BMI and body fat accumulation [8]. Physical inactivity and sedentary behavior are believed to be the main drivers of the increasing prevalence of overweight/obesity in Europe [9]. Higher PA levels were found to be associated with lower BMI values, better cardiorespiratory fitness, and lower body dissatisfaction [10].

Adolescent participation in team and individual sports is associated with better physical, mental, social, and emotional health [11]. In particular, adolescents participating in team sports appear to have better psychosocial health (lower anxiety, depression, and social problems) than those participating in individual sports [12].

Soccer is one of the most popular team sports, involving players of all ages at amateur and professional levels [13]. Therefore, it is important to understand possible physical and mental health risks for soccer players with a focus on youth groups. BID can lead to several negative health outcomes and the potential establishment of psychopathologies [14]. Another relevant element to evaluate in the sportsman is to consider not only body shape but also muscle development. In this specific case, a scale related to muscle development has been proposed for adolescents in addition to the one related to form [15]: the new figure-rating scales allow rapid and robust assessment of both aspects. In addition to the specific case of adolescents engaged in sports, the use of these differentiated scales is appropriate because eating disorders would arise as a response to dissatisfaction related to body fat and not muscularity [16]. In general, using the two scales makes it possible to distinguish whether most of the BID concerns the thin, lean ideal. While in the female sex there is generally a preference toward an ideal of thinness ('drive for thinness'), in the male sex there may be a preference for a lean and toned body ('drive for leanness') or a larger and more muscular body [15].

Despite the importance of these aspects for adolescents' good health, there is limited research on BI and body composition or weight status in adolescents [7,17], and, in particular, research is lacking regarding trends in body composition parameters and BID in soccer players. Although we could verify that pre-adolescent soccer players showed anthropometric and body image perception changes after a 12-week training program [18], with a positive effect of sports on BI, an analogous longitudinal study during late adolescence has not been conducted to date. The main aims of the present study were as follows: i. To examine changes in body composition and BI perception throughout the pre-season and in-season soccer training in adolescent soccer players; ii. To test for any relationship between BID and body composition parameters.

2. Materials and Methods

2.1. Participants and Procedure

This longitudinal study was carried out after the approval by the Bioethics Committee of the University of Bologna (approval code: 25027; 13 March 2017) on a sample of 28 soccer players corresponding to the minimum sample size expected through an a priori power analysis using the G*Power statistical program (version 3.1.9.6; Universitat Kiel, Kiel, Germany) for 80% power, medium effect size, and a 0.05 significance level for performing a primary analysis (repeated measures ANOVA) according to Andrade [19]. Participation in the research was bound by informed consent, which was signed directly by participants over 18 years of age and by parents in the case of minors.

The non-elite group consisted of players aged 16–19 who participated in the under-18 teams in the Emilia-Romagna second-tier soccer league in 2023–2024. All thirty boys aged $\geq$16 registered with a soccer club, selected on a convenience basis, were invited to participate in the three surveys conducted every three months (first: September 2023; second: December 2023; third: March 2024). Thirty players voluntarily agreed to participate, but two of them could not take part in all the surveys because they were injured. The final sample was therefore 28 players, and there were no missing values in the data collected during the surveys of this final sample. The average age of beginning organized soccer activities was 7.4 $\pm$ 3.3 years in the surveyed sample. All players trained for five hours a week (subdivided into three training days), plus a match at the end of the week during the soccer in-season period.

2.2. Anthropometric Traits

A trained operator performed all anthropometric surveys following standard procedures [20,21]. Anthropometric measurements were taken on players in light clothing and without shoes. The anthropometric traits measured were the following: stature, weight, and skinfold thicknesses (triceps, subscapular). Stature was measured to the nearest 0.1 cm using an anthropometer (Magnimeter, Raven Equipment Ltd., Dunmow, Essex, UK) on participants with their heads aligned with the Frankfurt plane in the standing position. Weight was measured to the nearest 0.1 kg using a digital scale (SECA, Basel, Switzerland). Skinfold thicknesses were taken at the triceps and subscapular points to the nearest 0.5 mm on the left side of the body (according to Weiner and Lourie [22]) by a Lange caliper (Beta Technology Inc., Houston, TX, USA). The operator's TEMs (assessed before the survey) were <5% for skinfolds and <1% for other measurements.

Some body composition parameters were calculated using the anthropometric measurements that were taken. Body Mass Index (BMI) was computed as weight (kg)/height^2 (m^2). Using BMI, we classified participants according to Cole's cut-offs [23,24] into four weight status categories: underweight, normal weight, overweight, and obese. The sum of triceps and subscapular thickness skinfolds was used to obtain body density by Durnin and Womersley's equation [25], and then the percentage of body fat (%Fat) by Siri's equation [26]. Subsequently, fat mass in kg (FM) and fat-free mass (FFM) in kg were obtained.

2.3. Body Image Perception and Dissatisfaction

The Male Body Scale (MBS) and Male Fit Body Scale (MFBS), developed and validated by Ralph-Nearman and Filik [15], were submitted to the participants to assess for each scale the perceived current body figure (actual) and ideal body figure (ideal) related to body fat (MBS) and muscularity (MFBS). MBS includes nine male figures that vary progressively from emaciated to obese; MFBS, in turn, includes nine male figures ranging from very lean to very muscular. Using the first scale to derive the current and ideal figures, it was possible to derive the adipose-related BID, while the second one allowed for indications of muscle-related BID.

Participants were successively presented with the two scales, with the figures numbered from 1 to 9, and had to indicate the number corresponding to the chosen figures in response to the following questions asked first on the MBS and then on the MFBS: 1—Which figure do you think best represents you? 2—Which figure would you like to resemble? 3—Which figure do you think corresponds to the ideal soccer player?

Answers to the first question were reported as perceived shape (S-Feel) and perceived muscularity (M-Feel), and those to the second question as ideal shape (S-Ideal) and ideal muscularity (M-Ideal). The answers to the third question were given as ideal soccer player shape (S-Soccer) and ideal soccer player muscularity (M-Soccer).

The degree of BID was determined through the discrepancy between the actual and ideal figure (FID or Feel minus Ideal Discrepancy): the index was obtained by subtracting the ideal figure score from the actual one, thus resulting in a discrepancy relative to shape (S-FID) and a discrepancy relative to muscularity (M-FID) according to the used scale. Finally, the discrepancy from the ideal soccer player figure (FIDsport, according to [27,28]) was calculated by subtracting the score of the ideal figure of the soccer player from the actual figure of the boy examined on both the shape and muscularity scales (S-FIDsport and M-FIDsport). Generally, the FID score will be positive when the actual figure is larger than the ideal one and negative when the actual figure is thinner/leaner than the ideal one. If there is no discrepancy (the figure chosen as the real one matches the ideal one), the FID score will be 0.

2.4. Statistical Analysis

Assumptions of normality were verified using Kolmogorov–Smirnov tests. Comparisons of skinfold thicknesses among surveys were performed after their log transformation.

Mean and standard deviation (SD) were used to describe continuous variables, and percentage frequencies to describe categorical variables.

Longitudinal anthropometric and BI perception changes were assessed by analysis of data collected in the three successive surveys (repeated measures ANOVA). A nonparametric Friedman rank-sum test was applied to three successive surveys for variables related to BI perception (not normally distributed). The effect size was calculated using partial eta–squared for repeated measures analysis of variance and using Kendall's W coefficient of concordance for the Friedman test.

Linear multiple regression analyses were conducted between the BID (dependent variable) and the independent variables of body composition (BMI, %Fat, FFM). The analyses were preceded by a check for multicollinearity using the variance inflation factor (VIF).

All statistical analyses were computed using STATISTICA software, version 11 (StatSoft, Tulsa, OK, USA).

3. Results

Table 1 shows the mean values and SD of anthropometric traits of the soccer player sample and the statistical comparisons among the three surveys.

Table 1. Anthropometric traits and body composition changes in adolescent soccer players during the soccer pre-season and in-season.

	First Survey		Second Survey		Third Survey		ANOVA		
Variables	**Mean**	**SD**	**Mean**	**SD**	**Mean**	**SD**	**F**	***p***	**Partial η^2**
Stature (cm)	175.0	6.8	176.0	6.7	177.0	6.8	83.30	0.001	0.755
Weight (kg)	69.9	9.9	70.7	9.75	71.7	11.2	6.64	0.003	0.197
Triceps skinfold (mm)	10.8	4.3	10.8	4.2	10.9	4.7	0.06	0.944	0.002
Subscapular skinfold (mm)	8.2	2.4	9.1	2.9	8.9	3.3	9.15	<0.001	0.253
Indices									
BMI (kg/m^2)	22.8	2.9	22.8	2.8	22.9	3.2	0.42	0.662	0.015
Density (g/cc)	1.067	0.009	1.065	0.009	1.066	0.011	1.50	0.222	0.054
%Fat	14.1	4.0	14.7	4.1	14.4	4.7	1.55	0.222	0.054
Fat Mass (kg)	10.1	4.0	10.7	4.2	10.7	5.0	2.10	0.132	0.072
Fat-Free Mass (kg)	59.8	7.1	60.1	6.8	61.0	7.4	6.33	0.003	0.190

Note: Comparisons among skinfold thicknesses were performed using log skinfolds.

Significant changes were observed for stature, weight, subscapular skinfold thickness, and FFM with increasing values during the period under consideration and a large effect size (partial eta-squared > 0.14). Some of these changes can be attributed to the normal growth process (e.g., statural increase); others are attributable to both the growth process and the effect of sports training (e.g., weight, FFM). Unlike FFM (+1.2 kg), the FM and %Fat showed non-significant changes during the soccer pre-season and in-season.

Considering the weight status, no differences in frequency were found over the period considered. The most represented category was normal weight (67.9%), followed by overweight (28.6%) and underweight (3.6%). There were no obese boys in the soccer player sample.

Regarding the perception of BI, the figures most frequently selected were No. 4 as actual figures for both body shape and muscularity and as ideal figures for both body shape and soccer player shape, and No. 5 as ideal figures for both personal and soccer player muscularity. Table 2 shows the changes in BI perception and dissatisfaction during the pre-season and in-season.

Table 2. BI perception and dissatisfaction changes in adolescent soccer players during the soccer pre-season and in-season.

	First Survey		Second Survey		Third Survey		Friedman Test		
MBS	**Mean**	**SD**	**Mean**	**SD**	**Mean**	**SD**	F_r	p	**Kendall's W**
S-Actual Figure	4.1	1.1	4.1	1.0	4.3	1.1	2.377	0.305	0.042
S-Ideal Figure	4.1	0.8	4.3	0.7	4.4	0.6	6.465	0.039	0.115
S-Soccer	4.3	0.9	4.1	0.8	4.6	0.6	8.818	0.012	0.157
S-FID	0.00	1.02	−0.21	0.92	−0.11	0.92	1.705	0.426	0.030
S-FIDsport	−0.18	1.25	−0.04	1.07	−0.29	1.15	3.127	0.209	0.056
MFBS									
M-Actual Figure	4.0	1.2	3.9	1.0	4.4	1.2	7.841	0.020	0.140
M-Ideal Figure	5.4	1.4	5.1	1.1	5.5	1.0	3.844	0.146	0.069
M-Soccer	5.5	1.1	5.1	1.1	5.3	0.9	5.746	0.057	0.103
M-FID	−1.46	1.04	−1.25	0.84	−1.07	0.98	4.415	0.110	0.079
M-FIDsport	−1.57	1.29	−1.18	1.16	−0.93	1.33	6.727	0.035	0.120

Note: MBS: Male Body Scale; MFBS: Male Fit Body Scale; S: shape; M: muscular; FID: feel minus ideal discrepancy.

BI perception variables changed significantly for the ideal body shape (S-Ideal), showing a desire for a more robust shape, and for current musculature (M-Actual), where there was a significant increase in perceived muscular figure. The figures of the ideal soccer player changed over time in both shape (S-Soccer) and musculature (M-Soccer): the former increased, the latter decreased.

Regarding BID indices, the adolescent soccer players examined generally showed low levels of dissatisfaction, with non-significant changes over the period. The only exception was the changes in FID related to the muscularity of the ideal soccer player (M-FIDsport). In general, M-FIDsport mean values were negative, indicating a desire to be more muscular to come closer to the ideal image of the soccer player. This index decreased significantly throughout repetitions in probable relation to the player's awareness of increased musculature with soccer training, consistent with the concomitant significant increase in M-Actual.

Table 3 shows the results obtained by the multiple regression analyses: the three independent variables analyzed were significant predictors of FID, indicating an increase in dissatisfaction as BMI and %Fat increased and a decrease in FID as FFM increased. The obtained models explain from more than 30 to more than 60% of the variance, emphasizing the importance of body composition parameters for BI purposes. In the S-FID, only BMI was a significant predictor in the third survey, implying an increase in BID as BMI increased.

Table 3. Body composition predictors of BID by multiple regressions during the soccer pre-season and in-season.

	First Survey			Second Survey			Third Survey		
	VIF	β	p	VIF	β	p	VIF	β	p
S-FID									
BMI (kg/m^2)	5.77	0.700	0.138	5.368	0.160	0.622	5.991	0.720	0.022
%Fat	3.40	−0.249	0.484	3.192	0.409	0.111	3.230	−0.098	0.653
Fat-Free Mass (kg)	2.50	−0.276	0.366	2.400	0.322	0.146	2.739	0.201	0.320
R^2		0.134			0.542			0.655	
R^2 adjusted		0.027			0.484			0.612	
p		0.315			<0.001			<0.001	
M-FID									
BMI (kg/m^2)	5.77	0.700	0.138	5.368	0.374	0.419	5.991	1.267	0.003
%Fat	3.40	−0.249	0.484	3.192	−0.059	0.867	3.230	−0.549	0.045
Fat-Free Mass (kg)	2.50	−0.276	0.366	2.40	−0.425	0.172	2.739	−0.348	0.190
R^2		0.134			0.090			0.416	
R^2 adjusted		0.027			0.024			0.343	
p		0.315			0.510			0.004	
S-FIDsport									
BMI (kg/m^2)	5.77	0.268	0.430	5.368	0.756	0.038	5.991	0.837	0.017
%Fat	3.40	0.357	0.176	3.192	−0.061	0.820	3.230	−0.083	0.732
Fat-Free Mass (kg)	2.50	0.232	0.302	2.40	−0.035	0.881	2.739	0.026	0.908
R^2		0.537			0.471			0.571	
R^2 adjusted		0.479			0.405			0.517	
p		<0.001			0.001			<0.001	
M-FIDsport									
BMI (kg/m^2)	5.77	0.876	0.067	5.368	0.546	0.219	5.991	1.741	<0.001
%Fat	3.40	−0.495	0.170	3.192	−0.179	0.596	3.230	−1.109	<0.001
Fat-Free Mass (kg)	2.50	−0.458	0.141	2.40	−0.028	0.924	2.739	−0.691	0.006
R^2		0.136			0.163			0.542	
R^2 adjusted		0.028			0.058			0.485	
p		0.311			0.226			<0.001	

Note: VIF: variance inflation factor; β: standardized coefficient; S-FID: feel minus ideal discrepancy relative to shape; M-FID: feel minus ideal discrepancy relative to muscularity; S-FIDsport: feel minus ideal soccer player figure discrepancy relative to shape; M-FIDsport: feel minus ideal soccer player figure discrepancy relative to muscularity.

Overall, the models constructed from the second and third surveys are highly significant, indicating that body composition parameters explain 48% and 61% of the variance in shape dissatisfaction, respectively. The models obtained by examining M-FID capture a significant effect only at the third survey: as BMI increases and %Fat decreases, dissatisfaction about muscularity increases. This model explains 34% of the variance. Dissatisfaction with the ideal soccer player shape has BMI as a significant predictor (in the second and third surveys): dissatisfaction increases as BMI increases. The three models obtained based on body composition were highly significant, going on to explain 48%, 40%, and 52% of the variance, respectively. Finally, dissatisfaction concerning muscle development of the ideal soccer player led in the third survey to a highly significant model based on the three predictors considered (BMI, %Fat, FFM): dissatisfaction increases as BMI increases and %Fat and FFM decrease. This model explains more than 48% of the variance.

4. Discussion

The current study aimed to detect any changes in body composition and BID perception of adolescent soccer players examined longitudinally, analyzing the associations between BID and body composition parameters.

The main findings showed significant changes in both anthropometric traits and BI perception in adolescent players over the year during the pre-season and in-season soccer training. Anthropometric changes are consistent with the process of growth and development, involving the completion of individual anthropometric characteristics with significant increases in stature and weight. In particular, as is well known, the statural growth, after the peak involving an average increase of 7 cm at puberty in the male sex reached on average at 14 years of age, goes down in speed, progressively decreasing from the year in which the spurt occurred and in the final stages of adolescence mainly implying an increase in the length of the trunk in comparison to the lower limb [29]. During this period, we observed the expected change in body composition with an increase in muscle mass in the male sex and a decrease in %Fat [29,30]. Indeed, although no significant changes were observed in FM and %Fat, soccer training of Italian adolescent players resulted in a significant increase in FFM, similar to what was found in Portuguese adolescents aged 15-16 years, for whom significant decreases in FM were also detected [31]. These anthropometric trends, partly related to the growth process, are connected to sports training and are consistent, in particular, with the expected effects on body composition due to soccer training [32]. Thus, a previous study comparing Italian adolescents practicing basketball versus those practicing soccer showed a lower %Fat and lower endomorphy in the latter [33]. A significant association between soccer performance and low-fat levels (particularly low in elite players) was also observed [34]. Soccer players have a slender body build compared to non-sporty individuals [33,35]. A recent systematic review and meta-analysis [32] highlighted that soccer improves children's body composition (FM and FFM).

More generally, the results obtained are satisfactory to the extent that the inclusion of soccer among health-enhancing physical activities is appropriate, as suggested by Hernandez-Martin et al. [32].

Considering anthropometric changes, the perception of BI in a body that is changing in shape and composition may lead to changes and eventual dissatisfaction. Adolescent players examined through the dual scale proposed by Ralph-Nearman et al. [15] showed a tendency to prefer slightly more robust shapes throughout the soccer season and a more muscular body ideal. Some apparent inconsistencies, such as increasing dissatisfaction as %Fat decreases, confirm the well-known male tendency to be less aware of their weight status than the female sex and to mistakenly believe that a more robust physique corresponds to greater muscularity [36,37]. Moreover, consistent with a study of 9- to 10-year-old Italian children practicing soccer [18], this research showed a positive effect of sports on the perception of BI: adolescent players were generally satisfied with their body shape (S-FID). This trend confirms the tendency of adolescents to improve BI with sport participation [37,38], with a more positive BI in athletes than in non-athletes [39]. The effect of PA reverberates on the perception of BI and, more generally, on wellbeing, so much so that PA among young people is suggested "to improve wellbeing and yield potential health benefits" [40]. Confirming this, previous studies have found that adolescents from various nationalities have higher levels of life satisfaction related to physical activities, such as team or non-team sports, probably because of the biological, psychological, and social benefits of physical activities on wellbeing [12,41,42].

Adolescence and the transition from adolescence to adulthood represent delicate periods for BID [43]. Several previous studies contributed to showing that BID can predict the tendency toward eating disorders, with possible public health implications (among others: [44–46]). PA generally results in decreased body dissatisfaction at all ages [37], and, in particular, athletes of both sexes are believed to have a lower level of dissatisfaction than non-athletes [39]. Distinguishing between non-esthetic/non-lean (e.g., ball sports) and esthetic/lean sports (e.g., gymnastics), a recent review showed that lean athletes have greater concerns about body image than non-lean athletes, although the meta-analysis found no significant differences between the athlete groups [39]. However, in addition to the background of sports training, individual/non-individual competition, and training intensity, other factors could intervene to influence BI in sports, such as pressures from other people (coaches, judges, parents, and peers) and training regimens [47]. Conversely, according to Webb et al. [43], athletes' BID with their body composition would be similar to that of the general population: their BI will be affected both by the ideal athlete practicing that sport and by the general social ideals. However, in our view, it is important to distinguish dissatisfaction with these different ideals, as the sporting ideal can be very different from the social ideal. Thus, for example, gymnasts were found to be satisfied with their BI in the social context even though they manifested dissatisfaction concerning their ideal of "gymnast" [27]. Young athletes face sport-specific pressures related to body shape and weight, increasing the risk of body dissatisfaction. These pressures include idealized body standards, critical comments, objectification, and sports regulations emphasizing physical appearance, particularly in weight-sensitive and esthetic sports (e.g., gymnastics, figure skating, long-distance running, and triathlons) [48–50]. Research on body image concerns has primarily focused on females, with studies on male athletes showing inconclusive results [39,51,52]. While lean athletes tend to report higher body image concerns, this pattern is observed in females, not males. Factors influencing body satisfaction in young athletes include motivation type: intrinsic and autonomous motivation are linked to better body satisfaction, while extrinsic motivation is associated with higher dissatisfaction [53–57]. Body satisfaction fluctuates during adolescence, with males showing an increase in satisfaction from ages 12 to 20, while females experience a decline between ages 10 and 16, which stabilizes and improves by age 20 [58]. Adolescent boys, unlike girls who typically pursue thinness, often have a desire for muscularity [36,59]. While this drive can encourage healthy habits, it can also lead to extreme, unhealthy behaviors when taken too far. Grieve [60] found that internalizing muscular ideals is strong in early adolescence. Around 25% of middle-school boys lift weights to increase muscle mass [61]. While sport can promote positive behaviors, certain sports may encourage unhealthy practices, such as weight-focused sports (e.g., wrestling, cross-country running) linked to eating disorders [62] or muscle-focused sports (e.g., football, weightlifting) associated with steroid use [63]. Symptoms of muscle dysmorphia often begin in adolescence [64].

In the sample of soccer players examined using longitudinal analysis, in the preseason, we were able to verify a complete satisfaction with the expected body shape in the social context (S-FID), while dissatisfaction was greater than in the ideal soccer player, especially for musculature (M-FID), which has been declining, however, as described, during the sports season. The aspiration toward greater muscle mass and body fat has also been reported in a previous cross-sectional study of 19-year-old soccer players [65]. In general, high self-confidence, self-esteem, and positive BI levels of athletes would depend on enhanced self-efficacy after completing physical tasks related to the sport played and better body consciousness [12]. The representation of body image across different eras reflects shifting ideals of beauty, strength, and human potential, shaped by cultural, social, and historical influences. In sports, the human body has symbolized the values of each time period. These ideals are dynamic, evolving with societal norms and cultural beliefs, and understanding them requires considering the broader historical and cultural context [66].

The main strengths of the study consist of its longitudinal design and the anthropometric measurements taken directly by a trained operator. This study also has several limitations. The absence of a control sample of non-athletes prevented us from testing for changes attributable with certainty to the normal growth process. Although the sample size analyzed meets the minimum required by the power analysis for a longitudinal study, it did not allow us to verify any differences in BI perception concerning the role of play. Differences in the competitive level and type of sport played are relevant aspects that could be addressed in future research. In addition, future studies should examine players of various age groups and both sexes and analyze differences with end-season and/or off-season periods. Furthermore, a more explicit discussion of the potential pedagogical implications of the findings could contribute to a deeper understanding of the psychological aspects involved in body image perception in adolescent athletes and help translate the research into practical strategies for sports educators, coaches, and practitioners.

Based on our study findings, some practical applications for coaches, trainers, and youth sport practitioners should be considered. First, regular monitoring of anthropometric traits and body image perception throughout the season can help tailor training loads and psychological support to the individual needs of adolescent players. Second, since adolescents may aspire to an unrealistic muscular ideal, practitioners should promote a healthy and functional body image, emphasizing strength and performance over appearance. Third, coaches and educators should be aware of the psychosocial impact of body image dissatisfaction and include BI education and body positivity strategies as part of youth development programs. It is important for those working with adolescent boys to recognize that body dissatisfaction can lead to harmful compensatory behaviors affecting their health and development. Promoting a positive body image, self-esteem, and early intervention is crucial. Coaches, educators, clinicians, and parents should be knowledgeable about identifying signs of muscle dysmorphia and discussing healthy ways to address body image concerns. Educating adolescents about natural body differences and the dangers of unhealthy body manipulation practices and providing healthier alternatives for weight management and fitness is key.

In conclusion, our findings confirm the importance of adolescents' participation in sports by showing physical and mental improvement resulting from playing soccer. Because improving BI and body composition parameters (increasing muscle mass and decreasing %Fat) is a relevant factor for both sports activity performance and good health perspectives, families and schools should make efforts to encourage children and adolescents' participation in sports activities.

Author Contributions: Conceptualization, L.Z., S.T., and E.G.-R.; methodology, L.Z.; software, M.R.; formal analysis, L.Z.; investigation, M.R.; resources, L.Z.; data curation, L.Z.; writing—original draft preparation, E.G.-R.; writing—review and editing, L.Z., M.R., S.T., and E.G.-R.; visualization, S.T.; supervision, L.Z.; project administration, L.Z. All authors have read and agreed to the published version of the manuscript.

Funding: This research received no external funding.

Institutional Review Board Statement: The study was conducted in accordance with the Declaration of Helsinki and approved by the Bioethics Committee of the University of Bologna (protocol code 25027; date of approval: 13 March 2017).

Informed Consent Statement: Informed consent was obtained from all subjects involved in the study.

Data Availability Statement: Data is available upon request due to ethical restrictions regarding participant privacy. Requests for the data may be sent to the corresponding authors.

Acknowledgments: The authors would like to thank the soccer club, coaches, and study participants.

Conflicts of Interest: The authors declare no conflicts of interest.

Abbreviations

The following abbreviations are used in this manuscript:

PA	Physical activity
BI	Body image
BID	Body image dissatisfaction
BMI	Body Mass Index
%Fat	percentage of body fat
FM	Fat mass
FFM	Fat-free mass
MBS	Male Body Scale
MFBS	Male Fit Body Scale
S-Feel	Perceived shape
M-Feel	Perceived muscularity
S-Ideal	Ideal shape
M-Ideal	Ideal muscularity
S-Soccer	Ideal soccer player's shape
M-Soccer	Ideal soccer player's muscularity
FID	Feel minus Ideal Discrepancy
S-FID	Discrepancy relative to shape
M-FID	Discrepancy relative to muscularity
S-FIDsport	Discrepancy from the ideal soccer player shape
M-FIDsport	Discrepancy from the ideal soccer player muscularity
VIF	Variance inflation factor

References

1. Logan, K.; Cuff, S.; Council on Sports Medicine and Fitness. Organized Sports for Children, Preadolescents, and 456 Adolescents. *Pediatrics* **2019**, *143*, e20190997. [CrossRef] [PubMed]
2. Allen, C.P.; Telford, R.M.; Telford, R.D.; Olive, L.S. Sport, physical activity and physical education experiences: Associations with functional body image in children. *Psychol. Sport Exerc.* **2019**, *45*, 101572. [CrossRef]
3. Fernández-Bustos, J.G.; Infantes-Paniagua, Á.; Cuevas, R.; Contreras, O.R. Effect of Physical Activity on Self-Concept: Theoretical Model on the Mediation of Body Image and Physical Self-Concept in Adolescents. *Front. Psychol.* **2019**, *10*, 1537. [CrossRef] [PubMed]
4. Davelaar, C.M.F. Body Image and its Role in Physical Activity: A Systematic Review. *Cureus* **2021**, *13*, e13379. [CrossRef]
5. Guthold, R.; Stevens, G.A.; Riley, L.M.; Bull, F.C. Global trends in insufficient physical activity among adolescents: A pooled 467 analysis of 298 population-based surveys with 1·6 million participants. *Lancet Child Adolesc. Health* **2020**, *4*, 23–35. [CrossRef]
6. UK Chief Medical Officers' Physical Activity Guidelines. 2019. Available online: https://assets.publishing.service.gov.uk/government/uploads/system/uploads/attachment_data/file/832868/uk-chief-medical-officers-physical-activity-guidelines.pdf (accessed on 15 April 2025).
7. Gualdi-Russo, E.; Rinaldo, N.; Zaccagni, L. Physical Activity and Body Image Perception in Adolescents: A Systematic Review. *Int. J. Environ. Res. Public Health* **2022**, *19*, 13190. [CrossRef]
8. Cabral, M.C.; Coelho, G.M.O.; Oliveira, N.; Canella, D.S.; Brasil, R.L.O.; Campos, T.A.M.; Faerstein, E.; Bezerra, F.F. Association of body image perception and (dis)satisfaction with adiposity in adults: The Pró-Saúde study. *PLoS ONE* **2024**, *19*, e0304987. [CrossRef] [PubMed]
9. Ruiz-Hermosa, A.; Sánchez-López, M.; Castro-Piñero, J.; Grao-Cruces, A.; Camiletti-Moirón, D.; Martins, J.; Mota, J.; Ceciliani, A.; Murphy, M.; Vuillemin, A.; et al. The Erasmus+ EUMOVE project-a school-based promotion of healthy lifestyles to prevent obesity in European children and adolescents. *Eur. J. Public Health* **2024**, *34*, 955–961. [CrossRef]

10. Moreno-Díaz, M.I.; Vaquero-Solís, M.; Tapia-Serrano, M.Á.; Sánchez-Miguel, P.A. Physical Activity, Body Composition, Physical Fitness, and Body Dissatisfaction in Physical Education of Extremadura Adolescents: An Exploratory Study. *Children* **2024**, *11*, 83. [CrossRef]
11. Collins, N.M.; Cromartie, F.; Butler, S.; Bae, J. Effects of Early Sport Participation on Self-esteem and Happiness. *Sport. J.* **2018**, *20*, 1–20.
12. Khan, A.; Ahmed, K.R.; Hidajat, T.; Kolbe-Alexander, T.; Edwards, E.J. Examining the Association between Sports Participation and Mental Health of Adolescents. *Int. J. Environ. Res. Public Health* **2022**, *19*, 17078. [CrossRef]
13. Staśkiewicz-Bartecka, W.; Zydek, G.; Michalczyk, M.M.; Kardas, M. Prevalence of Eating Disorders and Self-Perception Concerning Body Composition Analysis among Elite Soccer Players. *J. Hum. Kinet.* **2024**, *95*, 259–272. [CrossRef] [PubMed]
14. Neumark-Sztainer, D.; Paxton, S.J.; Hannan, P.J.; Haines, J.; Story, M. Does Body Satisfaction Matter? Five-year Longitudinal Associations between Body Satisfaction and Health Behaviors in Adolescent Females and Males. *J. Adolesc. Health* **2006**, *39*, 244–251. [CrossRef] [PubMed]
15. Ralph-Nearman, C.; Filik, R. New Body Scales Reveal Body Dissatisfaction, Thin-Ideal, and Muscularity-Ideal in Males. *Am. J. Mens Health* **2018**, *12*, 740–750. [CrossRef] [PubMed]
16. Smith, A.R.; Hawkeswood, S.E.; Bodell, L.P.; Joiner, T.E. Muscularity versus leanness: An examination of body ideals and predictors of disordered eating in heterosexual and gay college students. *Body Image* **2011**, *8*, 232–236. [CrossRef]
17. Anjum, A.; Hossain, S.; Hasan, M.T.; Uddin, M.E.; Sikder, M.T. Anxiety among urban, semi-urban and rural school adolescents in Dhaka, Bangladesh: Investigating prevalence and associated factors. *PLoS ONE* **2022**, *17*, e0262716. [CrossRef]
18. Rinaldo, N.; Zaccagni, L.; Gualdi-Russo, E. Soccer training programme improved the body composition of pre-adolescent boys and increased their satisfaction with their body image. *Acta Paediatr.* **2016**, *105*, e492–e495. [CrossRef]
19. Andrade, C. Sample Size and its Importance in Research. *Indian J. Psychol. Med.* **2020**, *42*, 102–103. [CrossRef]
20. Russo, E.G. Metodi Antropometrici: Generalità e principali caratteri antropometrici. In *Manuale di Antropologia. Evoluzione e Biodiversità Umana*; Sineo, L., Cecchi, J.M., Eds.; UTET: Torino, Italy, 2022; pp. 55–75.
21. Lohman, T.G.; Roche, A.F.; Martorell, R. *Anthropometric Standardization Reference Manual*; Human Kinetics Books: Champain, IL, USA, 1988.
22. Weiner, J.S.; Lourie, J.A. *Practical Human Biology*; Academic Press: Cambridge, MA, USA, 1981.
23. Cole, T.J.; Bellizzi, M.C.; Flegal, K.M.; Dietz, W.H. Establishing a standard definition for child overweight and obesity worldwide: International survey. *Br. Med. J.* **2000**, *320*, 1240–1243. [CrossRef]
24. Cole, T.J.; Flegal, K.M.; Nicholls, D.; Jackson, A.A. Body mass index cut offs to define thinness in children and adolescents: International survey. *Br. Med. J.* **2007**, *335*, 194–197. [CrossRef]
25. Durnin, J.V.; Womersley, J. Body fat assessed from total body density and its estimation from skinfold thickness: Measurements on 481 men and women aged from 16 to 72 years. *Br. J. Nutr.* **1974**, *32*, 77–97. [CrossRef] [PubMed]
26. Siri, W.E. Body composition from fluid spaces and density: Analysis of methods. In *Techniques for Measuring Body Compositions*; Brozek, J., Henschel, A., Eds.; National Academy of Sciences: Washington, DC, USA, 1961; p. 233.
27. Zaccagni, L.; Rinaldo, N.; Gualdi-Russo, E. Anthropometric Indicators of Body Image dissatisfaction and Perception Inconsistency in Young Rhythmic Gymnastics. *Asian J. Sports Med.* **2019**, *10*, 87871. [CrossRef]
28. Gualdi-Russo, E.; Rinaldo, N. Antropometria e percezione dell'immagine corporea. In *Manuale di Antropologia. Evoluzione e Biodiversità Umana*; Sineo, L., Cecchi, J.M., Eds.; UTET: Torino, Italy, 2022; pp. 485–493.
29. Tanner, J.M. *Auxologia Dal Feto All'uomo: La Crescita Fisica Dal Concepimento Alla Maurità*; UTET, Ed.; UTET: Torino, Italy, 1981.
30. Malina, R.M.; Bouchard, C.; Bar-Or, O. *Growth, Maturation and Physical Activity*, 2nd ed.; Human Kinetics: Champaign, IL, USA, 2004.
31. França, C.; Martinho, D.V.; Gouveia, É.R.; Martins, F.; Marques, A.; Ribeiro, T.; Nascimento, M.M.; Lopes, H.; Rodrigues, A.; Ihle, A. Changes in Estimated Body Composition and Physical Fitness of Adolescent Boys after One Year of Soccer Training. *Children* **2023**, *10*, 391. [CrossRef] [PubMed]
32. Hernandez-Martin, A.; Garcia-Unanue, J.; Martínez-Rodríguez, A.; Manzano-Carrasco, S.; Felipe, J.L.; Carvalho, M.J.; Gallardo, L.; Sanchez-Sanchez, J. The effects of football practice on nutritional status and body composition in children: A systematic review and meta-analysis. *Nutrients* **2021**, *13*, 2562. [CrossRef]
33. Toselli, S.; Campa, F.; Maietta Latessa, P.; Greco, G.; Loi, A.; Grigoletto, A.; Zaccagni, L. Differences in Maturity and Anthropometric and Morphological Characteristics among Young Male Basketball and Soccer Players and Non-Players. *Int. J. Environ. Res. Public Health* **2021**, *18*, 3902. [CrossRef]
34. Aurélio, J.; Dias, E.; Soares, T.; Jorge, G.; da Cunha Espada, M.A.; Filho, D.M.P.; Pereira, A.; Figueiredo, T. Relationship between Body Composition, Anthropometry and Physical Fitness in Under-12 Soccer Players of Different Positions. *Int. J. Sports Sci.* **2016**, *6*, 25–30.
35. Boraczyński, M.; Boraczyński, T.; Podstawski, R.; Wójcik, Z. Relationships between anthropometric traits, body composition and aerobic capacity in male soccer players aged 13–15 years. *J. Kinesiol. Exerc. Sci.* **2015**, *25*, 29–36.

36. Jones, D.C.; Crawford, J.K. Adolescent boys and body image: Weight and muscularity concerns as dual pathways to body dissatisfaction. *J. Youth Adolesc.* **2005**, *34*, 629–636. [CrossRef]
37. Gualdi-Russo, E.; Rinaldo, N.; Masotti, S.; Bramanti, B.; Zaccagni, L. Sex Differences in Body Image Perception and Ideals: Analysis of Possible Determinants. *Int. J. Environ. Res. Public Health* **2022**, *19*, 2745. [CrossRef]
38. Ouyang, Y.; Wang, K.; Zhang, T.; Peng, L.; Song, G.; Luo, J. The Influence of Sports Participation on Body Image, Self-Efficacy, and Self-Esteem in College Students. *Front. Psychol.* **2020**, *10*, 3039. [CrossRef]
39. Burgon, R.H.; Beard, J.; Waller, G. Body image concerns across different sports and sporting levels: A systematic review and meta-analysis. *Body Image* **2023**, *46*, 9–31. [CrossRef] [PubMed]
40. Sollerhed, A.-C.; Fransson, J.; Skoog, I.; Garmy, P. Physical Activity Levels, Perceived Body Appearance, and Body Functioning in Relation to Perceived Wellbeing Among Adolescents. *Front. Sports Act. Living* **2022**, *4*, 830913. [CrossRef]
41. Pierannunzio, D.; Spinelli, A.; Berchialla, P.; Borraccino, A.; Charrier, L.; Dalmasso, P.; Lazzeri, G.; Vieno, A.; Ciardullo, S.; Nardone, P. Physical activity among Italian adolescents: Association with life satisfaction, self-rated health and peer relationships. *Int. J. Environ. Res. Public Health* **2022**, *19*, 4799. [CrossRef]
42. Kvaskova, L.; Recka, K.; Jezek, S.; Macek, P. Time spent on daily activities and its association with life satisfaction among Czech adolescents from 1992 to 2019. *Int. J. Environ. Res. Public Health* **2022**, *19*, 9422. [CrossRef]
43. Webb, M.D.; Melough, M.M.; Earthman, C.P.; Katz, S.E.; Pacanowski, C.R. Associations between anthropometry, body composition, and body image in athletes: A systematic review. *Front. Psychol.* **2024**, *15*, 1372331. [CrossRef]
44. Calzo, J.P.; Sonneville, K.R.; Haines, J.; Blood, E.A.; Field, A.E.; Austin, S.B. The development of associations among body mass index, body dissatisfaction, and weight and shape concern in adolescent boys and girls. *J. Adolesc. Health* **2012**, *51*, 517–523. [CrossRef] [PubMed]
45. Mäkinen, M.; Puukko-Viertomies, L.R.; Lindberg, N.; Siimes, M.A.; Aalberg, V. Body dissatisfaction and body mass in girls and boys transitioning from early to mid-adolescence: Additional role of self-esteem and eating habits. *BMC Psychiatry* **2012**, *12*, 35. [CrossRef] [PubMed]
46. Morin, G.; Meilleur, D. Association Between Emotion Regulation and Body Image Concerns in a Group of Adolescent Boys: Interaction With the Internalization of the Sociocultural Body Ideal. *J. Emot. Behav. Disord.* **2024**, *32*, 213–225. [CrossRef]
47. Reel, J.J.; Petrie, T.A.; SooHoo, S.; Anderson, C.M. Weight pressures in sport: Examining the factor structure and incremental validity of the weight pressures in sport- females. *Eat. Behav.* **2013**, *14*, 137–144. [CrossRef]
48. Scott, C.L.; Haycraft, E.; Plateau, C.R. The impact of critical comments from teammates on athletes' eating and exercise psychopathology. *Body Image* **2022**, *43*, 170–179. [CrossRef]
49. Lundqvist, C.; Kolbeinsson, Ö.; Asratian, A.; Wade, T.D. Untangling the relationships between age, gender, type of sport, perfectionistic self-presentation and motivation on body satisfaction: A cross-sectional study on aesthetic and non-aesthetic female and male athletes aged 10 to 22 years. *BMJ Open Sport Exerc. Med.* **2024**, *10*, e001975. [CrossRef]
50. Matheson, E.L.; Schneider, J.; Tinoco, A.; Silva-Breen, H.; LaVoi, N.M.; Diedrichs, P.C. How can we help you? A global investigation into girls' body image experiences in sport and intervention preferences. *Body Image* **2023**, *46*, 265–279. [CrossRef] [PubMed]
51. Karrer, Y.; Halioua, R.; Mötteli, S.; Iff, S.; Seifritz, E.; Jäger, M.; Claussen, M.C. Disordered eating and eating disorders in male elite athletes: A scoping review. *BMJ Open Sport Exerc. Med.* **2020**, *6*, e000801. [CrossRef] [PubMed]
52. Zaccagni, L.; Gualdi-Russo, E. The impact of sports involvement on body image perception and ideals: A systematic review and meta-analysis. *Int. J. Environ. Res. Public Health* **2023**, *20*, 5228. [CrossRef] [PubMed]
53. Kopp, L.L.; Zimmer-Gembeck, M.J. Women's global self-determination, eating regulation, and body dissatisfaction: Exploring the role of autonomy support. *Eat. Behav.* **2011**, *12*, 222–224. [CrossRef]
54. Sheeran, P.; Wright, C.E.; Avishai, A.; Villegas, M.E.; Rothman, A.J.; Klein, W.M.P. Does increasing autonomous motivation or perceived competence lead to health behavior change? A meta-analysis. *Health Psychol.* **2021**, *40*, 706–716. [CrossRef]
55. Furnham, A.; Badmin, N.; Sneade, I. Body image dissatisfaction: Gender differences in eating attitudes, self-esteem, and reasons for exercise. *J. Psychol.* **2002**, *136*, 581–596. [CrossRef]
56. Alcaraz-Ibáñez, M.; Paterna, A.; Sicilia, Á.; Griffiths, M.D. A systematic review and meta-analysis on the relationship between body dissatisfaction and morbid exercise behaviour. *Int. J. Environ. Res. Public Health* **2021**, *18*, 585. [CrossRef]
57. Ruiz-Turrero, J.; Massar, K.; Kwasnicka, D.; Ten Hoor, G.A. The relationship between compulsive exercise, self-esteem, body image and body satisfaction in women: A cross-sectional study. *Int. J. Environ. Res. Public Health* **2022**, *19*, 1857. [CrossRef]
58. Lundqvist, C.; Schary, D.P.; Eklöf, E.; Zand, S.; Jacobsson, J. Elite lean athletes at sports high schools face multiple risks for mental health concerns and are in need of psychosocial support. *PLoS ONE* **2023**, *18*, e0284725. [CrossRef]
59. McCabe, M.P.; Ricciardelli, L.A. Body image dissatisfaction among males across the lifespan: A review of past literature. *J. Psychosom. Res.* **2004**, *56*, 675–685. [CrossRef] [PubMed]
60. Grieve, F.G. A conceptual model of factors contributing to the development of muscle dysmorphia. *Eat. Disord.* **2007**, *15*, 63–80. [CrossRef] [PubMed]

61. Smolak, L.; Murnen, S.K.; Thompson, J.K. Sociocultural influences and muscle building in adolescent boys. *Psychol. Men Muscularity* **2005**, *6*, 227–239. [CrossRef]
62. Carl, R.L.; Johnson, M.D.; Martin, T.J. Promotion of healthy weight-control practices in young athletes. *Pediatrics* **2017**, *140*, e20171871. [CrossRef]
63. Irving, L.M.; Wall, M.; Neumark-Sztainer, D.; Story, M. Steroid use among adolescents: Findings from the Project EAT. *J. Adolesc. Health* **2002**, *30*, 243–252. [CrossRef]
64. Leone, J.E.; Wise, K.A.; Mullin, E.M.; Harmon, W.; Moreno, N.; Drewniany, J. The effects of pubertal timing and alexithymia on symptoms of muscle dysmorphia and the drive for muscularity in men. *Psychol. Men Masculinity* **2015**, *16*, 67–77. [CrossRef]
65. Arroyo, M.; Gonzalez-de-Suso, J.M.; Sanchez, C.; Ansotegui, L.; Rocandio, A.M. Body image and body composition: Comparisons of young male elite soccer players and controls. *Int. J. Sport Nutr. Exerc. Metabolism* **2008**, *18*, 628–638. [CrossRef]
66. Cereda, F. Sporting bodies, societal norms in history: Examining body image and identity. *Sci. J. Sport Perform.* **2023**, *2*, 572–588. [CrossRef]

Journal of
Functional Morphology and Kinesiology

MDPI

Article

Positional Profiling of Anthropometric, Baropodometric, and Grip Strength Traits in Male Volleyball Players: Insights from a National Colombian Study

Adrián De la Rosa [1,2,3,*,†], María Alejandra Camacho-Villa [3,4,†], Fernando Millan-Domingo [1], Juan Carlos Saavedra [5], Marina Politi Okoshi [6] and Luana Urbano Pagan [6]

1 Freshage Research Group, Department of Physiology, Faculty of Medicine, University of Valencia, CIBERFES, Fundación Investigación Hospital Clínico Universitario/INCLIVA, 46010 Valencia, Spain; fernando.millan-domingo@uv.es
2 Body, Physical Activity and Sport Study Group (GECAFD), Sports Department, Universidad Industrial de Santander, Bucaramanga 680002, Colombia
3 Harpeer Research Group, Yumbo 760001, Colombia; alejacvilla@gmail.com
4 Pain Study Group (GED), Physical Therapy School, Universidad Industrial de Santander, Bucaramanga 680002, Colombia
5 GCED Research Group, Bucaramanga 688006, Colombia; juancs19@hotmail.com
6 Botucatu Medical School, São Paulo State University, Botucatu 18618-970, Brazil; marina.okoshi@unesp.br (M.P.O.); luana.pagan@unesp.br (L.U.P.)
* Correspondence: adrianmdegonz@outlook.com
† These authors contributed equally to this work.

Abstract: Background: In volleyball, upper limb dimensions, handgrip strength (HGS), and baropodometric parameters are critical for executing offensive and defensive actions during the match. These movements demand not only physical precision but also carry a significant risk of injury, varying by playing position. **Objectives:** This study aimed to determine the differences in specific upper limb anthropometric characteristics, HGS, and selected baropodometric variables among U-23 male volleyball players concerning playing position. **Methods:** The sample consisted of 92 U-23 male players who prepared for the U-23 Men's Volleyball National Championship 2022 (20.39 (1.74) years, 184 (8.46) cm, 75.52 (10.20) kg). Playing positions analyzed were setters (n = 12), outside (n = 18), opposites (n = 19), middle blockers (n = 16), and liberos (n = 12). **Results:** player position differences in HGS and several anthropometric upper limb variables were observed. Middle blockers, outsides, and opposites exhibited superior anthropometric traits in most of the measurements compared to liberos and setters ($p < 0.05$). Differences in baropodometric parameters were only found between feet and their zones when the entire sample was evaluated. Finally, regression analysis identified dominant hand breadth ($\beta = 3.42$, 95%CI [0.43, 6.40], upper arm muscle area ($\beta = 0.157$, 95%CI [0.02, 0.29]), and wrist diameter ($\beta = 3.59$, IC 95% [0.49, 6.68]) as associated variables of HGS. **Conclusions:** The study underscores the importance of positional profiling in volleyball, revealing key physical traits linked to performance. The observed differences are likely attributable to the specific role and physical demands inherent to each playing position. These findings can guide targeted training and injury prevention strategies to enhance performance.

Keywords: upper limb variables; positional differences; baropodometry; volleyball performance

1. Introduction

Volleyball players are expected to exhibit specific functional and structural characteristics that significantly influence their athletic success and performance [1–3]. Several anthropometric measurements, including height, body composition, and segmental length of body parts, particularly the upper limbs, play a crucial role in skill execution and team strategy [3,4].

In volleyball, as in other ball sports (i.e., basketball, softball, and handball) where hand–ball interaction is fundamental, numerous movements rely on the continuous use of wrist and digit flexors, particularly in attacking, blocking, and serving [5,6]. HGS emerges as another physical factor influencing these specific movements, serving as a key determinant of optimal performance and providing athletes with considerable advantages during gameplay [7–9].

Previous studies have examined [1,10–12] the anthropometric characteristics of the upper limbs and their relationship with HGS in volleyball players. However, most research has been conducted in general athlete populations (i.e., inter-university, elite, and female volleyball players) without considering differences across playing positions or providing clear links to functional performance metrics [13–15]. Consequently, these findings may not be directly applicable to U-23 national-level athletes, especially in the Colombian context, where talent selection processes may differ.

While upper limb strength is crucial for ball handling and attack efficiency, lower limb performance and static balance are equally essential for maintaining postural control and executing explosive movements [16–21]. Effective postural control in static positions and optimal plantar pressure distribution has been associated with enhanced neuromuscular efficiency, improved movement precision, and greater resilience to injury [22,23]. In volleyball athletes, deficits in static balance have been linked to an increased risk of lower extremity injuries (i.e., ankle sprains and functional ankle instability) [24] due to the sport's high demands for precise body positioning, dynamic control of the center of gravity, and a high degree of ankle stability [19,20].

In this sense, baropodometric analysis provides a detailed understanding of static postural control and plantar distribution by mapping pressure across the plantar surface, which reflects key postural variables [21,25,26]. This computerized system captures plantar imprints and ground reaction forces during quiet standing, dividing the measurements between the right and left feet and further subdividing them into the forefoot, midfoot, and rearfoot [27]. Moreover, it classifies foot types and provides stabilometric parameters based on the spatial and temporal behavior of the center of pressure [28]. These insights allow sports professionals to identify athletes at higher risk of injury early and support the development of targeted proprioceptive and neuromuscular training programs aimed at risk reduction [23,29,30]. Furthermore, monitoring static balance throughout the competitive season yields valuable data for fatigue management and rehabilitation progress, highlighting its relevance for both performance and promoting athlete health.

Despite its proven value in clinical and rehabilitation settings, the application of static baropodometric analysis in volleyball remains limited, and there is still a lack of sport-specific evidence exploring baropodometric characteristics across playing positions [31,32]. This aspect is particularly pivotal in a team sport like volleyball, where positional differences impose distinct physical demands due to the varying requirements of game phases, technical and tactical strategies, and risk of injury associated with each playing position [19,20].

Furthermore, research exploring these variables remains scarce among Latin American volleyball players, especially in the Colombian U-23 male population. This category represents a transitional phase toward elite competition, during which players consolidate technical and tactical skills and physical capacities [2,33]. In Colombia, this category represents a critical talent pool for national and professional teams yet remains understudied. Addressing this gap is essential for developing position-specific profiling to identify talent in the region, assess strengths and weaknesses based on playing position, and implement targeted training programs aimed at improving performance and reducing injury risk [34–36].

Given these considerations, the first purpose of the present study was to investigate the anthropometric, baropodometric, and HGS characteristics of male Colombian volleyball players, considering potential differences based on playing position. The second purpose was to assess the influence of upper limbs anthropometric parameters on HGS to better understand the relationship between segmental dimensions and functional performance in volleyball athletes.

2. Materials and Methods

2.1. Subjects

The present study employed a cross-sectional analytical design and was conducted during the qualification phase of the U-23 Men's National Volleyball Championship, held in July 2022 in Bucaramanga, Colombia. All eligible athletes participating in the competition were included in the study.

A total of ninety-two male volleyball players from the regional teams of Santander (n = 13), Antioquia (n = 14), Valle (n = 13), Caldas (n = 14), Nariño (n = 12), Atlántico (n = 14), and Cesar (n = 12) participated in the study (age: 20.39 (1.74) yrs; height: 184 (8.46) cm; weight: 75.52 (10.20) kg). Players were categorized according to their playing position as setters (n = 15), outside (n = 21), opposites (n = 22), middle blockers (n = 19), and liberos (n = 15).

All participants were healthy and free from any neuromuscular, orthopedic, or neurological conditions that might interfere with their sports performance, hand function, anthropometric characteristics, or activities of daily living. All participants were thoroughly informed about the study, including the risks and benefits of participation, and if, after this explanation, their decision was not to be included in the analysis, this did not adversely affect any current or future team selection. All included athletes provided written informed consent for testing and data. This study was conducted following the ethical principles outlined in the Declaration of Helsinki. The Ethics Committee for Human Subjects of the University approved this study research (0010-2022/2 May 2022)

2.2. Testing Procedures

Comprehensive details regarding the anthropometric parameters, baropodometric evaluations, and HGS tests are provided in the subsequent section. All assessments were performed before training sessions in a private setting at the Bicentenario Volleyball Coliseum during morning hours (between 8:00 and 11:00 a.m.).

The evaluations were carried out in an environment without climate control and followed a predefined order: body composition, upper limb anthropometric measurements, baropodometry, and finally, the HGS assessment. The dominance of the upper and lower limbs was determined by asking the athletes which arm/leg they would use to throw/kick a ball [37–39].

Assessments were conducted by two researchers, each with nine years of experience in sports research. Anthropometric measurements were collected by a level 2 anthropometrist. The intraclass correlation coefficient (ICC) values for intra-rater reliability ranged from 0.91 to 0.96, indicating excellent measurement consistency. HGS assessments were conducted by the other researcher, who underwent training that included protocols for the participant's position and verbal encouragement. The intra-rater reliability showed ICC values of 0.98 for the dominant hand and 0.97 for the non-dominant hand, exhibiting very high reliability (ICC > 0.90) [40].

2.3. Body Composition and Anthropometric Measurements

A certified level 2 anthropometrist conducted the measurements following the standardized protocols established by the International Society for the Advancement of Kinanthropometry (ISAK). The mean of two repeated measurements was calculated for each anthropometric variable for data analysis. Using a mechanical stadiometer platform (Seca® 274, Hamburg, Germany; Technical Error of Measurement = 0.019%), height measurements were obtained with participants standing barefoot. The adjustable headpiece was carefully lowered to contact the vertex of the head while the participant performed a deep inhalation. All measurements were taken in meters and rounded to the nearest 0.5 cm.

A bioelectrical impedance analysis (BIA) device (TANITA BC 240 MA, Arlington Heights, IL, USA) was used to assess body composition, with results rounded to the nearest 0.1 unit. Prior to the assessment, participants were instructed to remove any metal objects, abstain from caffeine or diuretics for at least three hours, and void their bladder within 30 min before the testing. The variables collected included weight and body fat percentage (BF%). The body mass index (BMI) was computed as weight (kg)/stature (m^2).

2.4. Upper Limbs Anthropometric Measurements

Participants wore minimal clothing and remained barefoot during the anthropometric measurements to ensure accuracy. A segmometer, a steel tape, a small bone anthropometer, and a skinfold caliper (Cescorf, Porto Alegre, Brazil) were employed to evaluate upper limb lengths, circumferences, and diameters and skinfold triceps thickness, respectively. All upper limb measurements were recorded to the nearest 0.1 cm. For each upper limb, the following parameters were assessed: arm and forearm length, hand breadth, hand length, first-to-fifth finger distance, as well as arm circumference, elbow, and wrist diameters according to the International Anthropometric Standardization Manual edited by ISAK [41].

Additionally, the triceps skinfold was measured following the standard technique [42], as illustrated in Figure 1.

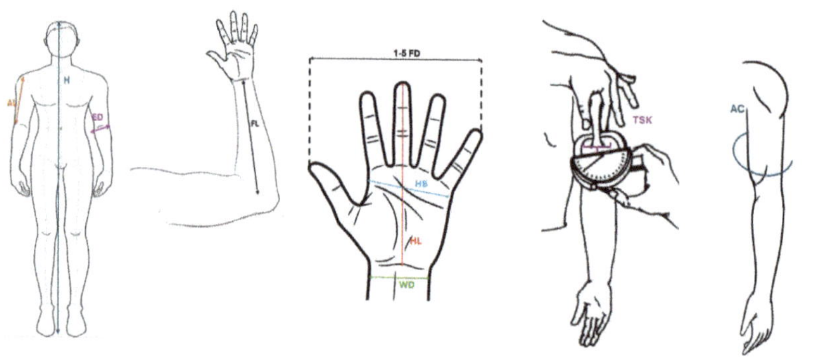

Figure 1. (AL) Arm length. (H) Height. (ED) Elbow length. (FL) Forearm length. (WD) Wrist diameter. (HL) Hand length. (HB) Hand breadth. First–to–fifth finger distance (1–5 FD). Triceps skinfold (TSK). Arm circumference (AC).

Arm length was assessed as the straight-line distance between the marked acromial point and the radiale, with the participant standing upright, arms relaxed at the sides, and palms resting against their thighs. *Forearm length* was determined by measuring the distance from the radiale to the stylion, with the elbow flexed and the tape positioned parallel to the longitudinal axis of the radius.

Hand length was assessed as the shortest distance from the marked mid-stylion line to the dactylion, while *hand breadth* was measured as the span between the radial side of the second metacarpal joint and the ulnar side of the fifth metacarpal joint [43]. The *first-to-fifth finger distance* was recorded as the linear measurement from the outer border of the tip of the thumb to the outer border of the little finger. The fingers and thumb are stretched as widely apart as the person finds comfortable [4].

Arm circumference was determined by identifying the midpoint of the distance between the acromial process and the radiale, ensuring proper alignment with the medial and lateral borders of the humerus. The participant stood in a relaxed, upright position with arms hanging naturally by their sides. In the same posture, the *triceps skinfold thickness* was measured by identifying the midpoint between the acromial and the olecranon processes. The skinfold caliper was positioned perpendicular to the fold, approximately 1 cm below the fingers holding the skin. The measurement was recorded 2 s after applying the caliper's pressure. If two consecutive measurements differed by more than 0.2 mm, a third measurement was taken, and the two closest values were averaged to enhance reliability.

Elbow diameter was assessed by measuring the distance between the medial and lateral epicondyles of the humerus. *Wrist diameter* was determined as the linear distance between the outer edges of the radial and ulnar styloid processes.

The arm muscle circumference, arm area, and arm muscle area were calculated according to the formulas outlined in Table 1.

Table 1. Formulas to estimate upper arm anthropometric variables.

Anthropometric Variables	Formula
Arm muscle circumference (cm)	$S = c - (T \times 3.14)$ [44]
Arm area (cm^2)	$A = c^2/12.56$ [45]
Arm muscle area (cm^2)	$M = S^2/12.56$ [44]

S: arm muscle circumference; c: arm circumference; T = triceps skinfold; A: arm area; M: arm muscle area.

2.5. Baropodometry Assessment

Baropodometric measurements were acquired using the electronic portable pressure platform Ecowalk (Ecosanit, Ecotechnology, Inc., Anghiari, Italy). The platform operates at a sampling frequency of 100 Hz and incorporates approximately 1–2 sensors per square centimeter, ensuring high spatial resolution during data acquisition. Data were processed using EcoFoot 4.0 software. Before measurement, participants stood quietly on the platform for approximately 60 s to perform the calibration.

During the measurement process, participants maintained a static bipedal stance on the platform for 20 s, maintaining a forward gaze and barefoot posture (Figure 2A).

Their feet were positioned side by side, and their arms were relaxed along the trunk. Gaze was fixed on a visual marker placed at eye level on the wall, approximately two meters away, to ensure standardized head and neck positioning. This controlled posture was employed to minimize measurement variability. Each athlete underwent two consecutive trials, separated by a one-minute rest interval. The mean value of the two trials was used for subsequent analysis [46].

The following parameters were evaluated for both feet: 95% confident ellipse area, percentage of load distribution, surface area, peak pressure, plantar arch index, and calcaneus angle (Figure 2B).

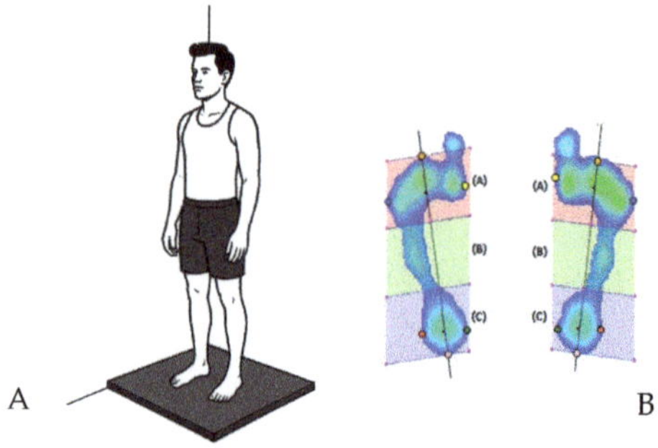

Figure 2. (**A**) Participant position during baropodometry assessment. (**B**) Static analysis of plantar pressure maps using EcoFoot 4.0 software. (A), (B), and (C) zones in (**B**) refer to forefoot, midfoot, and rearfoot zones.

2.6. *Handgrip Strength Assessment Protocol*

Maximal HGS was measured on both hands using a portable digital hand dynamometer (Takei 5401; Tokyo, Japan) with a precision of 0.1 kg. Participants stood upright during the assessment, with the test arm's shoulder adducted and the elbow flexed at a 90° angle. The forearm and wrist were maintained in a neutral position to maintain proper alignment of the hand with the forearm during grip assessment. The dynamometer was tailored to each participant's hand size to promote proper flexion of the metacarpophalangeal joints. Before testing, participants were provided with standardized verbal instructions and verbal encouragement was given throughout the procedure to guarantee maximal effort [2].

A total of three maximum voluntary contractions were performed per hand, with each trial lasting between 3 and 5 s. A 60 s rest interval was allowed between attempts to reduce the risk of fatigue. HGS values were expressed in kilograms (kg), and the highest value for the three trials for each hand was used for statistical analysis [47,48].

2.7. *Statistical Analysis*

All statistical analyses were conducted using the Statistical Package for Social Sciences (SPSS) software v.25 for Mac OS (IBM, Armonk, NY, USA). The assumptions of normality and homoscedasticity were assessed using the Shapiro–Wilk and Levene tests, respectively.

T-paired test, analysis of variance (ANOVA) with a Bonferroni post hoc test, or Kruskal–Wallis with multiple pairwise comparisons was used to determine the differences in anthropometric, baropodometric, and HGS measures between playing positions in volleyball. All data are presented as mean and standard deviation.

Pearson's and Spearman's correlation coefficients were used to analyze the relationship between dominant and non-dominant HGS and their corresponding side upper limb anthropometric variables. Based on the conventional approach to interpreting a correlation coefficient [49], coefficients are categorized as "negligible" (r = 0.00–0.10), "weak" (r = 0.10–0.39), "moderate" (r = 0.40–0.69), "strong" (r = 0.70–0.89), and "very strong" (0.90–1.00). Additionally, r-scores were used to identify multicollinearity and shared variance between the variables.

Effect sizes (ES) were reported to measure the magnitude of observed differences. For comparisons involving three or more independent groups, Epsilon Squared (ε^2) (derived from the Kruskal–Wallis test) and Eta Squared (η^2) (derived from the ANOVA test) were used to indicate the proportion of variance explained. Interpretation of ES was based on conventional thresholds: Epsilon Squared: small ($\varepsilon^2 < 0.01$), moderate ($\varepsilon^2 = 0.01$–0.06), and large ($\varepsilon^2 \geq 0.06$). Eta Squared: small ($\eta^2 = 0.01$), medium ($\eta^2 = 0.06$), and large ($\eta^2 = 0.14$) [50].

A stepwise multiple linear regression analysis was performed to identify the most relevant upper limb anthropometric variables associated with dominant HGS. To reduce the risk of overfitting given the sample size [51], the number of covariates was restricted to a maximum of nine, selecting those with the strongest correlation coefficients with HGS. The stepwise method followed entry and removal criteria based on significance levels ($p < 0.05$ and $p > 0.10$, respectively). Statistical significance was set at $p < 0.05$.

3. Results

Descriptive statistics for the sample, including anthropometric characteristics and HGS of both upper limbs in U-23 male volleyball players by playing position, are summarized in Table 2. One-way ANOVA revealed significant differences for player position for body height ($F_{(4, 87)} = 6.13$; $p < 0.001$) and body fat percentage ($F_{(4, 74)} = 2.76$; $p < 0.05$), with outsides, opposites, and middle blockers exhibiting the largest height and lower body fat percentage than liberos.

Table 2. Descriptive statistics of anthropometric and HGS variables in U-23 Colombian male volleyball players ($n = 92$).

Variable	Setter ($n = 15$) [1]	Outside ($n = 21$) [2]	Opposite ($n = 22$) [3]	Middle Blocker ($n = 19$) [4]	Libero ($n = 15$) [5]	Effect Size
Body weight (Kg)	74.03 (11.38)	76.42 (8.07)	76.84 (8.04)	77.34 (11.57)	73.66 (12.40)	0.22
Height (cm)	181.10 (10.28) [4]	183.82 (5.75)	185.30 (7.86)	189.43 (7.58)	174.48 (6.84) [2,3,4]	
Body fat (%)	8.27 (3.82)	10.29 (3.87)	7.95 (2.62)	7.66 (3.80)	11.28 (4.80)	0.13
Body water (%)	63.58 (2.96)	62.12 (2.61)	63.54 (2.48)	63.95 (5.00)	61.68 (2.84)	
Upper limbs anthropometric variables—Dominant side						
Upper arm length (cm)	34.58 (2.81)	35.89 (2.05)	35.51 (2.12)	36.81 (2.19) [1,3]	33.16 (2.04)	0.17
Arm circumference (cm)	29.58 (2.48)	30.14 (2.54)	29.58 (2.19)	29.54 (2.80)	30.42 (3.04)	
Forearm length (cm) †	25.66 (1.30)	27.11 (1.28)	26.68 (2.88)	28.37 (1.42) [1,3]	25.58 (1.24)	0.20
Elbow diameter (cm)	6.86 (0.40)	6.94 (0.43)	6.98 (0.35)	7.08 (0.29)	6.75 (0.26)	
Wrist diameter (cm) †	5.65 (0.41)	5.73 (0.27)	6.01 (0.77) [5]	5.85 (0.36) [5]	5.51 (0.19)	0.16
Hand length (cm) †	19.70 (1.13)	19.22 (4.30)	20.13 (1.07) [5]	20.93 (1.19) [5]	19.00 (0.64)	0.15
Hand breadth (cm)	8.34 (0.50)	8.59 (0.48) [5]	8.50 (0.56) [5]	8.73 (0.35) [5]	8.20 (0.32)	0.14
1–5 finger distance (cm) †	21.79 (1.42)	22.50 (1.49)	22.19 (1.21)	22.69 (1.81) [5]	21.08 (1.14)	0.08
Upper arm muscle area (cm^2)	58.01 (11.39)	59.37 (9.51)	59.93 (10.05)	58.99 (10.09)	61.10 (9.78)	
Upper arm muscle circumference (cm) †	26.88 (2.48)	27.2 (2.17)	27.3 (2.32)	27.1 (23.10)	27.6 (21.74)	
Upper arm area (cm^2)	70.12 (12.05)	72.80 (12.09)	70.02 (10.21)	70.07 (13.24)	74.37 (15.29)	
Handgrip strength (kgf)	41.39 (6.82) [3]	44.87 (7.32)	48.65 (8.58)	47.10 (5.05)	43.81 (4.86)	0.12
Upper limbs anthropometric variables—Non-dominant side						
Upper arm length (cm)	34.33 (2.80)	35.58 (1.63)	35.60 (1.93)	36.34 (2.20)	32.96 (2.13) [1–4]	0.19
Arm circumference (cm)	29.46 (2.51)	30.07 (2.58)	29.44 (2.25)	28.90 (2.90)	30.00 (3.00)	
Forearm length (cm) †	25.54 (1.86) [2–4]	26.77 (1.46)	28.26 (4.05)	28.12 (1.76)	25.08 (1.58) [2–4]	0.23
Elbow diameter (cm)	6.75 (0.64)	6.95 (0.47)	7.09 (0.26)	7.11 (2.78)	6.77 (0.28)	
Wrist diameter (cm) †	5.71 (0.84)	5.69 (0.27)	5.73 (0.34)	5.73 (0.39)	5.33 (0.14) [1–4]	0.18
Hand length (cm) †	19.58 (1.01)	20.39 (1.05)	20.47 (1.54) [5]	20.97 (1.21)	19.25 (1.38)	0.09
Hand breadth (cm)	8.30 (0.41)	8.52 (0.50)	8.50 (0.44) [5]	8.63 (0.33)	8.05 (0.30)	0.11
1–5 finger distance (cm) †	21.91 (1.37)	23.25 (1.94)	21.93 (5.22)	23.03 (1.84)	20.91 (1.25) [2–4]	0.15
Upper arm muscle area (cm^2)	57.51 (11.56)	59.23 (9.48)	59.61 (10.29)	55.99 (9.78)	58.75 (9.88)	
Upper arm muscle circumference (cm)	26.88 (2.48)	27.19 (2.17)	27.26 (2.48)	26.42 (2.29)	27.07 (2.27)	
Upper arm area (cm^2)	69.55 (12.17)	72.50 (12.25)	69.40 (10.68)	67.12 (13.42)	72.35 (14.69)	
Handgrip strength (kgf)	43.03 (8.17)	45.52 (6.64)	49.42 (8.70)	48.61 (6.14)	44.13 (4.46)	

Significantly different from the following: [1]—setter; [2]—outside; [3]—opposite; [4]—middle blocker; [5]—libero; † differences were assessed with Kruskal–Wallis test. All data are presented as mean (standard deviation).

Similarly, the analysis also showed significant differences on the dominant side for several variables. For HGS ($F_{(4, 87)} = 2.93$; $p < 0.05$), opposites outperformed setters. For arm length ($F_{(4, 87)} = 4.53$; $p < 0.01$) and forearm length ($\chi^2_{(4)} = 22.65$; $p < 0.01$), middle blockers outdid setters and liberos in both variables. Significant differences were also found for wrist diameter ($\chi^2_{(4)} = 18.59$; $p < 0.01$) and hand length ($\chi^2_{(4)} = 18.02$; $p < 0.01$), where opposites and middle blockers outperformed liberos. Regarding hand breadth ($F_{(4, 87)} = 3.66$; $p < 0.01$), outsides, middle blockers, and opposites exhibited greater values than those of the liberos. Finally, for the 1–5 finger distance ($\chi^2_{(4)} = 11.40$; $p < 0.05$), middle blockers outperformed liberos.

On the non-dominant side, significant differences were found for several anthropometric variables. Arm length ($F_{(4,87)} = 5.08$; $p < 0.01$) and wrist diameter ($\chi^2_{(4)} = 20.06$; $p < 0.01$) were greater in opposites, outsides, and middle blockers compared to liberos. Differences were also observed in forearm length ($\chi^2_{(4)} = 24.54$; $p < 0.01$), with opposites, outsides, and middle blockers outperforming both liberos and setters. In hand length ($\chi^2_{(4)} = 12.94$; $p < 0.05$) and hand breadth ($F_{(4, 87)} = 2.72$; $p < 0.05$), middle blockers and opposites outperformed liberos, respectively. Lastly, the 1–5 finger distance was significantly larger ($\chi^2_{(4)} = 17.67$; $p < 0.01$) in opposites, outsides, and middle blockers than in liberos.

Analysis of load percentage distribution, which refers to which foot each player carries the most weight, showed no differences between feet ($t_{(65)} = 1.07$, $p > 0.05$) (Figure 3A) or between positions in the non-dominant foot ($F_{(4, 61)} = 0.115$; $p > 0.05$) or the dominant foot ($F_{(4, 61)} = 0.113$; $p > 0.05$) (Figure 3B). After that, we analyzed the behavior of peak pressure on the feet. T-paired test revealed higher levels on the non-dominant foot ($t_{(65)} = 2.67$, $p < 0.01$) (Figure 3C). Then, when we analyzed peak pressure levels according to the zone on foot, which revealed greater levels in the forefoot and rearfoot zones of the non-dominant ($F_{(2, 195)} = 142.17$; $p < 0.01$) and dominant ($F_{(2, 195)} = 135.20$; $p < 0.01$) feet than in the midfoot zone. Likewise, the rearfoot zone exhibited higher pressure levels than those found in the forefoot zone ($p < 0.01$) in both feet, as shown in Figure 3D. Finally, we also analyzed the peak pressure levels according to the position of players, with no differences detected in either foot.

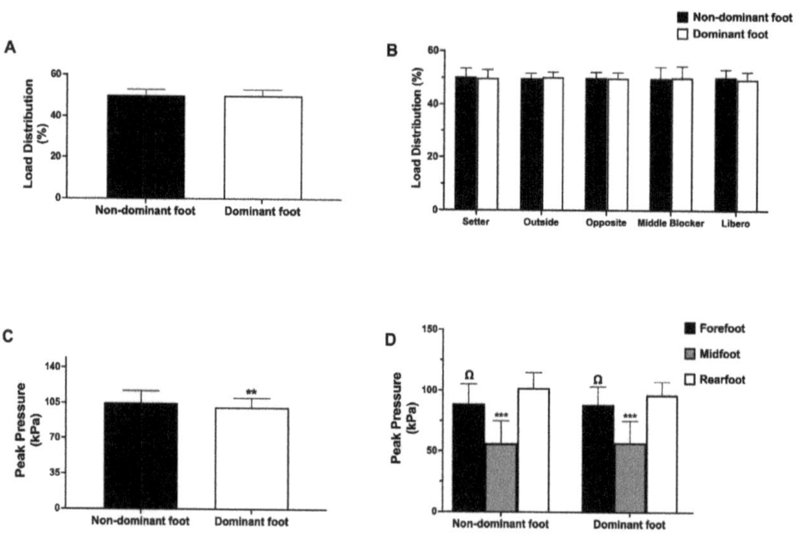

$^{**}p < 0.01$ from the non-dominant foot
$^{\Omega/***}$ denotes $p < 0.01$/ $p < 0.001$ from the rearfoot

Figure 3. Analysis of load percentage and peak pressure distribution. (**A**) Distribution of load percentage between the feet. (**B**) Load percentage distribution by position. (**C**) Distribution of peak pressure between the feet. (**D**) Distribution of peak pressure according to foot zones. Statistical significance was assessed using a T-paired test and one-way ANOVA for (**A**), (**C**), and (**B**), (**D**), respectively.

Additionally, the analysis of other baropodometric parameters, including the ellipse area, calcaneus angle, and total foot surface, showed no statistically significant differences in the whole sample, even when the between-group position factor was considered (Table S1).

Table 3 presents the correlations between selected upper limb anthropometric variables measured in the study with dominant and non-dominant HGS in U-23 male volleyball players. We found moderate and weak positive correlations between most anthropometric variables and HGS on both sides. Furthermore, no significant correlations were identified with the hand shape index for the dominant ($r_{(91)} = 0.09$; $p > 0.05$) and non-dominant side ($r_{(91)} = -0.02$; $p > 0.05$).

Table 3. Pearson's and Spearman correlation coefficients between all anthropometric variables and HGS in U-23 male volleyball players.

Variables	Dominant HGS	Non-Dominant HGS
Height (cm)	0.33 **	0.43 ***
Upper arm length (cm)	0.30 **	0.47 ***
Arm circumference (cm)	0.33 **	0.41 ***
Forearm length (cm)	0.37 ***	0.43 ***
Elbow diameter (cm)	0.24 *,†	0.37 ***
Wrist diameter (cm)	0.42 ***	0.42 ***
Hand length (cm)	0.44 ***	0.47 ***
Hand breadth (cm)	0.47 ***,†	0.60 ***
1–5 finger distance (cm)	0.29 **,†	0.36 ***
Upper arm muscle area (cm)	0.40 ***	0.49 ***
Upper arm muscle circumference (cm)	0.40 ***	0.48 ***
Upper arm area (cm^2)	0.33 **	0.42 ***

*/**/***/ denotes $p < 0.05/p < 0.01/p < 0.001$; † denotes Spearman correlations. All the anthropometric variables correspond to the same side as the HGS measurement (dominant or non-dominant).

Finally, a multivariate analysis was conducted to evaluate whether selected anthropometric parameters could explain dominant HGS (Table 4). Based on the stepwise multiple linear regression applied to the nine variables with the strongest correlation with dominant HGS (Table 3), three variables were retained in the final model. Specifically, the analysis showed that dominant hand breadth ($\beta = 3.42$, 95% CI [0.43, 6.40]), dominant upper arm muscle area ($\beta = 0.157$, 95% CI [0.02, 0.29]), and dominant wrist diameter ($\beta = 3.59$, 95% CI [0.49, 6.68]) entered the regression equation and explained 27.5% ($R^2 = 0.30$; adjusted $R^2 = 0.27$) of the variance in dominant HGS ($F_{(3,88)} = 12.53$; $p < 0.001$).

Table 4. Stepwise multiple regression analysis between dominant HGS-associated variables in U-23 volleyball players.

Independent Variables	β	t	p	95% CI	VIF
Dominant hand breadth (cm)	3.42	2.27	0.025	[0.43, 6.40]	1.43
Dominant upper arm muscle area (cm^2)	0.157	2.32	0.023	[0.02, 0.29]	1.22
Dominant wrist diameter (cm)	3.59	2.30	0.023	[0.49, 6.68]	1.31

VIF: Variance Inflation Factor.

4. Discussion

This cross-sectional study provides a comprehensive analysis of the morphological characteristics, baropodometric variables, and HGS performance of U-23 male Colombian volleyball players. Notably, significant differences ($p < 0.05$) were observed in all measurements of lengths and breadths of both the dominant and non-dominant upper limbs among players, according to their position. Furthermore, an association was found between all hand, forearm, and arm dimensions with HGS on both sides, identifying selected upper limb dimensions as moderately associated variables of HGS. Significant differences ($p < 0.05$) were also found in some baropodometric measures between feet and across different foot regions in both feet.

Anthropometric characteristics have been extensively investigated in volleyball, with most studies focusing on describing players' somatotypes and body fat percentage. However, specific upper limb dimensions, such as the lengths, diameters, and breadth of arms, forearms, and hands, have received less attention [2].

Research has shown position-specific descriptions of anthropometric and physical performance in volleyball players [19,52]. Thus, middle blockers have been described as the tallest, most ectomorphic, least mesomorphic, and endomorphic. At the same time, liberos are more likely to be shorter, less ectomorphic, more mesomorphic, and endomorphic than players in other positions [53]. In this sport, certain anthropometric and morphological characteristics (e.g., height, weight, body composition, arm, forearm, and hand dimensions) are essential for success during the game. For instance, serves, blocks, and spikes are performed more efficiently when players possess larger upper limb dimensions [53].

Results of the present study showed that significant differences exist among volleyball players of different playing positions. Therefore, opposites, middle blockers, and outsides were taller, had a lower body fat percentage than liberos, and exhibited higher values in most of the selected upper limb variables (upper arm length, forearm length, wrist diameter, hand length, and hand breadth) than the liberos and setters (Table 2). Similar results to those reported in our study have been found by other researchers [19,52,54]. For instance, the latest findings of Milić et al. (2024) [19] on female volleyball players described liberos and setters as the smaller players in the team. These findings were also reported by Toselli et al. (2018) [52] in a sample of elite male volleyball players. Researchers also observed great differences in upper limb dimensions (arm's length and humerus width) among the players of the different roles, with opposites and middle blockers outdoing setters and liberos, respectively.

Although height is a critical factor in volleyball success, as it enhances a player's ability to compete effectively over the net, liberos are not required to be taller. Moreover, height does not seem critical for them since they play in the back row and are not allowed to spike or block. Instead, having a low center of mass is critical for effectively handling low balls during landings, receiving, and defense, which suits shorter players [13]. Thus, the shorter height of libero players in this study is aligned with these roles in volleyball. Conversely, other positions (opposites, middle blockers, and outsides) require taller players with larger upper limb lengths to be successful during spiking and blocking [55]. These anthropometric characteristics enable players to extend their reach, allowing them to contact the ball over the top of the opponent's block and from a higher point, thereby increasing the ball's downward trajectory and speed [56].

While a greater height provides better reach above the net, facilitating easier control of defensive and offensive actions, a broader wrist allows for more forceful hits during attacks. Additionally, longer arms have been related to better performance over the net in both attacking and defensive actions [55].

Liberos and setters are key volleyball positions that require agility, precision, and technical skills rather than power or extended reach. Liberos are defensive specialists whose primary responsibilities demand quick lateral movements and a low body position to respond to the ball effectively. On the contrary, extending upward or outward for blocks or spikes is not a function of this position [57,58]. To succeed in most of the actions, liberos must often maintain a low center of gravity, which could be supported, in part, by having shorter arm spans and smaller upper limb dimensions [19,59].

Setters, meanwhile, focus on ball distribution and coordination, requiring fast, accurate hand movements rather than the extended reach needed for offensive or defensive net play. Because decision-making is the main ability expected from athletes in this position [60], the setter plays an essential role in a volleyball team to effectively organize offensive strategies [61,62]. In this regard, having smaller upper limb dimensions could facilitate rapid arm motions, which are necessary to set the ball accurately under time constraints. Along with this, players in this position rarely spike during a match [63,64], so high levels of upper body power may also not be necessary.

Here, setters also showed the lowest values in the dominant HGS among the different positions, being statistically outperformed by the opposites (Table 2). As previously mentioned, players in this position do not rely heavily on upper limb strength as their roles emphasize ball control, agility, and decision-making. For that reason, it would be reasonable for these players to have lower HGS values compared to those in offensive positions. Despite several studies focusing on strength in male volleyball players, there is little published research evaluating upper limb strength. Consistent with the findings of our study, Toselli et al. (2018) [52] also observed lower HGS scores in setters compared to opposite players. Similarly, the González-Badillo group also described worse strength performance in setters compared with opposites and middle blockers in a 4RM bench press test [65].

Our findings can be partially explained by the geometric scaling paradigm [65,66], which has been successfully used to examine the influence of body size on athletic performance. According to this paradigm, strength is closely linked to the muscle cross-sectional area, which increases with increases in body height. This suggests that individuals with greater limb dimensions would be likely to excel in activities that require a strength component [65]. Thus, this theory is supported in the current study by the significant positive correlations observed between height and selected upper limb dimensions with the dominant HGS (Table 3).

In addition, hand breadth, upper arm muscle area, and wrist diameter were identified as anthropometric variables that explain 27% of the variance of HGS (Table 4). These findings are consistent with the existing literature, which underscores the critical role of larger hand dimensions and upper arm musculature in determining HGS, a key factor influencing various technical movements in volleyball [53,67]. Nevertheless, this association should be interpreted with caution due to the explained variance being modest, indicating that other factors, such as neuromuscular control or sport-specific strength, likely play an important role.

Fallahi et al. (2011) [53] suggest that larger hand anthropometric variables lead to a reduced finger spread, thereby enhancing grasp efficiency and reducing fatigue during ball manipulation. Additionally, wrist diameter, which reflects bone structure and muscular support, may contribute to greater force application in repeated gripping actions. This biomechanical advantage allows players to exert better control over the ball during overhead passes and serves, improving precision and consistency [2,68]. Furthermore, the upper arm muscle area serves as an indirect marker of overall upper limb strength. This characteristic is essential for executing blocks and spikes with maximal power while also enabling athletes to withstand high-impact forces [69].

Building upon this evidence, these parameters could be valuable for identifying talent and guiding coaching strategies in volleyball. Integrating HGS-related parameters into scouting and training processes may provide a more objective approach to assessing players' physical potential and skill development.

Regarding the baropodometric profile results, no significant differences were observed in the percentage of load distribution between the feet or across playing positions

(Figure 3A,B). Similarly, when analyzing balance-related variables such as the ellipse area, no positional differences were found (Table S1), which aligns with the results of load percentage distribution. This finding may be attributed to the fact that when load distribution is balanced, oscillations in the center of pressure (CoP) are minimized, resulting in a reduced ellipse area, which reflects greater postural efficiency [25,70]. In our study, the sample consisted of high-level players in each position, suggesting a relatively homogeneous postural control and stability.

Interestingly, when the entire sample was analyzed, significant differences in peak pressure were observed between feet, with the non-dominant foot experiencing higher pressure levels (Figure 3C). However, no significant differences were found when peak pressure was analyzed across playing positions. One possible explanation for this asymmetry could be variation in the plantar surface area among players, as pressure is inversely proportional to this variable [71]. Nevertheless, our analysis revealed no significant differences in plantar surface area between feet (Table S1), suggesting that other biomechanical or neuromuscular factors may contribute to this discrepancy, as has been suggested in previous biomechanical studies [72,73]. However, this hypothesis requires confirmation through dynamic or kinematic assessments.

As reported in the literature [71,74], volleyball-specific movements may contribute to plantar pressure asymmetry. The frequent jumps, lateral movements, and unbalanced landings could explain the higher pressure observed in the non-dominant foot, given its stabilizing role, particularly when setting up for a spike or block. This repeated exposure to sport-specific loading patterns may reinforce the asymmetrical pressure distribution during movements, as previously reported by our research group [46,75]. Moreover, our findings regarding pressure distribution across foot regions (Figure 3D) align with previous evidence indicating that, in a normal foot, pressure is typically distributed as 60% in the rearfoot, 30% in the forefoot, and 10% in the midfoot [76,77]. This distribution is supported by the fact that 80% of our sample exhibited this kind of foot, reinforcing the consistency between expected biomechanical pressure patterns and the observed data.

Finally, the identification of upper limb-associated anthropometric variables of HGS and the baropodometric profile across playing positions could be integrated into an athlete's screening protocol and talent scouting, serving as practical applications of these findings. Additionally, these results may contribute to the development of injury prevention strategies by identifying potential asymmetries in plantar pressure distribution. Such insights may be crucial for managing overuse injuries and biomechanical inefficiencies.

This study provides a comprehensive assessment of anthropometric, baropodometric, and HGS characteristics across playing positions in a homogeneous sample of high-level U-23 Colombian male volleyball players. Consequently, this research offers a broad and in-depth perspective on the physical attributes that contribute to volleyball success.

However, these results may not be directly applicable to younger players, female athletes, or recreational volleyball players, as the participants in the present study are highly trained individuals. The sample size of the study was influenced by the availability of participants in the competition, which may have limited the detection of subtle effects and affected the generalizability of our findings. Furthermore, the exclusive use of static baropodometric assessment limits the generalizability of postural control findings to dynamic, sport-specific contexts. Additionally, the modest associations observed between anthropometric variables and HGS should be interpreted with caution, as they indicate that other physiological or biomechanical factors likely contribute to strength performance. Future research should employ longitudinal studies to explore how these variables evolve with training and their impact on injury risk and performance over time.

5. Conclusions

This study provides the first characterization of anthropometric, HGS, and baropodometric parameters in U-23 Colombian male volleyball players, highlighting position-specific differences. Positional specific differences were observed in upper limb morphology, with opposites, outsides, and middle blockers showing greater segmental dimensions and HGS compared to setters and liberos. Selected upper limb parameters such as hand breadth, upper arm muscle area, and wrist diameter were moderately associated with HGS. Although baropodometric parameters did not differ by playing position, foot dominance and specific zones were linked to changes in load distribution and peak pressure.

These findings suggest that positional profiling in volleyball provides critical insights into the physical attributes necessary for optimizing performance and injury prevention. The identification of moderately associated anthropometric variables with HGS, along with baropodometric pressure patterns, might contribute to the development of position-specific training and screening strategies.

Supplementary Materials: The following supporting information can be downloaded at https://www.mdpi.com/article/10.3390/jfmk10020197/s1, Table S1: Descriptive statistics of baropodometric variables in Colombian male volleyball players.

Author Contributions: Conceptualization, A.D.l.R., M.A.C.-V., F.M.-D. and M.P.O.; methodology, A.D.l.R., M.A.C.-V. and M.P.O.; software, A.D.l.R. and M.A.C.-V.; validation, L.U.P., F.M.-D. and J.C.S.; formal analysis, A.D.l.R., M.A.C.-V. and L.U.P.; investigation, A.D.l.R., M.A.C.-V. and F.M.-D.; resources, A.D.l.R. and L.U.P.; data curation, J.C.S., F.M.-D. and L.U.P.; writing—original draft preparation, A.D.l.R., M.A.C.-V., F.M.-D. and L.U.P.; writing—review and editing, A.D.l.R., M.A.C.-V. and M.P.O.; visualization, M.A.C.-V. and F.M.-D.; supervision, A.D.l.R. and M.A.C.-V.; project administration, L.U.P. and M.P.O.; funding acquisition, L.U.P. and M.P.O. All authors have read and agreed to the published version of the manuscript.

Funding: This research received no external funding.

Institutional Review Board Statement: The study was conducted in accordance with the Declaration of Helsinki and approved by the Ethics Committee for Human Beings from the Unidades Tecnológicas de Santander (no. 0010-2022/2 May 2022).

Informed Consent Statement: Informed consent was obtained from all subjects involved in the study.

Data Availability Statement: The datasets used and/or analyzed during the current study are available from the corresponding author upon reasonable request.

Acknowledgments: The authors are thankful to the coaches and player participants.

Conflicts of Interest: The authors declare no conflicts of interest.

References

1. Khanna, A.; Koley, S. Comparison of Anthropometric Profile and Handgrip Strength between Inter-University Volleyball Players and a Reference Group. *Biomed. Hum. Kinet.* **2020**, *12*, 82–90. [CrossRef]
2. Camacho-Villa, M.A.; Hurtado-Alcoser, J.; Jerez, A.S.; Saavedra, J.C.; Paredes Prada, E.T.; Merchán, J.A.; Millan-Domingo, F.; Silva-Polanía, C.; De la Rosa, A. Handgrip Strength and Upper Limb Anthropometric Characteristics among Latin American Female Volleyball Players. *J. Funct. Morphol. Kinesiol.* **2024**, *9*, 168. [CrossRef] [PubMed]
3. Koley, S.; Kaur, S.P. Correlations of Handgrip Strength with Selected Hand-Arm-Anthropometric Variables in Indian Inter-University Female Volleyball Players. *Asian J. Sports Med.* **2011**, *2*, 220. [CrossRef]
4. Pizzigalli, L.; Cremasco, M.M.; Torre, A.L.; Rainoldi, A.; Benis, R. Hand Grip Strength and Anthropometric Characteristics in Italian Female National Basketball Teams. *J. Sports Med. Phys. Fit.* **2017**, *57*, 521–528. [CrossRef]
5. Obana, K.K.; Singh, P.; Namiri, N.K.; Levine, W.N.; Parsons, B.O.; Trofa, D.P.; Parisien, R.L. The Finger, Wrist, and Shoulder Are the Most Commonly Injured Areas in Youth Volleyball Players but the Incidence of Injuries Decreased Overall Between 2012 and 2022. *Arthrosc. Sports Med. Rehabil.* **2024**, *6*, 100862. [CrossRef]

6. Sattler, T.; Hadžic, V.; Derviševic, E.; Markovic, G. Vertical Jump Performance of Professional Male and Female Volleyball Players: Effects of Playing Position and Competition Level. *J. Strength Cond. Res.* **2015**, *29*, 1486–1493. [CrossRef]
7. Almashaqbeh, S.F.; Al-Momani, S.; Khader, A.; Qananwah, Q.; Marabeh, S.; Maabreh, R.; Al Badarneh, A.A.B.; Abdullah, K. The Effect of Gender and Arm Anatomical Position on the Hand Grip Strength and Fatigue Resistance during Sustained Maximal Handgrip Effort. *J. Biomed. Phys. Eng.* **2022**, *12*, 171. [CrossRef]
8. Subramani Manoharan, V.; Ganesh Sundaram, S.; Isaac Jason, J. Factors Affecting Hand Grip Strength And Its Evaluation: A Systemic Review. *Int. J. Physiother. Res.* **2015**, *3*, 1288–1293. [CrossRef]
9. Patel, D.J.; Bhatt, K.; Ghevariya, K.; Joshi, P. Comparison of Hand Grip Strength and Hand–Arm Anthropometric Measures among Volleyball and Basketball Players—A Cross Sectional Study. *Sport Sci. Health* **2024**, *21*, 407–414. [CrossRef]
10. ÖcalK, D. Evaluating the Relation between Dominant and Non-Dominant Hand Perimeters and Handgrip Strength of Basketball, Volleyball, Badminton and Handball Athletes. *Int. J. Environ. Sci. Educ.* **2016**, *11*, 3297–3309.
11. Carvalho, A.; Roriz, P.; Duarte, D. Comparison of Morphological Profiles and Performance Variables between Female Volleyball Players of the First and Second Division in Portugal. *J. Hum. Kinet.* **2020**, *71*, 109–117. [CrossRef] [PubMed]
12. Gaurav, V.; Sıngh, A. Anthropometric Characteristics of Indian Volleyball Players in Relation to Their Performance Level. *Turk. J. Sport Exerc.* **2014**, *16*, 87–89. [CrossRef]
13. Milić, M.; Grgantov, Z.; Chamari, K.; Ardigò, L.P.; Bianco, A.; Padulo, J. Anthropometric and Physical Characteristics Allow Differentiation of Young Female Volleyball Players According to Playing Position and Level of Expertise. *Biol. Sport* **2016**, *34*, 19. [CrossRef]
14. Wiliam, C.; Hamlet, B.; Sofia, L.; Yanel, D.; Miriam, M.; Ivis, E.; María, C.; Noemí, S. Kinanthropometric Profile of Cuban Women Olympic Volleyball Champions. *MEDICC Rev.* **2012**, *14*, 16. [CrossRef]
15. Mielgo-Ayuso, J.; Calleja-González, J.; Clemente-Suárez, V.J.; Zourdos, M.C. Influence of Anthropometric Profile on Physical Performance in Elite Female Volleyballers in Relation to Playing Position. *Nutr. Hosp.* **2015**, *31*, 849–857. [CrossRef]
16. Hadzic, V.; Sattler, T.; Topole, E.; Jarnovic, Z.; Burger, H.; Dervisevic, E. Risk Factors for Ankle Sprain in Volleyball Players: A Preliminary Analysis. *Isokinet. Exerc. Sci.* **2009**, *17*, 155–160. [CrossRef]
17. Cronin, J.; Lawton, T.; Harris, N.; Kilding, A.; Mcmaster, D.T. A Brief Review of Handgrip Strength and Sport Performance. *J. Strength Cond. Res.* **2017**, *31*, 3187–3217. [CrossRef]
18. Ángel Campo-Ramírez, M.; Enrique Hernández-Oñate, G.; Eliecer López-Salamanca, D.; Lucía Hincapié-Gallón, O.; Mosquera, W.; Marcela Paz Sánchez, G.; de Citar, F.; Ramírez, C.M.; Oñate, H.G.; de Salamanca, L.; et al. Caracterización Del Equilibrio Dinámico y La Tipología de Pie En Futbolistas Juveniles. *Salud UIS* **2022**, *54*, e:22030. [CrossRef]
19. Milić, M.; Vukadinović, N.; Grgantov, Z.; Nikolić, K.; Savić, Z.; Majstorović, N.; Nešić, G. Basic Anthropological Characteristics of Elite U17 Youth Serbian Female Volleyball Players According to Playing Position. *Appl. Sci.* **2024**, *14*, 11197. [CrossRef]
20. Pocek, S.; Milosevic, Z.; Lakicevic, N.; Pantelic-Babic, K.; Imbronjev, M.; Thomas, E.; Bianco, A.; Drid, P. Anthropometric Characteristics and Vertical Jump Abilities by Player Position and Performance Level of Junior Female Volleyball Players. *Int. J. Environ. Res. Public Health* **2021**, *18*, 8377. [CrossRef]
21. Farivar, M.; Harris, S.; Agana, A.; King, A.C. Lower Limb Cross-Over Effects on Postural Control: Impact of Proximal and Distal Muscle Fatigue. *Biomechanics* **2024**, *4*, 50–62. [CrossRef]
22. Kilic, O.; Maas, M.; Verhagen, E.; Zwerver, J.; Gouttebarge, V. Incidence, Aetiology and Prevention of Musculoskeletal Injuries in Volleyball: A Systematic Review of the Literature. *Eur. J. Sport Sci.* **2017**, *17*, 765–793. [CrossRef] [PubMed]
23. Guzmán-Muñoz, E.; Daigre-Prieto, M.; Soto-Santander, K.; Concha-Cisternas, Y.; Méndez-Rebolledo, G.; Sazo-Rodríguez, S.; Valdés-Badilla, P. Efectos de un entrenamiento neuromuscular sobre el control postural de voleibolistas universitarios con inestabilidad funcional de tobillo: Estudio piloto. *Arch. Med. Deporte* **2019**, *5*, 283–287.
24. Chan, L.Y.T.; Sim, Y.T.N.; Gan, F.K.; Bin Abd Razak, H.R. Effect of Chronic Ankle Instability on Lower Extremity Kinematics, Dynamic Postural Stability, and Muscle Activity during Unilateral Jump-Landing Tasks: A Systematic Review and Meta-Analysis. *Phys. Ther. Sport* **2022**, *55*, 176–188. [CrossRef]
25. Rosário, J.L.P. A Review of the Utilization of Baropodometry in Postural Assessment. *J. Bodyw. Mov. Ther.* **2014**, *18*, 215–219. [CrossRef]
26. Bunnell, E.; Stratton, M.T.; Bunnell, E.; Stratton, M.T. The Impact of Functional Training on Balance and Vestibular Function: A Narrative Review. *J. Funct. Morphol. Kinesiol.* **2024**, *9*, 251. [CrossRef]
27. Molinaro, L.; Taborri, J.; Rossi, S. Baropodometric Analysis in Different Feet Positions: Reliability and Repeatability Evaluation. In Proceedings of the 2021 IEEE International Workshop on Metrology for Industry 4.0 and IoT, MetroInd 4.0 and IoT, Rome, Italy, 7–9 June 2021; pp. 295–300. [CrossRef]
28. Lidiane Teles de Menezes, A.; Henrique Ferreira de Araujo Barbosa, P.; Souza Costa, A.; Castro Mundim, A.; Craveiro Ramos, G.; Cardoso dos Santos Couto Paz, C.; Fachin Martins, E. Baropodometric Technology Used to Analyze Types of Weight-Bearing during Hemiparetic Upright Position. *Fisioter. Mov.* **2012**, *25*, 583–594. [CrossRef]

29. Agostini, F.; de Sire, A.; Furcas, L.; Finamore, N.; Farì, G.; Giuliani, S.; Sveva, V.; Bernetti, A.; Paoloni, M.; Mangone, M. Postural Analysis Using Rasterstereography and Inertial Measurement Units in Volleyball Players: Different Roles as Indicators of Injury Predisposition. *Medicina* **2023**, *59*, 2102. [CrossRef]
30. Zarei, M.; Soltani, Z.; Hosseinzadeh, M. Effect of a Proprioceptive Balance Board Training Program on Functional and Neuromotor Performance in Volleyball Players Predisposed to Musculoskeletal Injuries. *Sport Sci. Health* **2022**, *18*, 975–982. [CrossRef]
31. Borozan, I.-S.; Grădinaru, S.; Miron, P.; Puta, T.; Bota, E. Postural Differences of Volleyball Players. *Timis. Phys. Educ. Rehabil. J.* **2016**, *9*, 42–46. [CrossRef]
32. Kumar, A.R.M. Pressure Analysis of the Dominant Foot and Non Dominant Foot of Volleyball Players. *Int. J. Res. Pedagog. Technol. Educ.* **2018**, *7*, 1–6. Available online: https://ijems.net/index.php/ijem/article/view/178 (accessed on 16 February 2025).
33. Hanley, B. The Role of the European U23 Championships in the Development of Elite Athletes. *New Stud. Athl.* **2014**, *5*, 41–56.
34. Fuchs, P.X.; Menzel, H.J.K.; Guidotti, F.; Bell, J.; von Duvillard, S.P.; Wagner, H. Spike Jump Biomechanics in Male versus Female Elite Volleyball Players. *J. Sports Sci.* **2019**, *37*, 2411–2419. [CrossRef]
35. Reeser, J.C.; Fleisig, G.S.; Bolt, B.; Ruan, M. Upper Limb Biomechanics During the Volleyball Serve and Spike. *Sports Health* **2010**, *2*, 368–374. [CrossRef]
36. Young, W.K.; Briner, W.; Dines, D.M. Epidemiology of Common Injuries in the Volleyball Athlete. *Curr. Rev. Musculoskelet. Med.* **2023**, *16*, 229–234. [CrossRef]
37. De, L.; Mendonça, M.; Alia, N.; Netto Bittencourt, F.; Lima Freire, R.; Campos, V.C.; Ferreira, T.V.; Silva, P.L. Hip External Rotation Isometric Torque for Soccer, Basketball, and Volleyball Athletes: Normative Data and Asymmetry Index. *Braz. J. Phys. Ther.* **2022**, *26*, 100391. [CrossRef]
38. Almeida, G.P.L.; Silveira, P.F.; Rosseto, N.P.; Barbosa, G.; Ejnisman, B.; Cohen, M. Glenohumeral Range of Motion in Handball Players with and without Throwing-Related Shoulder Pain. *J. Shoulder Elb. Surg.* **2013**, *22*, 602–607. [CrossRef]
39. Almeida, G.P.L.; Rodrigues, H.L.D.N.; De Freitas, B.W.; De Paula Lima, P.O. Reliability and Validity of the Hip Stability Isometric Test (HipSIT): A New Method to Assess Hip Posterolateral Muscle Strength. *J. Orthop. Sports Phys. Ther.* **2017**, *47*, 906–913. [CrossRef]
40. Koo, T.K.; Li, M.Y. A Guideline of Selecting and Reporting Intraclass Correlation Coefficients for Reliability Research. *J. Chiropr. Med.* **2016**, *15*, 155. [CrossRef]
41. Marfell-Jones, M.; Olds, T.; Stewart, A.; Carter, L. *International Standards for Anthropometric Assessment*; ISAK: Potchestroom, South Africa, 2006.
42. Quintero, A.M.; Fonseca, S.; Chagnaud, C.A.; De la Rosa, A. Comparison between Plethysmography and Body Fat Equations in Elite Taekwondo Athletes. *Ido Mov. Cult. J. Martial Arts Anthropol.* **2022**, *22*, 14–22. [CrossRef]
43. Uhrová, P.; Beňuš, R.; Masnicová, S.; Obertová, Z.; Kramárová, D.; Kyselicová, K.; Dörnhöferová, M.; Bodoriková, S.; Neščáková, E. Estimation of Stature Using Hand and Foot Dimensions in Slovak Adults. *Leg. Med.* **2015**, *17*, 92–97. [CrossRef] [PubMed]
44. Frisancho, R. New Norms of Upper Limb Fat and Muscle Areas for Assessment of Nutritional Status. *Am. J. Clin. Nutr.* **1981**, *11*, 2540–2545. [CrossRef] [PubMed]
45. Heymsfield, S.B.; McManus, C.; Smith, J.; Stevens, V.; Nixon, D.W. Anthropometric Measurement of Muscle Mass: Revised Equations for Calculating Bone-Free Arm Muscle Area. *Am. J. Clin. Nutr.* **1982**, *36*, 680–690. [CrossRef]
46. Monterrosa Quintero, A.; Fuentes-Garcia, J.P.; Poblete-Valderrama, F.; Pereira de Andrade, G.A.; De la Rosa, A. Body Composition, Power Muscle, and Baropodometric Assessment in Elite Muay Thai Athletes. *Ido Mov. Cult. J. Martial Arts Anthropol.* **2024**, *24*, 12–22. [CrossRef]
47. Mathiowetz, V.; Wiemer, D.M.; Federman, S.M. Grip and Pinch Strength: Norms for 6- to 19-Year-Olds. *Am. J. Occup. Ther.* **1986**, *40*, 705–711. [CrossRef]
48. Pereira, H.M.; De Oliveira Menacho, M.; Takahashi, R.H.; Cardoso, J.R.; Paulo -Sp, S. Handgrip Strength Evaluation on Tennis Players Using Different Recommendations. *Rev. Bras. Med. Esporte* **2011**, *17*, 184–188. [CrossRef]
49. Schober, P.; Schwarte, L.A. Correlation Coefficients: Appropriate Use and Interpretation. *Anesth. Analg.* **2018**, *126*, 1763–1768. [CrossRef]
50. Cohen, J. *Statistical Power Analysis for the Behavioral Sciences*; Routledge: New York, NY, USA, 2013. Available online: https://www.taylorfrancis.com/books/mono/10.4324/9780203771587/statistical-power-analysis-behavioral-sciences-jacob-cohen (accessed on 11 February 2025).
51. Groenwold, R.H.H.; Dekkers, O.M. Is It a Risk Factor, a Predictor, or Even Both? The Multiple Faces of Multivariable Regression Analysis. *Eur. J. Endocrinol.* **2023**, *188*, E1–E4. [CrossRef]
52. Toselli, S.; Campa, F. Anthropometry and Functional Movement Patterns in Elite Male Volleyball Players of Different Competitive Levels. *J. Strength Cond. Res.* **2018**, *32*, 2601–2611. [CrossRef]
53. Fallahi, A.A.; Jadidian, A.A. The Effect of Hand Dimensions, Hand Shape and Some Anthropometric Characteristics on Handgrip Strength in Male Grip Athletes and Non-Athletes. *J. Hum. Kinet.* **2011**, *29*, 151. [CrossRef]

54. Bekele, B.A.; Adane, A.K.; Kabiso, T.A. Anthropometric characteristics of volleyball players with respect to playing positions: In ethiopian female premier league. *Cent. Eur. J. Sport Sci. Med.* **2022**, *39*, 65–73. [CrossRef]
55. Pawar, N. Development Of Discriminant Model For Classifying The Volleyball Players On The Basis Of Selected Anthropometric Variables According To Their Playing Position. *Eur. J. Phys. Educ. Sport Sci.* **2018**, *5.*, 142–153. [CrossRef]
56. Tilp, M. The Biomechanics of Volleyball. In *Handbook of Sports Medicine and Science*; John Wiley & Sons: Newark, NJ, USA, 2017; pp. 29–37. [CrossRef]
57. Serge, E.; Vincent, L.; Kendall, J. The Role of the Libero in Volleyball as a Paradoxical Influence on the Game: Logical Debate and the Proposal for a Rule Change. *Sport J.* **2015**, *19*, 1–21.
58. Bobula, G.; Piech, J.; Płonka, A.; Król, P.; Czarny, W.; Pinto, R.; Pawlik, D.; Rydzik, Ł.; Bajorek, W. Evaluation of Lower Extremities Power, Movement, Position and Effectiveness in Volleyball. *Appl. Sci.* **2024**, *14*, 10065. [CrossRef]
59. Yumi, O.; Suzune, F.; Shinichi, D. Anthropometric Profile and Physical Fitness Characteristics in the Difference of Position of Japanese Female High-School Volleyball Players. *Asian Exerc. Sport Sci. Assoc.* **2022**, *6*, 2588–4832. [CrossRef]
60. Fortin-Guichard, D.; Laflamme, V.; Julien, A.S.; Trottier, C.; Grondin, S. Decision-Making and Dynamics of Eye Movements in Volleyball Experts. *Sci. Rep.* **2020**, *10*, 17288. [CrossRef]
61. Nikos, B.; Karolina, B.; Elissavet, N.M. Performance of Male and Female Setters and Attackers on Olympic-Level Volleyball Teams. *Int. J. Perform. Anal. Sport* **2009**, *9*, 141–148. [CrossRef]
62. Matias, C.J.A.D.S.; Greco, P.J. Offensive Organization Assessment of Winners of Brazilian Volleyball Superleague Setters. *Rev. Bras. De. Ciências Do Esporte* **2011**, *33*, 1007–1028. [CrossRef]
63. Mori, Y.; Yamada, Y.; Umezaki, S.; Kida, N.; Nomura, T.; Mori, Y.; Yamada, Y.; Umezaki, S.; Kida, N.; Nomura, T. A Study on the Number of Jumps and Jump Height in Volleyball: From a Mock Game of College Men Players. *Adv. Phys. Educ.* **2022**, *12*, 1–10. [CrossRef]
64. Lima, R.F.; Silva, A.F.; Matos, S.; de Oliveira Castro, H.; Rebelo, A.; Clemente, F.M.; Nobari, H. Using Inertial Measurement Units for Quantifying the Most Intense Jumping Movements Occurring in Professional Male Volleyball Players. *Sci. Rep.* **2023**, *13*, 5817. [CrossRef]
65. Marques, M.C.; Van Den Tillaar, R.; Gabbett, T.J.; Reis, V.M.; González-Badillo, J.J. Physical Fitness Qualities of Professional Volleyball Players: Determination of Positional Differences. *J. Strength Cond. Res.* **2009**, *23*, 1106–1111. [CrossRef] [PubMed]
66. Challis, J.H. The Relationship between Ball Mass and Throw Distance: Implications for Coaching Practice. *Int. J. Sports Sci. Coach.* **2024**, *19*, 1940–1949. [CrossRef]
67. Abe, T.; Counts, B.R.; Barnett, B.E.; Dankel, S.J.; Lee, K.; Loenneke, J.P. Associations between Handgrip Strength and Ultrasound-Measured Muscle Thickness of the Hand and Forearm in Young Men and Women. *Ultrasound Med. Biol.* **2015**, *41*, 2125–2130. [CrossRef]
68. Masanovic, B.; Vukasevic, V. Differences in Anthropometric Characteristics between Junior Handball and Volleyball Players. *J. Anthropol. Sport Phys. Educ.* **2020**, *4*, 9–14. [CrossRef]
69. Hadadnezhad, M.; Letafatkar, A.; Moradi, M.; Shah Saheb, M.S.; Norouzi, G.; Salehi, S. The Relationship Between Upper Extremity Functional Tests Scores With Strength, Proprioception, and Posture in Male Volleyball Players. *Phys. Treat.-Specif. Phys. Ther. J.* **2024**, *14*, 115–124. [CrossRef]
70. Álvarez Solano, C.; González Camacho, L.A.; Castaño Duque, S.P.; Cortés Velosa, T.; Vanoy Martin, J.A.; Chambrone, L. To Evaluate Whether There Is a Relationship between Occlusion and Body Posture as Delineated by a Stabilometric Platform: A Systematic Review. *CRANIO®* **2023**, *41*, 368–379. [CrossRef]
71. Wang, J.; Fu, W. Asymmetry between the Dominant and Non-Dominant Legs in the Lower Limb Biomechanics during Single-Leg Landings in Females. *Adv. Mech. Eng.* **2019**, *11*, 1687814019849794. [CrossRef]
72. Padrón Cabrera, L.; López-López, D.; Becerro-de-Bengoa-Vallejo, R.; Losa-Iglesias, M.; Gómez-Salgado, J.; Bayod-López, J. Implications of Plantar Static Pressure Distribution among Bilateral Flatfoot Patients: An Analytical Case-Control Research. *Biomed. Signal Process Control* **2024**, *96*, 106578. [CrossRef]
73. Wilzman, A.R.; Tenforde, A.S.; Troy, K.L.; Hunt, K.; Fogel, N.; Roche, M.D.; Kraus, E.; Trikha, R.; Delp, S.; Fredericson, M. Medical and Biomechanical Risk Factors for Incident Bone Stress Injury in Collegiate Runners: Can Plantar Pressure Predict Injury? *Orthop. J. Sports Med.* **2022**, *10*, 23259671221104790. [CrossRef]
74. Afonso, J.; Ramirez-Campillo, R.; Lima, R.F.; Laporta, L.; Paulo, A.; Castro, H.D.O.; Costa, G.D.C.T.; García-De-alcaraz, A.; Araújo, R.; Silva, A.F.; et al. Unilateral versus Bilateral Landing after Spike Jumps in Male and Female Volleyball: A Systematic Review. *Symmetry* **2021**, *13*, 1505. [CrossRef]
75. Monterrosa Quintero, A.; De La Rosa, A.; Arc-Chagnaud, C.; Quintero Gómez, J.M.; Pereira Moro, A.R. Morphology, Lower Limbs Performance and Baropodometric Characteristics of Elite Brazilian Jiu-Jitsu Athletes. *Ido Mov. Cult.* **2023**, *23*, 58–69. [CrossRef]

76. Cavanagh, P.R.; Rodgers, M.M.; Liboshi, A. Pressure Distribution under Symptom-Free Feet during Barefoot Standing. *Foot Ankle Int.* **1987**, *7*, 262–278. [CrossRef] [PubMed]
77. Alexander, I.J.; Chao, E.Y.S.; Johnson, K.A. The Assessment of Dynamic Foot-to-Ground Contact Forces and Plantar Pressure Distribution: A Review of the Evolution of Current Techniques and Clinical Applications. *Foot Ankle Int.* **1990**, *11*, 152–167. [CrossRef] [PubMed]

Journal of
Functional Morphology and Kinesiology

MDPI

Article

Anthropometric Characteristics and Body Composition Changes in a Five-Time Olympic Champion in Greco-Roman Wrestling: A Longitudinal Case Study Towards the Paris 2024 Olympic Games

Wiliam Carvajal-Veitía [1,2], Carlos Abraham Herrera-Amante [2,3,*], Rodrigo Yáñez-Sepúlveda [4], Vladimir Gainza-Pérez [5], Yanell Deturnell-Campos [5], Carlos Cristi-Montero [6], Guillermo Cortés-Roco [7] and César Octavio Ramos-García [2,3,*]

1 Subdirectorate of Teaching and Research, Institute of Sports Medicine (IMD), Havana 10800, Cuba; wiliamcarvajal790@gmail.com
2 Ibero-American Network of Researchers in Applied Anthropometry (RIBA2), 04120 Almería, Spain
3 Nutritional Assessment and Nutritional Care Laboratory (LECEN), Division of Health Sciences, Tonalá University Center, University of Guadalajara (UdeG), Tonalá 45425, Mexico
4 Faculty of Education and Social Sciences, Universidad Andres Bello, Viña del Mar 2520000, Chile; rodrigo.yanez.s@unab.cl
5 Subdirectorate of Medical Control, Institute of Sports Medicine (IMD), Havana 10800, Cuba; prefimed@gmail.com (V.G.-P.); yanell201074@gmail.com (Y.D.-C.)
6 IRyS Group, Physical Education School, Pontificia Universidad Católica de Valparaíso, Valparaíso 2530388, Chile; carlos.cristi.montero@gmail.com
7 Facultad de Ciencias de la Vida, Universidad Viña del Mar, Viña del Mar 2520000, Chile; guillermo.cortes@uvm.cl
* Correspondence: carlos.amante@academicos.udg.mx (C.A.H.-A.); cesar.ramos@cutonala.udg.mx (C.O.R.-G.)

Abstract: Purpose: This case study examines the anthropometric characteristics and body composition changes of a 41-year-old Cuban Greco-Roman 130 kg wrestler, a five-time Olympic gold medalist (2008–2024). To optimize his preparation for the Paris 2024 Olympic Games, another athlete participated in the qualifying process, allowing him to train without competition gear. **Methods**: The study monitored changes in body composition using anthropometry and bioelectrical impedance analysis (BIA) at three key time points in 2024: January, June, and July. The final assessment occurred 25 days before the Olympic event, coinciding with the final phase of his preparation. **Results**: The analysis revealed a significant reduction in total body mass, from 150 kg in January to 138.5 kg in July, with fat mass decreasing from 37.06 kg (24.11%) to 29.7 kg (21.5%). Muscle mass decreased slightly (77.41 kg to 72.3 kg), while bone mass remained stable. The somatotype classification was endomorphic–mesomorphic at all assessments, with slight shifts in its components (4.6–10.4–0.1 in January to 4.4–10.3–0.1 in July), reflecting an improved muscle–fat ratio. Notably, hydration levels and cellular integrity remained stable, as indicated by BIVA analysis. **Conclusions**: This study provides insight into the anthropometric characteristics and body composition of an elite Greco-Roman wrestler, as well as the changes observed during his preparation for his final Olympic participation. These data serve as a valuable reference for wrestlers and sports professionals, highlighting the physical profile of one of the most emblematic figures in Olympic history.

Keywords: wrestling; elite athlete; anthropometry; body composition; somatotype; sports nutritional sciences; athletic performance

1. Introduction

Greco-Roman wrestling is a weight-classified Olympic event that requires exceptional upper-body strength, motor control, and technical precision because all action is confined above the waist. Athletic performance relies on a combination of explosive, high-intensity efforts—such as throws and lifts—and sustained aerobic capacity to endure two physically demanding rounds [1–3]. A common and controversial practice in this sport is weight cutting, which involves a rapid reduction in body mass (BM) to meet the demands of competition [4,5]. Techniques such as dehydration, caloric restriction, and fluid deprivation pose significant physiological and psychological challenges, including glycogen depletion, hormonal imbalances (e.g., cortisol and testosterone), and increased psychological stress, which can impair strength, endurance, and recovery [6–14].

When body mass loss exceeds 5%, the negative impact on athletic performance becomes more pronounced [15,16]. In contrast, gradual weight management strategies, such as those recommended by the American College of Sports Medicine, aim to minimize these risks by preserving muscle mass and ensuring adequate hydration [17]. Accordingly, an assessment of body composition is critical in evaluating the effects of different weight management strategies on both performance and athlete health.

Wrestlers often begin reducing BM at least two weeks prior to competition [18–20]. However, longitudinal data documenting prolonged periods of significant BM reduction—greater than 10%—remain scarce. Notably, Kordi et al. [21] found that 5% of 198 Iranian wrestlers lost more than 10% of their BM prior to competition, while Zhong et al. [22] reported losses of up to 17.5% in martial artists within a 29-day window.

Although wrestling is one of the most studied sports in terms of weight management and BM loss [18,23–26], longitudinal studies evaluating the effects of prolonged BM reduction on anthropometric profiles, fractional body composition, and bioelectrical impedance vector analysis (BIVA) are still lacking, despite the widespread acceptance of these methods in applied sports science [27–30].

The need for such research is particularly pressing in the case of heavyweight wrestlers, whose unique physiological characteristics and competitive demands are underrepresented in the literature. This study addresses this gap by investigating the effects of an extended weight management intervention in an elite athlete: a five-time Olympic heavyweight Greco-Roman wrestler, aged 41, who returned to competition at the Paris 2024 Games after an extended hiatus following Tokyo 2021. Over an eight-month preparation period, the athlete achieved a body mass reduction of over 10% while maintaining an exceptional 31-year career in elite sport. Through a multidimensional approach—including whole-body, tissue, cellular, and molecular assessments—this study provides novel insights into the physiological adaptations associated with prolonged weight management in elite sport and offers practical implications for evidence-based coaching and health management of athletes in heavyweight wrestling. In addition, the study describes the anthropometric and morphofunctional characteristics of the athlete, including body composition, somatotype, and human proportionality scores, that make him unique and distinguish him from others in his category.

2. Materials and Methods

2.1. Athlete and Case Study Background

A case study in which the athlete was a Cuban Olympic wrestler, aged 41.4 decimal years, with a stature of 194.4 cm. Now retired from elite competition, he competed in the 130 kg Greco-Roman wrestling category at the Paris 2024 Olympic Games. He previously won gold medals at the 2008 Summer Olympics in Beijing, 2012 in London, 2016 in Rio de Janeiro, and 2020 in Tokyo.

The athlete took a break after the Tokyo Olympics until January 2024, when he began a six-month preparation period. He did not compete for the Olympic qualification in the 130 kg category for the Paris Games in 2024; instead, another athlete secured the qualification at the 2023 World Championships in Belgrade [31]. This strategy allowed the Athlete to prepare for the Paris 2024 Games without competitive demands, given his age.

The athlete has read, approved, and provided written consent for this publication, which meets the ethical standards for case studies. The procedure followed the ethical standards of the Institute of Sports Medicine (IMD) Ethics Committee on Human Experimentation (code: CEI-IMD-01-11-2023) and complied with the 2013 updated Declaration of Helsinki [32]. This case study adheres to the CARE (for CAse RE-ports) guidelines [33].

As part of the confidentiality protocols, data on body mass loss during the last 25 days before the competition, when 8.5 kg (6.1%) remained to reach the competition weight, will not be disclosed. Protection of this information is an integral part of the procedures established in the Kinanthropometry Laboratory of the Institute of Sports Medicine.

2.2. Nutritional Intervention Strategy

The body mass management strategy, designed according to the recommendations of Reale et al. [34], was structured in three progressive phases. The goal was to achieve controlled fat reduction, preserve muscle mass, and maintain cellular integrity to ensure that the athlete reached peak condition to compete in the Paris 2024 Olympic Games after a hiatus since Tokyo 2021.

The first phase, known as the initial stabilization phase, ran from January 9 to June 6, 2024, and aimed to restore the athlete's physical readiness without immediate competitive pressure. The 9 January and 6 June assessments used anthropometry and impedance methods to monitor the athlete's condition. This phase focused on a controlled energy balance with an adjusted macronutrient intake: protein (1.6–2 g/kg/day), carbohydrates distributed throughout training sessions (4–6 g/kg/day), and healthy fats accounting for 25–30% of total caloric intake. Hydration protocols were optimized to ensure cellular functionality and water balance stability.

The second phase, termed the active strategy phase, ran from 6 June to 11 July 2024, and implemented progressively aggressive nutritional strategies to induce significant body mass loss. Protein intake was increased to 2–2.2 g/kg/day, carbohydrates were adjusted to 3.5–5 g/kg/day, and fats were reduced to 20–25% of total caloric intake. A final assessment on July 11th evaluated the adjustments made, with hydration remaining a priority to maintain cell quality throughout the body mass loss process.

Although the third phase, the cutting phase, was not specifically evaluated, the athlete successfully reached the competition weight of 130 kg on August 5 and 6, following the same principles established by Reale et al. [34]. This phase followed progressive caloric deficit strategies, maintaining consistency with the nutritional and hydration framework of the previous phases to ensure readiness for competition.

The experimental design included training sessions at high performance camps in Bulgaria and Croatia, with regular evaluations at the Anthropometry Laboratory of the Institute of Sports Medicine in Havana, under the supervision of an ISAK-certified specialist. This meticulous approach to monitoring and progressive adjustments at each stage addressed a critical gap in the literature on weight management in combat sports. Figure 1 illustrates the complete flow of this experimental design, detailing the sequence and objectives of each phase.

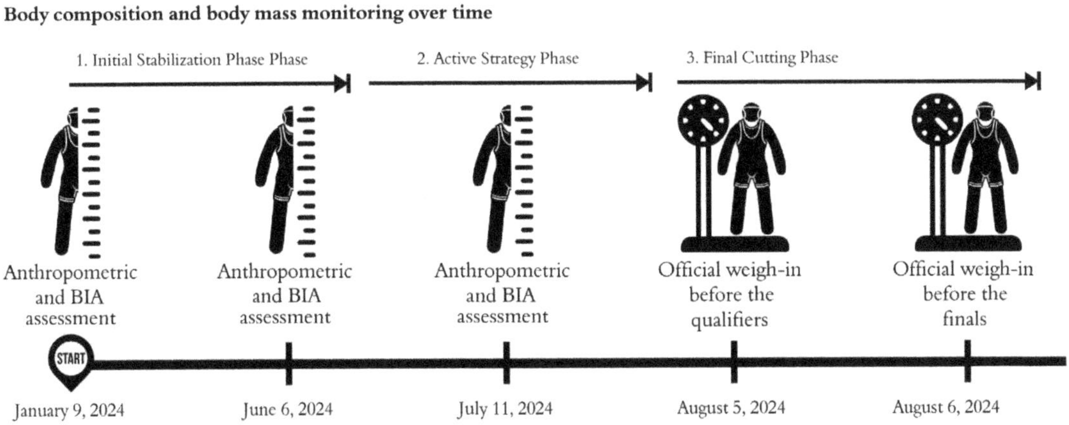

Figure 1. Flowchart of the experimental design for the weight management strategy in preparation for Paris 2024.

At the beginning of the athlete's preparation process (9 January 2024), his minimum wrestling weight (MWW) was certified at 132.8 kg. The projection for reaching the MWW was set by the American College of Sports Medicine [17]: a loss of 1.5% BM per week starting on 6 June (Figure 2).

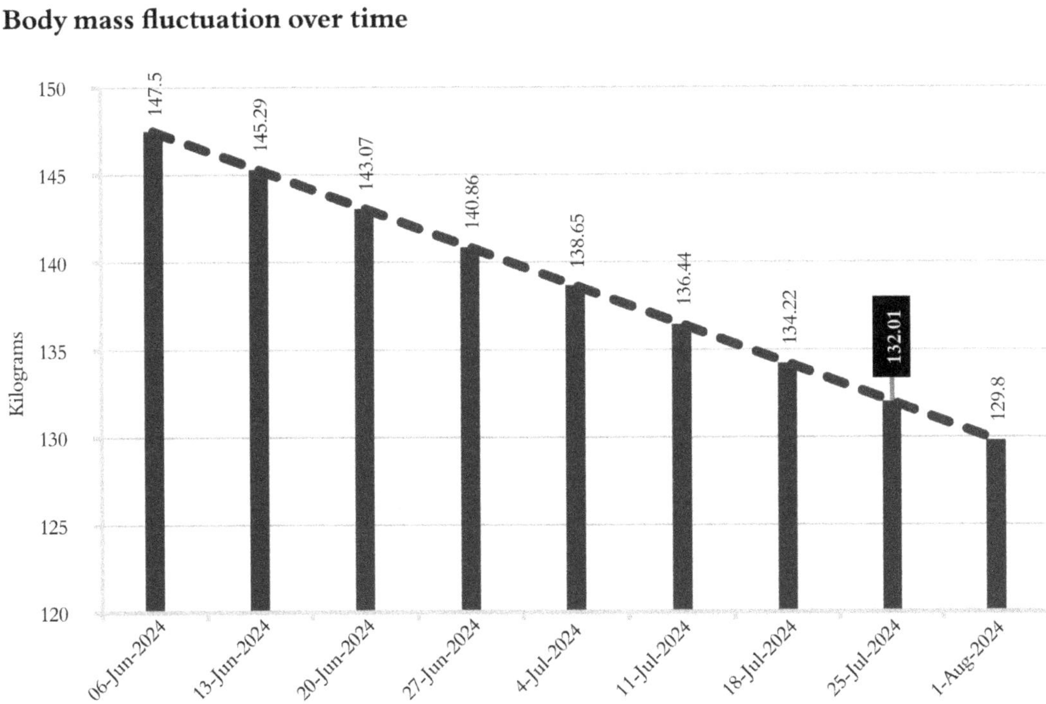

Figure 2. Estimated body mass on the measurement dates. The estimated body mass was reached on 25 July 2024.

2.3. Experimental Procedures, Measurements, and Data Analysis

Body composition was assessed using two different methods: (i) anthropometry and (ii) bioelectrical impedance analysis (BIA). All data from the developmental assessment were presented at three points prior to the Olympic event (January, June, and July). The final evaluation was performed 25 days before his participation in the Paris 2024 Olympic Games, coinciding with the final stage of his preparation.

2.4. Anthropometric Profile: Body Composition, Somatotype, Indices, and Phantom Proportionality

The assessments were performed by the same level III anthropometrist according to the protocols established by the International Society for the Advancement of Kinanthropometry (ISAK) standards [35]. During 2024, 27 anthropometric variables were measured based on the phantom stratagem [36], as shown in Figure 3. Analyses were performed using Holtain instruments (Holtain Ltd., Crymych, UK). The technical measurement error was less than 5% for skinfolds and less than 1% for other anthropometric variables. Body composition was determined using the five-fold fractionation method of body mass [27], which includes an estimation of fat mass. However, fat mass was also estimated using Lohman's equation [37], which has been validated for the calculation of minimum wrestling weight (MWW = fat-free mass/0.87) in wrestlers [17]. A reference value of 13% was used for MWW as this was the lowest fat percentage achieved by this athlete during his peak performance at the London 2012 Games. Somatotype was determined using the Heath–Carter anthropometric method [38].

2.5. Body Composition Assessed by BIA

Body composition was assessed by BIA as a complementary method using the Seca medical Body Composition Analyzer (mBCA) 214/215 (Seca GmbH and Co., KG, Hamburg, Germany). The indicators evaluated included fat mass and fat-free mass indices, phase angle, total body water, extracellular and intracellular water, resistance, reactance, and qualitative indicators such as bioelectrical impedance vector analysis (BIVA) and body composition chart (BCC).

The Seca mBCA device used in this study is considered a reliable instrument, having been validated against established reference methods, including the four-compartment model (4C), deuterium dilution (D2O), and sodium bromide dilution (NaBr). Empirical evidence has consistently demonstrated correlations greater than 95% for critical parameters such as fat-free mass and total body water, underscoring its accuracy and applicability in tracking anthropometric and compositional changes [39–41].

From a reliability perspective, the device exhibits minimal intra-device variability, ensuring consistent measurements when performed under standardized conditions [42]. In addition, the study protocol included an ISAK-certified operator, which minimized intra-operator variability and increased the reproducibility of results.

2.6. Analysis Approach

The anthropometric and bioimpedance results in this case study were expressed as percentage change over the measurement series. Bar charts, somatochart, and line graphs were used to illustrate trends in body mass fractionation, somatotype, and proportionality profile. Somatotype analysis included trend comparisons with two reference groups: world championship medalists and Olympic athletes in the same competitive category from Cuba.

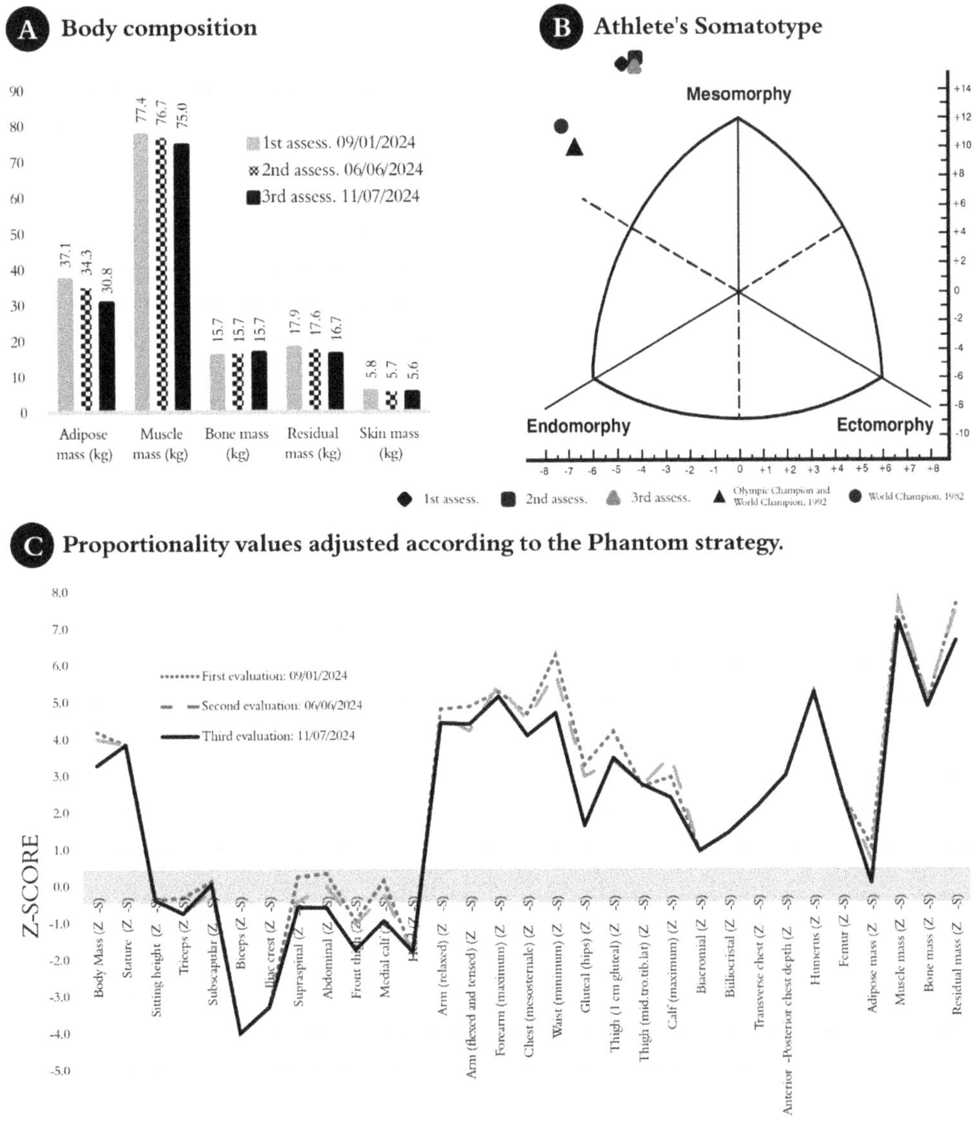

Figure 3. Evolution of body composition, somatotype, and human proportionality based on anthropometric variables (anthropometric profile) across three key evaluations conducted on 2024: 9 January, 6 June, and 11 July, prior to the Paris 2024 Olympic Games (charts (**A**)–(**C**)).

3. Results

3.1. Anthropometric Profile: Body Composition, Somatotype, Indexes, and Phantom Proportionality

The monitoring of the athlete's anthropometric profile is shown in Table 1. From the first to the last evaluation, absolute and relative values of body mass, fat mass, bone mass, body fat (Lohman), endomorphy, WHR, AMR, BMI, and ∑6SKF decreased more significantly than muscle mass, FFM (Lohman), mesomorphy, and ∑4CG. Residual mass, ectomorphy, bone mass, thorax, HBBI, HBiiL, and BBI indices remained unchanged.

Table 1. Body composition corrected girths and anthropometric indices over time.

Variables	Eval 1 (9 January 24)	Eval 2 (6 June 24)	Eval 3 (11 July 24)	Change 1→2 (%)	Change 2→3 (%)	Change 1→3 (%)
Body mass (kg)	150	147.5	138.5	−1.6	−6.1	−7.6
Body composition						
Adipose mass (kg)	37.1	34.3	30.8	−7.5	−10.2	−17.0
Adipose (%)	24.1	22.9	21.4	−5.0	−6.6	−11.2
Muscle mass (kg)	77.4	76.7	75	−0.9	−2.2	−3.1
Muscle (%)	50.3	51.1	52.2	1.6	2.2	3.8
Bone mass (kg)	15.7	15.7	15.7	0.0	0.0	0.0
Bone (%)	10.2	10.5	10.9	2.9	3.8	6.9
Residual mass (kg)	17.9	17.6	16.7	−1.7	−5.1	−6.7
Residual (%)	11.6	11.7	11.6	0.9	−0.9	0.0
Skin mass (kg)	5.8	5.7	5.6	−1.7	−1.8	−3.4
Skin (%)	3.8	3.8	3.9	0.0	2.6	2.6
Body Fat Lohman (kg)	34.4	32.2	27.0	−6.4	−16.1	−27.4
Body Fat Lohman (%)	22.2	21.8	19.5	−1.8	−10.6	−13.8
FFM Lohman (kg)	115.6	115.3	111.5	−0.25	−3.29	−3.54
Somatotype						
Endomorphy	4.9	4.3	4.3	−12.2	0.0	−12.2
Mesomorphy	10.4	10.4	10.1	0.0	−2.9	−2.9
Ectomorphy	0.1	0.1	0.1	0.0	0.0	0.0
X	−4.8	−4.2	−4.2	−12.5	0.0	−12.5
Y	15.9	16.4	15.8	3.1	−3.7	−0.6
Corrected girths (cm)						
Arm	38.5	38.7	38.2	0.5	−1.3	−0.8
Chest	121.6	121	118.2	−0.5	−2.3	−2.8
Thigh	77.2	73.6	76	−4.7	3.3	−1.6
Calf	42	44.2	42.4	5.2	−4.1	1.0
Indexes						
BMI (kg/m^2)	39.7	39	36.6	−1.8	−6.2	−7.8
AKS (g/cm^3)	1.55	1.57	1.52	1.3	−3.2	−1.9
WHR	0.6	0.6	0.5	0.0	−16.7	−16.7
HBBI	23.4	23.4	23.4	0.0	0.0	0.0
HBiiL	5.4	5.4	5.4	0.0	0.0	0.0
BBI	1.3	1.3	1.3	0.0	0.0	0.0
MBR	4.9	4.9	4.8	0.0	−2.0	−2.0
AMR	0.5	0.4	0.4	−20.0	0.0	−20.0
Cormic index	51.9	51.9	51.9	0.0	0.0	0.0
Thoracic index	68.2	68.2	68.2	0.0	0.0	0.0
Other						
∑6SKF (mm) †	128	115.2	99.6	−10.0	−13.5	−22.2
∑4CG (mm) ‡	316.5	315	311.8	−0.5	−1.0	−1.5

FFM, fat-free mass; BMI, body mass index; WHR, waist-to-height ratio; HBBI, height-to-biacromial breadth index; HBiiL, height-to-biiliocristal breadth index; BBI, biacromial–biiliocristal breadth index; MBR, muscle–bone ratio; AMR, adipose–muscular ratio; ∑6SKF, sum of 6 skinfolds; ∑4CG, sum of 4 correct girths. † Sum of triceps, subscapular, supraspinal, abdominal, front thigh, and medial calf skinfold thicknesses. ‡ Sum of correct girths: Arm, chest, thigh, and calf.

In the first 150 days, from January to July, the changes were less pronounced than those observed in the following 36 days, from 6 June to 24 July 2024, 25 days before the tournament.

With the most significant change in body mass, from 147.5 to 138.5 kg, the athlete lost 6.1% of BM, 10% of adipose tissue mass, and only 2.2% of muscle mass. Lipid mass decreased by 16.1%, whereas FFM decreased by only 3.29%. The sum of skinfold thicknesses (∑6SKF) changed by 13.5%, while corrected girths (∑4CG) decreased by only 1%.

3.2. Anthropometric Parameters: Body Composition, Somatotype, and Proportionality

Figure 3A shows a significant reduction in total body mass, from 150 kg in January to 138.5 kg in July. In terms of absolute body mass fractionation, this change was most pronounced in fat mass (37.06 kg to 29.7 kg), while muscle mass showed a less pronounced reduction (77.41 kg to 75.0 kg). Other compartments decreased slightly.

The athlete's somatotype was categorized as endomorphic mesomorph (4.9–10.4–0.1), indicating a pronounced musculoskeletal development over relative adiposity. By July, this profile had shifted to a lower endomorph (4.3–10.1–0.1), as shown in Table 1. Compared to World Championship medalists from the 1980s and 1990s, this athlete had greater mesomorphy and lower relative adiposity (Figure 3B).

Figure 3C shows the changes in body proportionality using the phantom strategy, expressed as z-scores. The most pronounced changes occurred in waist girth (minimum, Z-S), from 6.3 to 4.7. Skinfold thickness measurements such as triceps (Z-S) and subscapular (Z-S) also showed slight reductions, with triceps (Z-S) decreasing most significantly from −0.3 to −0.7. BM (Z-S) decreased from 4.2 in January to 3.3 in July. Fat mass (Z-S) decreased from 7.7 to 7.2, while muscle mass (Z-S) remained stable at high values (~5.0).

3.3. Body Composition Assessed by BIA

Table 2 provides a detailed summary of the evolution of body composition at the cellular and molecular levels using bioelectrical impedance analysis (BIA). At the cellular level, phase angle (PA) showed a slight decrease from 6.1° in January to 5.6° in June and July. The absolute increase in TBW was most pronounced from January to June, coinciding with greater increases in ICW and IWC/EWC than in ECW, resulting in a relative decrease in ECW/ICW. During the June–July rapid loss period, changes in TBW (0.0 L), extracellular water (+0.3 L), and intracellular water (−1.9 L) were not substantial. The resistance decreased continuously, but the reactance remained stable since June.

Table 2. Evolutionary assessment of body composition from Bioelectrical Impedance Analysis (BIA): Estimates of the different levels of body composition studied (molecular and cellular).

Variables	Eval 1 (9 January 4)	Eval 2 (6 June 4)	Eval 3 (11 July 24)	Change 1→2 (%)	Change 2→3 (%)	Change 1→3 (%)
Cellular level						
PA (°)	6.1	5.6	5.6	8.2	0	8.2
ECW (L)	31.8	34.3	34.4	7.9	0.3	8.2
ECW (%)	21	23.5	24.7	11.9	5.1	17.6
ICW (L)	39.1	47.8	46.9	22.3	−1.9	19.9
ICW (%)	29.3	31.6	34.1	7.8	7.9	16.4
ICW/ECW	1.22	1.39	1.36	13.9	−2.2	11.5
ECW/ICW	0.81	0.72	0.73	−0.11	0.01	−0.09
Xc (Ω)	40.9	32.4	32.4	−20.8	0	−20.8
Molecular level						
TBW (L)	70.9	82.1	81.3	15.8	−1	14.7
TBW (%)	50.3	50.3	58.8	0	16.9	16.9
R (Ω)	380.7	342.4	333.1	−10.1	−2.7	−11.5
FMI (kg/Ht^2)	12.68	9.85	7.59	−22.3	−22.9	−40.1
FFMI (kg/Ht^2)	27.01	29.19	27.45	8.1	6	−1.6

PA, phase angle; TBW, total body water; ECW, extracellular water; ICW, intracellular water; FMI, fat mass index; FFMI, fat-free mass index; Xc, reactance; R, resistance.

At the molecular level, the fat mass index (FMI) decreased significantly by 40.1%, while the fat-free mass index (FFMI) initially increased before decreasing (27.01 to 29.45 kg/Ht^2).

Figure 4A illustrates the evolution of the Bioelectrical Impedance Vector Analysis (BIVA), highlighting changes in the athlete's cellular and molecular state throughout the evaluated period. A high proportion of body water relative to cell count was observed from the first measurement. Figure 4B shows an FFMI above the population average, with migration reflecting a decrease in FMI, while FFMI remained stable.

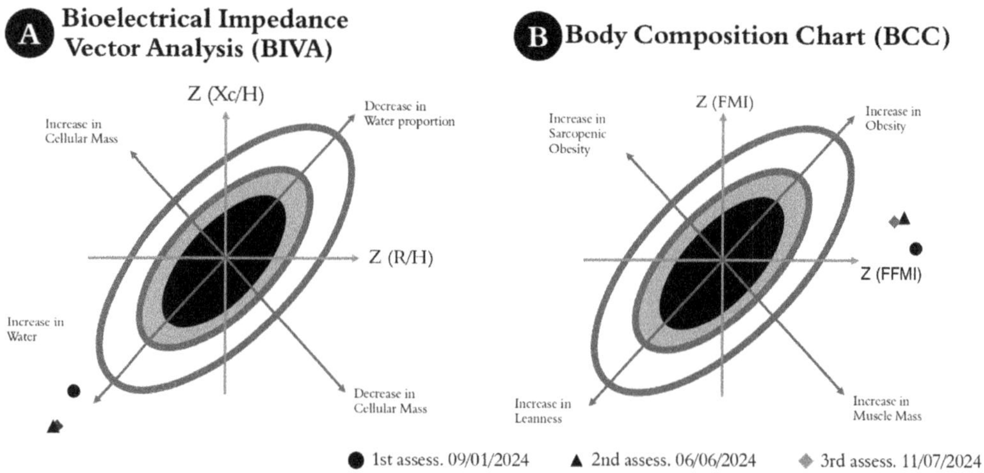

Figure 4. Analysis of the evolution of the bioelectrical impedance vector (BIVA) and body composition (BCC) in three key evaluations conducted in 2024, prior to the athlete's participation in the Paris 2024 Olympic Games (**A**,**B**). Dark markers represent the athlete's body composition at the cellular and molecular levels: black circles for January, black triangles for June, and gray diamonds for July 2024.

4. Discussion

4.1. Anthropometric Characteristics of Anthropometric Profile

This study represents a significant advancement in sports science as no previous research has specifically characterized the anthropometric profile of wrestlers competing in the 130 kg category across wrestling styles. Most existing studies tend to group super-heavyweight wrestlers with lighter categories or exclude them altogether to avoid increasing morphological variability, thus failing to account for their distinct structural and functional characteristics [43–45]. Consequently, the findings presented in this study offer a new perspective on performance analysis and talent identification in Greco-Roman wrestling.

Body composition and somatotype are fundamental pillars in determining competitive success in this discipline [46–48]. The elite wrestler analyzed in this study exhibits strong mesomorphic dominance, characterized by a high fat-free mass (FFM) and optimized structural proportions, establishing a new benchmark for evaluating athletes in the super-heavyweight category.

The analyzed wrestler showed superior values compared to the Cuban average for the same competitive category in several anthropometric parameters [48]. He demonstrated notable advantages in stature (194.4 cm vs. 188.0 cm), mesomorphy (10.1 vs. 7.0), the sum of skinfold thickness (99.6 mm vs. 79.6 mm), and body fat percentage (19.5% vs. 11.7%), reinforcing his structural profile optimized for performance. However, in terms of active body substance index (1.52 vs. 1.59 g/cm^3), the wrestler showed slightly lower values than the category average, which could indicate differences in the distribution of functional mass. In terms of somatotype, the subject was classified as endomorphic–mesomorphic (4.3–10.1–0.1), similar to the Cuban average for the 130 kg category (3.0–7.0–2.4), but with a greater predominance of endomorphy and mesomorphy. This suggests an evolutionary trend towards a more compact and structurally efficient morphology, adapted to the increasing physical demands of Greco-Roman wrestling.

In terms of somatotype evolution, the current Cuban champion shows a marked difference from previous Cuban medalists, as shown in Figure 3B. He is significantly more mesomorphic than a world champion who competed in the 120+ kg category in the 1980s and significantly more mesomorphic than another Olympic and world medalist who competed in the 100 and 120 kg categories in the late 1990s. This trend suggests that the Cuban 130 kg wrestler has evolved toward a more compact somatotype, likely as an adaptive response to the increasing physical demands of the sport.

While the primary focus of this study is on the individual anthropometric characteristics of the wrestler, a comparison with Azerbaijani wrestlers provides additional context for evaluating his physical advantages. Rahmani et al. [47] reported that Azerbaijani wrestlers in this category had an average stature of 183.83 cm, a BMI of 32.9 kg/m^2, a %BF of 14.8, an FFM of 102.86 kg, and somatotype values of 3.13–8.76–0.6. In contrast, the Cuban athlete exceeded these parameters with a stature of 194.4 cm, a BMI of 36.6 kg/m^2, an FFM of 111.5 kg, and somatotype values of 4.3–10.1–0.1. Due to methodological differences in body fat measurement, %BF values cannot be directly compared, but the marked variation in somatotype and FFM reinforces the structural advantages of the Cuban wrestler.

This study fills a critical gap by providing a focused examination of an elite competitor at the highest level of the sport. While direct comparisons remain limited, the reference to Azerbaijani wrestlers is highly relevant as Azerbaijan consistently produces world-class athletes in Greco-Roman wrestling, solidifying its status as a dominant force in the discipline.

Regarding anthropometric indices, although previous studies—such as the research by Škugor et al. [43] on Croatian wrestlers—have shown that anthropometric indices do not show significant differences between medalists and non-medalists, the results of this study should not be disregarded in talent selection. On the contrary, these indices remain essential for identifying structural patterns associated with performance outcomes, allowing coaches to make strategic decisions regarding athlete development.

An example of this can be seen by examining the proportionality profile between the Polish wrestlers and the subject of the study (Figure 3C). The femur width, humerus width, and flexed arm girth of the Cuban athlete reflected greater skeletal robustness and muscular development (Z-S > 0 and Z-S < 5). In contrast, Polish wrestlers had more balanced proportions across weight classes (Z-S from 1.76 to 2.69) [45].

When analyzing the proportionality profiles of the subscapular, triceps, biceps, and calf skinfolds, the Cuban athlete presented higher skinfold thickness values, indicating greater energy reserves (Z-S > −4 and Z-S < 0). Polish wrestlers, on the other hand, maintained lower values, emphasizing a leaner structure adapted to their performance strategies (Z-S from −1.82 to −2.19). These results confirm how super-heavyweight athletes develop specific anthropometric adaptations, differentiating the structural advantages of the Cuban athlete in biomechanical leverage from his Polish counterparts.

An innovative application within this study is the estimation of minimal wrestling weight (MWW) based on a 13% body fat (BF) threshold, which provides a more realistic reference for super-heavyweight wrestlers. Contrary to the 5% BF threshold recommended by the American College of Sports Medicine [17,49], which applies mainly to light- and medium-weight categories, long-term analyses of more than 500 combat sports athletes (wrestling, judo, boxing, and taekwondo) at the Cuban Institute of Sports Medicine (IMD) have shown that this lower threshold is unattainable for heavier athletes, even under optimal conditions.

Although a limitation of this research, the study lacks body composition data at competition weight as the athlete underwent an additional 8.5 kg reduction in the last 25 days prior to the event. However, his initial %BF of 19.5 provided sufficient physiological reserves to complete the body mass loss process without compromising muscle integrity. In contrast, in lower-weight-class wrestlers, FFM becomes the primary limiting factor during body mass loss. In such cases, athletes often experience significant muscle wasting during the acute body mass reduction phase, particularly in the days leading up to competition [43,50].

This finding suggests that super-heavyweight wrestlers may exhibit distinct patterns of body composition management, particularly in their ability to maintain higher initial fat percentages while maintaining functional muscle mass throughout the body mass loss process. Unlike their lighter counterparts, who often approach critical physiological thresholds in terms of fat-free mass availability, athletes in the 130 kg category appear to possess greater metabolic flexibility, allowing them to progressively reduce body mass while maintaining performance capacity. This adaptation may be essential for maintaining competitive readiness in the super-heavyweight divisions, where absolute strength, structural integrity, and endurance play a critical role in performance outcomes.

4.2. Analysis of Body Mass Monitoring

The process of body mass reduction in this elite athlete provides critical insights into the physiological and structural adaptations required for high-performance competition. One of the key findings of this study is the athlete's ability to achieve a controlled and balanced reduction of 6.1% body mass over 36 days, primarily through targeted fat loss while maintaining muscle mass and phase angle (PhA). These results are consistent with those of Reale et al. [37] and confirm that gradual and structured body mass loss strategies are essential in elite sports.

A key methodological innovation in this study was the use of non-traditional anthropometric techniques, such as corrected girth and five-way fractionation of body mass, which provided a more detailed perspective on composition shifts. In contrast to conventional methods, this approach allowed precise differentiation between reductions in adipose tissue and muscle mass, providing a deeper understanding of the physiological adaptations of the athlete during body mass loss.

Fluid balance played a fundamental role in this process. Previous studies using dilution techniques have linked fluctuations in the ICW/ECW ratio to cellular damage, inflammation, and dehydration [31]. In this case, however, the athlete exhibited acute hyperhydration due to increased fluid intake and optimized sodium levels during the rehydration phase. This strategy, combined with a controlled reduction in sodium intake in the days prior to competition, is consistent with the hydration protocols recommended by Reale et al. [37].

The stability of phase angle and reactance throughout this body mass loss process underscores the effectiveness of hydration and nutritional management in maintaining cellular integrity, confirming the findings of Campa et al. [29]. The absence of significant changes in intracellular and extracellular water ratios, bioelectrical impedance vector analysis (BIVA), and body cell mass further highlights the athlete's resilience to fluid balance shifts—in contrast to amateur boxers who often experience dehydration effects with similar body mass losses [28].

Regarding the relationship between phase angle, intracellular water (ICW) and ECW/ICW ratio, this study contradicts previous findings by Marini et al. [51] and Campa et al. [52]. Marini et al. [51] reported a positive correlation between PhA and ICW ($r = 0.327$, $p < 0.001$) and a negative correlation with ECW/ICW ($r = -0.493$, $p < 0.001$), whereas Campa et al. [52] found that PhA was positively correlated with TBW and ICW ($r = 0.458$, $p < 0.01$

and r = 0.564, $p < 0.01$, respectively) and negatively correlated with ECW/ICW (r = −0.436, $p < 0.01$). In contrast, ICW and TBW increased in this study, while PhA decreased from 6.1° to 5.6°, reversing previous trends.

Despite this discrepancy, the BIVA vector shift and the increase in the ICW/ECW ratio from 1.22 to 1.39 suggest improved cellular stability and metabolic efficiency. These observations are consistent with findings in Chinese and European athletic populations. Zhang et al. [53] demonstrated that ICW/ECW values between 1.35 and 1.45 were indicative of optimal cellular stability and performance, supporting the physiological adaptations observed in this study.

Similarly, Canda et al. [54] studied Spanish athletes and concluded that moderate fluctuations in PhA could be associated with hydration adaptations without compromising cellular integrity. Their research found a mean PhA of 7.3° in male athletes, whereas the PhA values recorded in this study ranged from 6.1° to 5.6°, suggesting a sport-specific adaptation to body mass loss rather than a decline in cellular function.

These findings support the hypothesis that super-heavyweight athletes may exhibit a differential regulation of cellular conductivity, consistent with previous observations by Marini et al. [51] and Campa et al. [52], where bioelectrical parameters varied according to body composition and body mass loss history. In this context, the shift of the BIVA vector towards higher total water content, together with the sustained increase in ICW, confirms that the fluid management strategy implemented in this athlete favored optimal fluid redistribution without compromising muscle function.

In terms of phase angle interpretation, values between 5° and 7° indicate good cell structure and membrane integrity, while lower values are associated with fluid accumulation and a loss of membrane integrity, and higher values are associated with dehydration [55]. The range of 6.1° to 5.6° recorded in this study falls within these physiological parameters. However, comparisons with reference values established by Campa et al. [56] for Italian athletes using the BIA 101, Akern (Florence, Italy), are not applicable to this Greco-Roman wrestler due to technological and methodological differences.

Although phase angle estimates do not require biological assumptions [30], research has shown that raw impedance parameters—resistance, reactance, and phase angle at 50 kHz—vary between devices due to technical factors [57]. As a result, data from this study can only be compared to results obtained using similar bioimpedance technologies, specifically those that incorporate mBCA formulas.

In addition to tracking body mass loss, phase angle monitoring has been shown to be a valuable tool in assessing somatic maturation over the course of a competitive season [51]. Additionally, PhA is positively correlated with vertical jump performance [58], a critical variable influencing the competitive success of Greco-Roman wrestlers [1,46].

In conclusion, it is important to emphasize that the case of this athlete demonstrates the benefits of structured preparation without exposure to competitive stress and body mass cycling, a strategy that may have optimized his performance and preserved his metabolic functionality. As emphasized by Lebron et al. [59] in their review on physiological disturbances in combat sports, repetitive cycles of body mass loss can lead to dysfunctions in insulin and leptin regulation, negatively affecting the energy efficiency of athletes. The training plan allowed this athlete to minimize cumulative metabolic stress, ensuring a more efficient adaptation during the Olympic qualifying period.

In addition, given his 41 years of decimal age and extensive competitive career, avoiding extreme body mass fluctuations may have reduced the impact of endocrine alterations that often affect aging athletes. Lebron et al. [59] emphasized that aggressive body mass reduction processes may impair post-event recovery and increase the risk of long-term metabolic dysfunction. The strategy adopted by the coaching staff in this case not only optimized physical preparation but also preserved metabolic homeostasis, reflecting a comprehensive approach grounded in both professional experience and scientific evidence to support the sustainability of long-term performance.

4.3. Limitations

One of the major limitations of this study is the lack of body composition data at competition weight as the athlete underwent an additional body mass reduction in the final days before the event. This prevents an accurate assessment of how the body mass reduction may have affected muscle integrity, hydration levels, and overall physiological adaptations. Additionally, while the anthropometric profile of a super-heavyweight wrestler has been thoroughly characterized, the findings are based on a single elite athlete, limiting generalizability to the broader population of wrestlers in this category. Methodological differences in body fat measurement between studies also challenge direct comparisons with international competitors. Despite these limitations, the study fills a critical gap by providing a reference framework for talent identification and performance optimization in the 130 kg Greco-Roman wrestling category.

Additionally, another limitation of this study is the lack of psychological assessment, despite the athlete's significant motivation in his pursuit of a fifth Olympic gold medal. The literature highlights that mental advantage, motivational factors, and psychological responses to body mass change are often overlooked in studies of wrestlers [60]. Given the significant mental and emotional demands placed on elite athletes, future research could explore the psychological component of body mass management in Greco-Roman wrestling to better understand its influence on performance.

4.4. Implications for Coaches and Practitioners

This study demonstrates the usefulness of non-traditional methods in the assessment and management of body mass in super-heavyweight wrestlers, highlighting anthropometric fractionation, corrected girths, and bioelectrical impedance vector analysis (BIVA) as essential tools for accurate athlete monitoring. Fractionation enables differentiation of body compartments, allowing for adjustment of body mass reduction without compromising functional muscle mass, while corrected girths facilitate the assessment of muscle structure in key areas for performance. In addition, BIVA provides insight into cellular stability and fluid balance, ensuring that body mass adjustments do not compromise metabolic function. For coaches, the integration of these techniques enhances athletic planning, enabling more precise training and nutritional adjustments while ensuring that the body mass loss process maintains the athlete's competitive ability.

One of the most valuable tools for coaches is the monitoring of phase angle (PhA) as an indicator of cellular integrity and hydration status. While PhA has been extensively studied in endurance and power sports [51,56,58], its application in combat sports remains underutilized. In this study, PhA remained relatively stable despite body mass loss, highlighting the effectiveness of proper hydration and nutritional protocols. Coaches should incorporate PhA tracking into athlete assessments to ensure that body mass loss does not compromise physiological resilience.

In addition, coaches can use intracellular and extracellular water (ICW/ECW) monitoring to assess hydration levels and fluid balance [28,52]. The maintenance of ECW/ICW ratios without significant changes in body composition suggests that careful hydration strategies can help super-heavyweight wrestlers avoid dehydration-induced performance declines. Ensuring proper sodium regulation and adjustment of fluid intake is critical in the final stages before competition [28].

This study also challenges conventional estimates of minimal wrestling weight (MWW) by demonstrating that 13% body fat is a more realistic threshold for super-heavyweight wrestlers, as opposed to the 5% BF suggested by the American College of Sports Medicine [17,49]. Given the unique physiological requirements of heavy and super-heavyweight combat athletes, coaches should reconsider rigid MWW standards and adapt them to practical observations and athlete-specific metrics.

Finally, the findings encourage coaches to take a holistic approach to body mass management, considering both physiological and psychological factors. Although this study did not directly assess mental resilience, the athlete's high motivation to achieve a fifth Olympic gold medal undoubtedly contributed to his ability to withstand the physical and strategic demands of body mass reduction. Future coaching strategies should integrate mental conditioning alongside physiological monitoring to ensure optimal preparation and peak performance in competition [60].

5. Conclusions

This study makes a valuable contribution to sport science by presenting a detailed anthropometric profile of an Olympic champion, a finding with direct implications for talent identification and athlete development in elite wrestling. Understanding the structural characteristics of a world-class competitor enables coaches and researchers to refine selection criteria and optimize training strategies aimed at achieving high performance. Moreover, the study underscores the practical value of non-traditional methods for assessing and managing body mass in super-heavyweight wrestlers. Anthropometric fractionation, corrected girths, and bioelectrical impedance vector analysis (BIVA) emerge as key tools for precise athlete monitoring. While fractionation distinguishes between body compartments to guide body mass reduction without compromising functional muscle mass, corrected girths offer insights into muscle structure in performance-relevant areas. BIVA, in turn, provides information on cellular integrity and fluid balance, ensuring that weight loss strategies do not impair metabolic function. For coaches and nutritionists, integrating these techniques into training planning allows for more accurate adjustments and safeguards the athlete's competitive capacity. Altogether, the combination of benchmark anthropometric data from an elite wrestler with advanced monitoring techniques offers a robust framework for athlete selection, performance optimization, and individualized intervention design.

Author Contributions: Conceptualization, W.C.-V., C.A.H.-A., R.Y.-S. and C.O.R.-G.; methodology, W.C.-V., C.A.H.-A., V.G.-P. and Y.D.-C.; software, C.A.H.-A., R.Y.-S., Y.D.-C. and G.C.-R.; validation, W.C.-V., C.C.-M., G.C.-R. and C.O.R.-G.; formal analysis, W.C.-V., C.A.H.-A. and R.Y.-S.; investigation, W.C.-V., C.A.H.-A., V.G.-P. and Y.D.-C.; resources, C.C.-M., G.C.-R. and C.O.R.-G.; data curation, W.C.-V., C.A.H.-A. and R.Y.-S.; writing—original draft preparation, W.C.-V., C.A.H.-A., R.Y.-S., V.G.-P. and Y.D.-C.; writing—review and editing, C.C.-M., G.C.-R. and C.O.R.-G.; visualization, W.C.-V. and C.O.R.-G.; supervision, W.C.-V. and C.O.R.-G.; project administration, W.C.-V., C.A.H.-A., R.Y.-S. and C.O.R.-G.; funding acquisition, C.C.-M., G.C.-R. and C.O.R.-G. All authors have read and agreed to the published version of the manuscript.

Funding: This research received no external funding.

Institutional Review Board Statement: The study was conducted in accordance with the Declaration of Helsinki and approved by the Institutional Ethical Committee of the Institute of Sports Medicine (IMD) (code: CEI-IMD-01-11-2023 on 24 November 2023).

Informed Consent Statement: The athlete has read, approved, and provided written consent for this publication, which complies with the ethical standards for case studies.

Data Availability Statement: The original contributions presented in this study are included in the article. Further inquiries can be directed to the corresponding authors.

Acknowledgments: We would like to thank the athlete who participated in this study, whose cooperation was essential to its completion. We also thank the coaching and support team, especially Raúl Trujillo Díaz, and Susel Suárez Armas, for their invaluable assistance. We would also like to thank Antonio Córdova for all his support in developing this and other projects using SECA mBCA in Cuba.

Conflicts of Interest: The authors declare that they have no conflicts of interest.

Abbreviations

The following abbreviations are used in this manuscript:

BM	Body Mass
BIA	Body Electrical Impedance Analysis
MWW	Minimum Wrestling Weight
BF	Body Fat
FFM	Fat-Free Mass

References

1. Chaabene, H.; Negra, Y.; Bouguezzi, R.; Mkaouer, B.; Franchini, E.; Julio, U.; Hachana, Y. Physical and Physiological Attributes of Wrestlers: An Update. *J. Strength Cond. Res.* **2017**, *31*, 1411–1442. [CrossRef]
2. Škugor, K.; Karnincic, H.; Zugaj, N.; Stajer, V.; Gilic, B. Repetition of the Exhaustive Wrestling-Specific Test Leads to More Effective Differentiation between Quality Categories of Youth Wrestlers. *Appl. Sci.* **2024**, *14*, 3677. [CrossRef]
3. Cieśliński, I.; Gierczuk, D.; Sadowski, J. Identification of Success Factors in Elite Wrestlers—An Exploratory Study. *PLoS ONE* **2021**, *16*, e0247565. [CrossRef] [PubMed]
4. Levy, J.J.; Boyd, C. Getting Small to Feel Big: The Psychology of Weight Cutting in Combat Sports. *Front. Sports Act. Living* **2025**, *7*, 1495612. [CrossRef] [PubMed]
5. Lakicevic, N.; Mani, D.; Paoli, A.; Roklicer, R.; Bianco, A.; Drid, P. Weight Cycling in Combat Sports: Revisiting 25 Years of Scientific Evidence. *BMC Sports Sci. Med. Rehabil.* **2021**, *13*, 154. [CrossRef] [PubMed]
6. Lakicevic, N.; Reale, R.; D'Antona, G.; Kondo, E.; Sagayama, H.; Bianco, A.; Drid, P. Disturbing Weight Cutting Behaviors in Young Combat Sports Athletes: A Cause for Concern. *Front. Nutr.* **2022**, *9*, 842262. [CrossRef]
7. Sarıakçalı, B.; Şahin, F.N.; Başoğlu, B.; Ceylan, L.; Güler, Ö.; Yamak, B.; Arıkan, G.; Acar, G.C.; Sekban, G.; Durmuşoğlu, M.V.; et al. The Dual Impact: Physiological and Psychological Effects of Rapid Weight Loss in Wrestling. *Front. Psychol.* **2025**, *15*, 1513129. [CrossRef]
8. Hammer, E.; Sanfilippo, J.L.; Johnson, G.; Hetzel, S. Association of In-Competition Injury Risk and the Degree of Rapid Weight Cutting Prior to Competition in Division I Collegiate Wrestlers. *Br. J. Sports Med.* **2023**, *57*, 160–165. [CrossRef]
9. Marten, A.D.; Mattin, M.D.; Rumps, M.V.; Saraf, S.M.; Mulcahey, M.K. Effects of Rapid Weight Loss in Male and Female Wrestlers. *J. Strength Cond. Res.* **2025**, *39*, e721–e729.
10. Ghaemi, J.; Rashidlamir, A.; Hosseini, R.A. A Comparison of Effects of Rapid and Gradual Weight Loss Methods on Body Composition, Aerobic Capacity, and Anaerobic Power in Trained Wrestlers. *Int. J. Wrestl. Sci.* **2014**, *4*, 79–94. [CrossRef]
11. Kuźmicki, S.; Kruszewski, A.; Kruszewski, M. The effects of body mass reduction on the anaerobic power and selected somatic characteristics of Greco-Roman wrestlers. *Biomed. Hum. Kinet.* **2023**, *15*, 35–42. [CrossRef]
12. Kondo, E.; Sagayama, H.; Yamada, Y.; Shiose, K.; Osawa, T.; Motonaga, K.; Ouchi, S.; Kamei, A.; Nakajima, K.; Higaki, Y.; et al. Energy Deficit Required for Rapid Weight Loss in Elite Collegiate Wrestlers. *Nutrients* **2018**, *10*, 536. [CrossRef]
13. Türkyilmaz, R.; Yarar, H. The Investigation of the Effects of Short-Term Body Weight Loss on Elite Wrestlers on Anaerobic Performance and Reaction Time in Tournament Conditions. *J. Sport Sci. Res.* **2024**, *9*, 1–16. [CrossRef]
14. Ayar, M.; Kucuk-Yetgin, M.; Agopyan, A.; Elmacioglu, F. The Effect of a Nutrition Program for Weight Loss during the Pre-competition Period on the Body Composition, Hydration, and Mood Profile of Elite Greco-Roman Wrestlers. *Sport Sci. Health* **2023**, *19*, 1245–1256. [CrossRef]
15. Okudan, B.; Isik, O.; Akyurek, Z.; Karakullukcu, O.F.; Talaghir, L.G.; Nanu, L. Physiological and Psychological Effects of Weight Loss-Induced Stress before a Competition in Senior Wrestlers. *Front. Psychol.* **2025**, *16*, 1568284. [CrossRef]
16. Trivic, T.; Roklicer, R.; Zenic, N.; Modric, T.; Milovancev, A.; Lukic-Sarkanovic, M.; Maksimovic, N.; Bianco, A.; Carraro, A.; Drid, P. Rapid Weight Loss Can Increase the Risk of Acute Kidney Injury in Wrestlers. *BMJ Open Sport Exerc. Med.* **2023**, *9*, e001617. [CrossRef]
17. Oppliger, R.A.; Case, H.S.; Horswill, C.A.; Landry, G.L.; Shelter, A.C. American College of Sports Medicine position stand. Weight loss in wrestlers. *Med. Sci. Sports Exerc.* **1996**, *28*, 135–138. [CrossRef] [PubMed]

18. Buğrahan, C.; Ahmet, S. The investigation of the weight loss methods and effects on the elite U23 wrestlers. *Prog. Nutr.* **2020**, *22*, 88–93.
19. Drid, P.; Figlioli, F.; Lakicevic, N.; Gentile, A.; Stajer, V.; Raskovic, B.; Vojvodic, N.; Roklicer, R.; Trivic, T.; Tabakov, S.; et al. Patterns of rapid weight loss in elite sambo athletes. *BMC Sports Sci. Med. Rehabil.* **2021**, *13*, 39. [CrossRef]
20. Roklicer, R.; Rossi, C.; Bianco, A.; Stajer, V.; Ranisavljev, M.; Todorovic, N.; Manojlovic, M.; Gilic, B.; Trivic, T.; Drid, P. Prevalence of rapid weight loss in Olympic-style wrestlers. *J. Int. Soc. Sports Nutr.* **2022**, *19*, 593–602. [CrossRef]
21. Kordi, R.; Ziaee, V.; Rostami, M.; Wallace, W.A. Patterns of weight loss and supplement consumption of male wrestlers in Tehran. *Sports Med. Arthrosc. Rehabil. Ther. Technol.* **2011**, *3*, 4. [CrossRef] [PubMed]
22. Zhong, Y.; Song, Y.; Artioli, G.G.; Gee, T.I.; French, D.N.; Zheng, H.; Lyu, M.; Li, Y. The Practice of Weight Loss in Combat Sports Athletes: A Systematic Review. *Nutrients* **2024**, *16*, 1050. [CrossRef]
23. Bagheri, R.; Negaresh, R.; Motevalli, M.S.; Wong, A.; Ashtary-Larky, D.; Kargarfard, M.; Rashidlamir, A. Spirulina supplementation during gradual weight loss in competitive wrestlers. *Br. J. Nutr.* **2022**, *127*, 248–256. [CrossRef]
24. Kondo, E.; Shiose, K.; Osawa, T.; Motonaga, K.; Kamei, A.; Nakajima, K.; Sagayama, H.; Wada, T.; Nishiguchi, S.; Takahashi, H. Effects of an overnight high-carbohydrate meal on muscle glycogen after rapid weight loss in male collegiate wrestlers. *BMC Sports Sci. Med. Rehabil.* **2021**, *13*, 96. [CrossRef] [PubMed]
25. Milovancev, A.; Miljkovic, T.; Ilic, A.; Stojšić-Milosavljević, A.; Petrović, M.; Roklicer, R.; Lakićević, N.; Trivić, T.; Drid, P. Disrupting arrhythmia in a professional male wrestler athlete after rapid weight loss and high-intensity training—Case report. *Front. Cardiovasc. Med.* **2023**, *10*, 1091603. [CrossRef]
26. Yu, L.; Lei, L.; Cheng, L. Influence of slow and rapid weight loss periods on physiological performance, mood state and sleep quality in male freestyle wrestlers: A study from Sichuan Province, China. *Front. Psychol.* **2024**, *15*, 1445810. [CrossRef] [PubMed]
27. Ross, W.; Kerr, D.A. Fraccionamiento de la masa corporal: Un nuevo método para utilizar en nutrición clínica y medicina deportiva. *Apunt. Med. l'Esport* **1991**, *28*, 175–187.
28. Reijic, D.; Hässler, E.; Jost, J.; Friedmann-Bette, B. Rapid weight loss and the body fluid balance and hemoglobin mass of elite amateur boxers. *J. Athl. Train.* **2013**, *48*, 109–117.
29. Campa, F.; Toselli, S.; Mazzilli, M.; Gobbo, L.A.; Coratella, G. Assessment of Body Composition in Athletes: A Narrative Review of Available Methods with Special Reference to Quantitative and Qualitative Bioimpedance Analysis. *Nutrients* **2021**, *13*, 1620. [CrossRef]
30. De la Cruz Marcos, S.; Redondo del Río, M.P.; de Mateo Silleras, B. Applications of Bioelectrical Impedance Vector Analysis (BIVA) in the Study of Body Composition in Athletes. *Appl. Sci.* **2021**, *11*, 9781. [CrossRef]
31. Marantz, K. Paris 2024 Preview: Day 1 at GR 60 kg, 130 kg & WW 68 kg. *United World Wrestling*. 22 July 2024. Available online: https://uww.org/article/paris-2024-preview-day-gr-60kg-130kg-ww-68kg (accessed on 4 February 2025).
32. World Medical Association. World Medical Association Declaration of Helsinki: Ethical Principles for Medical Research Involving Human Participants. *JAMA* **2025**, *333*, 71–74. [CrossRef] [PubMed]
33. Gagnier, J.J.; Kienle, G.; Altman, D.G.; Moher, D.; Sox, H.; Riley, D.; CARE Group. The CARE guidelines: Consensus-based clinical case reporting guideline development. *BMJ Case Rep.* **2013**, *53*, 1541–1547.
34. Reale, R.; Slater, G.; Burke, L.M. Individualised dietary strategies for Olympic combat sports: Acute weight loss, recovery and competition nutrition. *Eur. J. Sport Sci.* **2017**, *17*, 727–740. [CrossRef] [PubMed]
35. Esparza Ros, F.; Vaquero-Cristóbal, R.; Marfell-Jones, M. *International Standards for Anthropometric Assessment*; International Society for the Advancement of Kinanthropometry (ISAK): Murcia, Spain, 2019; 114p.
36. Ross, W.; Wilson, N.C. A stratagem for proportional growth assessment. *Acta Paediatr. Belg.* **1974**, *28*, 169–182.
37. Lohman, T.G. Skinfolds and body density and their relation to body fatness: A review. *Hum. Biol.* **1981**, *53*, 181–225. [PubMed]
38. Carter, J. *The Heath-Carter Anthropometric Somatotype Instruction Manual*; Department of Exercise and Nutritional Sciences, San Diego State University: San Diego, CA, USA, 2002.
39. Bosy-Westphal, A.; Schautz, B.; Later, W.; Kehayias, J.J.; Gallagher, D.; Müller, M.J. What makes a BIA equation unique? Validity of eight-electrode multifrequency BIA to estimate body composition in a healthy adult population. *Eur. J. Clin. Nutr.* **2013**, *67*, S14–S21. [CrossRef]
40. Stratton, M.T.; Smith, R.W.; Harty, P.S.; Rodriguez, C.; Johnson, B.A.; Dellinger, J.R.; Williams, A.D.; White, S.J.; Benavides, M.L.; Tinsley, G.M. Longitudinal agreement of four bioimpedance analyzers for detecting changes in raw bioimpedance during purposeful weight gain with resistance training. *Eur. J. Clin. Nutr.* **2021**, *75*, 1060–1068. [CrossRef] [PubMed]
41. Lahav, Y.; Goldstein, N.; Gepner, Y. Comparison of body composition assessment across body mass index categories by two multifrequency bioelectrical impedance analysis devices and dual-energy X-ray absorptiometry in clinical settings. *Eur. J. Clin. Nutr.* **2021**, *75*, 1275–1282. [CrossRef]
42. Kushner, R.F.; Gudivaka, R.; Schoeller, D.A. Clinical characteristics influencing bioelectrical impedance analysis measurements. *Am. J. Clin. Nutr.* **1996**, *64*, 423S–427S. [CrossRef]

43. Škugor, K.; Gilić, B.; Karninicic, H.; Jokai, M.; Babszky, G.; Ranisavljev, M.; Štajer, V.; Roklicer, R.; Drid, P. What determines the competitive success of young Croatian wrestlers: Anthropometric indices, generic or specific fitness performance? *J. Funct. Morphol. Kinesiol.* **2023**, *8*, 90. [CrossRef]
44. Reale, R.; Burke, L.M.; Cox, G.R.; Slater, G. Body composition of elite Olympic combat sport athletes. *Eur. J. Sport Sci.* **2020**, *20*, 147–156. [CrossRef] [PubMed]
45. Sterkowicz-Przybycień, K.L.; Sterkowicz, S.; Żarów, R.T. Somatotype, body composition and proportionality in Polish top Greco-Roman wrestlers. *Hum. Kinet.* **2011**, *28*, 141–154. [CrossRef] [PubMed]
46. Baranauskas, M.; Kupčiūnaitė, I.; Lieponienė, J.; Stukas, R. Dominant somatotype development in relation to body composition and dietary macronutrient intake among high-performance athletes in water, cycling and combat sports. *Nutrients* **2024**, *16*, 1493. [CrossRef]
47. Rahmani, F.; Mirzaei, B.; Faradjzadeh-Mevaloo, S. Anthropometric profile of elite Azerbaijani senior Greco-Roman wrestlers. *Pedagog. Psychol. Med. Biol. Probl. Phys. Train. Sports* **2019**, *23*, 196–201. [CrossRef]
48. Carvajal-Veitía, W.; León-Pérez, S.; González-Revuelta, M.E.; Deturnel-Campo, Y.; Echavarría-García, Y. Anthropometrical characteristics of Cuban sporting population: Reference data from a high performance national teams, 1992–2014. *Apunt. Med. Esport* **2018**, *53*, 129–137. [CrossRef]
49. Jagim, A.R.; Tinsley, G.M.; Merfeld, B.R.; Ambrosius, A.; Khurelbaatar, C.; Dodge, C.; Carpenter, M.; Luedke, J.; Erickson, J.L.; Fields, J.B.; et al. Validation of skinfold equations and alternative methods for the determination of fat-free mass in young athletes. *Front. Sports Act. Living* **2023**, *5*, 1240252. [CrossRef] [PubMed]
50. Donmez, E.; Isik, O.; Baic, M. Fluctuation in body composition and urine specific gravity of Turkish wrestlers in a top-level official wrestling competition. *Front. Physiol.* **2025**, *15*, 1516149. [CrossRef]
51. Marini, E.; Campa, F.; Buffa, R.; Stagi, S.; Matias, C.N.; Toselli, S.; Sardinha, L.B.; Silva, A.M. Phase angle and bioelectrical impedance vector analysis in the evaluation of body composition in athletes. *Clin. Nutr.* **2020**, *39*, 447–454. [CrossRef]
52. Campa, F.; Matias, C.N.; Marini, E.; Heymsfield, S.B.; Toselli, S.; Sardinha, L.B.; Silva, A.M. Identifying athlete body fluid changes during a competitive season with bioelectrical impedance vector analysis. *Int. J. Sports Physiol. Perform.* **2020**, *15*, 361–367. [CrossRef]
53. Zhang, J.; Zhang, N.; Li, Y.; He, H.; Song, G.; Chen, J.; Yan, Y.; Ma, G. Habitual water intake impacted the body composition of young male athletes in free-living conditions: A cross-sectional study. *Front. Sports Act. Living* **2024**, *6*, 1458242. [CrossRef]
54. Canda, A. Muscle mass index estimated by anthropometry vs bioelectrical impedance: Study in athletes competing by weight categories. *Apunt. Sports Med.* **2021**, *56*, 100360. [CrossRef]
55. Caparello, G.; Galluccio, A.; Ceraudo, F.; Pecorella, C.; Buzzanca, F.; Cuccomarino, F.; Bonofiglio, D.; Avolio, E. Evaluation of body composition changes by bioelectrical impedance vector analysis in volleyball athletes following Mediterranean diet recommendations during Italian Championship: A pilot study. *Appl. Sci.* **2023**, *13*, 2794. [CrossRef]
56. Campa, F.; Thomas, D.M.; Watts, K.; Clark, N.; Baller, D.; Morin, T.; Toselli, S.; Koury, J.C.; Melchiorri, G.; Andreoli, A.; et al. Reference percentiles for bioelectrical phase angle in athletes. *Biology* **2022**, *11*, 264. [CrossRef] [PubMed]
57. Bennett, J.P.; Cataldi, D.; Liu, Y.E.; Kelly, N.N.; Quon, B.K.; Gonzalez, M.C.; Heymsfield, S.B.; Shepherd, J. Variations in bioelectrical impedance devices impact raw measures comparisons and subsequent prediction of body composition using recommended estimation equations. *Clin. Nutr. ESPEN* **2024**, *63*, 540–550. [CrossRef]
58. Cirillo, E.; Pompeo, A.; Cirillo, F.T.; Vilaça-Alves, J.; Costa, P.; Ramirez-Campillo, R.; Dourado, A.C.; Afonso, J.; Casanova, F. Relationship between bioelectrical impedance phase angle and upper and lower limb muscle strength in athletes from several sports: A systematic review with meta-analysis. *Sports* **2023**, *11*, 107. [CrossRef]
59. Lebron, M.A.; Stout, J.R.; Fukuda, D.H. Physiological perturbations in combat sports: Weight cycling and metabolic function—A narrative review. *Metabolites* **2024**, *14*, 83. [CrossRef]
60. Karnincic, H.; Žugaj, N.; Škugor, K. Advice on regulating body mass in wrestling from the most cited combat sport literature—A systematic review. *J. Funct. Morphol. Kinesiol.* **2024**, *9*, 264. [CrossRef]

Journal of
Functional Morphology and Kinesiology

MDPI

Article

Comparative Analysis of Maturation Prediction Methods (Moore, Mirwald, BAUSport™): Croatian Female Volleyball Youth Team Example

Drazen Cular [1,2,3], Tea Beslija [3], Marijana Cavala [1], Matej Babic [1,4] and Ana Kezic [1,*]

1 Faculty of Kinesiology, University of Split, 21000 Split, Croatia; drazen.cular@kifst.eu (D.C.); marijana.cavala@kifst.eu (M.C.); matej.babic@kifst.eu (M.B.)
2 Einstein, Startup for Research, Development, Education, Trade and Services, 21000 Split, Croatia
3 European Institute for Talents, Education, Research & Development, 21000 Split, Croatia; tea.beslija@hotmail.com
4 Faculty of Kinesiology, University of Zagreb, 10000 Zagreb, Croatia
* Correspondence: ana.kezic@kifst.eu; Tel.: +385-21-302-440

Abstract: Objectives: The study aims to compare three distinct protocols—Moore, Mirwald, and the new BAUSport™ SonicBone system—for predicting somatic maturation in youth athletes. **Methods**: The participants were female members of the Croatian national volleyball youth team (U-17) (n = 16). The study involved comprehensive measurements, including height, weight, sitting height, leg length, wrist diameter, hand joint diameter, hand grip strength, and ultrasound measurements for skeletal age assessment. **Results**: Correlation analysis showed moderate to strong correlations between the Moore and Mirwald skeletal age estimates, but both showed weaker correlations with the BAUSport™ skeletal age. Repeated-measures ANOVA showed no significant difference between the Moore and Mirwald methods ($p > 0.05$); significant differences between both the Moore and Mirwald methods and the BAUSport™ method ($p < 0.05$). Regression analysis revealed that height, weight, sitting height, leg length, wrist diameter, and hand joint diameter explained 69% of BAUSport™, with wrist diameter being the only significant predictor. While the Moore and Mirwald methods remain useful tools for estimating the timing of an athlete's growth spurt, BAUSport™ represents a potential advancement in skeletal age assessment. Further research is needed to validate BAUSport™ across diverse populations and optimize its calibration to accommodate anatomical variations. **Conclusions**: The findings suggest that with further refinement, BAUSport™ could become a new standard for monitoring skeletal development in youth athletes. Additionally, studies should explore comparative analyses with other emerging technologies, such as genetic markers, hormonal assessments, and MRI, for further understanding of biological maturation in talent identification.

Keywords: somatic maturation; skeletal age; ultrasound technology; youth athletes; skeletal age; growth; talent identification

1. Introduction

The development of national youth teams is a critical component of talent identification in sports, as it sets the foundation for future athletic success. A key aspect of this process is understanding how biological maturation intersects with the selection of athletes. Considering biological maturation in talent selection processes is crucial to avoid disadvantaging late-maturing athletes and potentially reduce the relative age effect in sports [1].

Athletes who experience later maturation may face disadvantages in selection processes due to temporarily lower physical development compared to their earlier-maturing peers. In youth sports, coaches and evaluators often favor athletes with greater strength, speed, and height—traits typically associated with those who have entered puberty earlier. Without considering biological maturation, late-maturing athletes may be unfairly excluded despite their high long-term potential. As a result, promising talent may be overlooked, potentially hindering future excellence once full maturity is achieved. In addition to the challenges faced by late-maturing athletes, it is important to note that early maturation can offer benefits, such as improved initial athletic performance. However, while early maturing athletes may have an advantage in the short term, this can sometimes result in a plateau in development once their peers catch up in terms of maturity, highlighting the importance of considering maturation in long-term talent identification. By identifying an athlete's maturity stage, performance evaluations can be adjusted accordingly, ensuring a more accurate and equitable selection process. Current scientific research on biological maturation in young athletes employs a variety of methodologies for its assessment that reflect the non-linear nature of the process. These methods range from non-invasive anthropometric approaches, such as predicting maturation status with height or secondary sexual characteristics, to imaging and biomarker analysis. While each of these approaches has its strengths and is continually evolving, no single method has been considered the gold standard, and there are known sex-, age-, and population-specific variations.

The Mirwald equation predicts years from peak height velocity using anthropometric measurements, showing high reliability in cross-validation studies [2]. However, while widely used, it has been shown to misclassify a significant portion of athletes [3]. Additionally, anthropometric measures have been used to develop prediction models for somatic maturity. Moore and colleagues refined existing equations, demonstrating good fit and calibration in external samples [4]. These models provide alternatives to commonly used methods and can be applied without specialized equipment. Various equations have been developed to predict maturity status, including those by Mirwald and Moore, which have demonstrated varying degrees of accuracy in Chilean children [5]. Fransen and colleagues developed an improved equation for estimating age at peak height velocity (APHV) using a maturity ratio, which showed better accuracy than previous models for both general and athletic populations [6]. Moraes Macêdo and colleagues created equations to predict skeletal age and sexual maturation index using anthropometric measurements in Brazilian children [7]. Malina and associates validated a non-invasive maturity estimate based on the percentage of predicted mature height against skeletal age in youth football players, finding moderate concordance between the two methods [8]. These studies demonstrate the ongoing efforts to develop and refine non-invasive techniques for assessing biological maturity, which can be valuable for talent identification and development in sports, as well as for medical diagnostics and disease prevention. Visual evaluation of individual growth curves demonstrated the highest concordance ($\approx$80%) with maturity status classifications based on longitudinal data [3]. The percentage of predicted adult height method using Khamis–Roche or Tanner–Whitehouse 2 equations provides a reasonably valid alternative to maturity offset prediction equations, which tend to misclassify players [3].

Skeletal maturity refers to the development and maturation of bones, which can be assessed through methods such as bone age estimation. It is an important indicator of an athlete's physical development and maturity stage, as it directly influences body mass, strength, flexibility, and cardiorespiratory fitness [9]. In sports performance, skeletal maturity can significantly affect an athlete's abilities, particularly during periods of rapid growth, as those with more advanced skeletal maturity may have an advantage in strength and power. Understanding skeletal maturity helps in adjusting training loads, preventing injuries, and optimizing performance, which is critical for effective talent identification and long-term development in sports. Skeletal age assessment is crucial for evaluating growth, predicting final height, and guiding talent selection in youth sports [10]. While magnetic resonance imaging (MRI) is considered the gold standard, more pragmatic and cost-effective alternatives have been explored [11]. Ultrasound-based methods, such as BAUSport™ SonicBone, have shown a high correlation with traditional radiographic techniques like the Fels method [12]. Various ultrasound imaging methods for assessing biological maturity have been developed, with the hand and wrist being the most commonly analyzed regions [13]. These methods generally demonstrate high reliability, but require further development to become a new gold standard. Understanding maturity-related changes is essential for managing training load, injury risk, and physical performance in youth soccer, particularly around the age of peak height velocity [14]. Quantitative ultrasound (QUS) techniques, such as the BAUSport™ SonicBone device, have shown comparable results to traditional X-ray-based methods for skeletal age assessment [15]. Similarly, broadband ultrasonic attenuation (BUA) measurements of the calcaneus have been used to evaluate skeletal maturation in Japanese youth [16]. Such complementary non-invasive methods might provide advantages through reduced exposure to ionizing radiation when assessing particularly pediatric populations. Ultrasonographic techniques are also relatively inexpensive and readily available, and can be repeated for longitudinal studies. Additionally, they enable real-time imaging and dynamic evaluations, yielding detailed information on growth plate development and other biological maturation indicators without requiring specialized radiological infrastructure. Similarly, the Sunlight BonAge ultrasound device demonstrated good correlations with radiographic methods, offering a non-invasive and quick assessment for children aged 5–15 years [17]. Automated simplifications of the Eklof and Ringertz method, analyzing 3–5 ossification centers in carpal images, have shown high agreement with classical methods and offer a reliable, fast, and objective approach to skeletal age estimation [18].

Until now, no broad-scale studies have been conducted comparing ultrasound methods with traditional anthropometric methods at height, sitting height, and weight within algorithmic prediction. This study attempts to illustrate two different methodological approaches to obtain better information on the accuracy and applicability of various methods in sports selection and talent development contexts. The importance of accurately assessing the maturation of high-level athletes, particularly members of national teams, is crucial for ensuring fair competition, optimizing training regimens, and guiding future athletic development. National team athletes represent the pinnacle of talent, and their maturation status can have significant implications for long-term performance and injury prevention. This makes it all the more critical to accurately measure and track their progress, since the selection process for elite sports often favors athletes with specific physical and maturational characteristics, particularly in aesthetic sports [19]. The main aim of the study was to compare and evaluate three distinct protocols—Moor, Mirwald, and BAUSport™—in terms of their methodologies, accuracy, and practical applications in predicting somatic maturation in youth athletes. The study seeks to understand how these methodologies correlate with each other and their practical implications in the fields of sports science and youth talent detection and athletic development.

2. Materials and Methods

The overall sample consisted of 16 female Croatian national volleyball youth team members (U-17), aged between 14 and 16 years (mean $\pm$ SD = 15.89 $\pm$ 0.58 years). All participants were of Caucasian origin and were actively involved in competitive volleyball at the national level. The participants were selected based on their membership in the Croatian national volleyball team, with all having a minimum of 4 years of training experience in the sport. However, no specific inclusion or exclusion criteria were applied beyond their national team status and training background. The selection was made by the national team coach and technical staff based on the players' performances observed in their club games. All players were pre-selected by national-level scouts and coaching staff based on their technical and tactical performance, as well as club-level match assessments. This ensured a homogenous, high-performance sample representative of elite youth volleyball athletes. On average, participants trained 5 times per week. The inclusion criteria for this study were age and national team membership. Ethical approval for the study was obtained from the University of Split, Faculty of Kinesiology Research Ethics Board (003-08/20-04/00121818-205-02-05-20-006), and informed consent was provided by both the participants and their legal guardians.

The sample of variables was composed of chronological age, training experience, height (cm), weight (kg), sitting height (cm), leg length (cm), wrist diameter (cm), hand joint diameter (cm), hand grip strength (kg), and ultrasound measurements for skeletal age assessment. All measurements were performed by the same investigator who had extensive training and experience, and each measure was repeated three times, with the average used for statistical analysis. The participants wore the same training attire (shorts, shirt, socks) and removed footwear for all measures. All tests were carried out between 8:00 a.m. and 10:00 a.m. in a standardized indoor laboratory setting to minimize circadian variation.

Body weight (MC-780, Tanita Corporation, Tokyo, Japan) was recorded with electronic scales to the nearest 0.1 kg. Height was recorded to the nearest 0.1 cm using a portable stadiometer (Holtain, Harpenden, UK). Sitting height was assessed as the distance from the vertex to the base sitting surface. Results were recorded to the nearest 0.1 cm using a sitting height stadiometer (SHstad; Harpenden Sitting Table, Holtain, Harpenden, UK). The leg length was measured in a standing position, without shoes, with the feet slightly apart. The distance between the anterior iliac crest (upper part of the hip bone) and the medial malleolus (bony protrusion on the inner side of the ankle) was measured using a Martin Anthropometer. The result was displayed with a precision of 0.1 cm. Wrist and hand joint diameters were recorded to the nearest 0.01 cm using a sliding caliper (GPM Martin type Sliding Caliper, Bachenbülach, Switzerland). Hand diameter was measured across the metacarpophalangeal joints and wrist diameter across the styloid processes of the radius and ulna. All mentioned measures were obtained in one day. All anthropometric measurements were repeated three times, and average results were taken for statistical purposes. The grip strength of the dominant hand was measured using a standard adjustable digital hand grip dynamometer (Takei Scientific Instruments Co., Ltd., Tokyo, Japan) at standing position with shoulder adducted and neutrally rotated and elbow in full extension. The subjects were asked to put maximum force on the dynamometer thrice. The test was repeated three times, with adequate rest between attempts to avoid fatigue (1 min). The maximal value was recorded in kilograms.

Collected anthropological data were inserted into two algorithms to estimate three variables: age at peak height velocity (APHV), maturity offset (OFFSET) (measures the time from the peak height velocity), and skeletal age for each method (Mirwald and Moore method).

The BAUSport^{TM} SonicBone instrument system with accompanying software, produced by SonicBone Medical Ltd., Rishon LeZion, Israel, was used to estimate skeletal age (BAUSport^{TM} Skeletal Age) based upon ultrasound assessment of three skeletal locations on the left hand-wrist. Assessments were conducted by a professional who was trained in the use of the BAUSport^{TM} device. BAUSport^{TM} system device was placed on a stable table to avoid vibration or displacement during the test. The participants' chair height was adjusted to ensure natural and comfortable hand placement. Participants removed all jewelry, rings, bracelets, watches, etc., before the measurement. A washable marker was used to mark the contact spots on the patient's wrist and metacarpal sites. An even layer of ultrasound gel was applied at each designed location on the hand to establish acoustic contact between the ultrasound probes and the hand. During the measurement, the participants' elbow touched the top side of the device surface. The angle between the arm and hand during measurement was between 130 and 140 degrees, and the hand was placed parallel to the device. In order to obtain a valid measurement result, all three measurements were performed in the following order: 1—wrist, 2—phalanx, 3—metacarpal (Figure 1). Information, based upon the speed at which high-frequency waves of an ultrasound pulse propagate through bone and distance attenuation factors (i.e., decay rate), is fed into an integrated algorithm using the scoring method designed by Tanner and Whitehouse [20]. The algorithm then provides an estimate of skeletal age and future adult stature. The time durations for the scans at each of the various sites were 12 s for the radius and ulna and 4 s for the proximal phalange and distal metacarpal. Total time for completing the assessment was approximately five to ten minutes per participant. The BAUSport™ system has previously demonstrated high levels of repeatability and validity in young athletes and in the general population [12,15,21]. However, one of the technical limitations of the BAUSport™ device is that its skeletal age estimates do not fully align with traditional radiographic methods, such as the Fels method [12]. While it provides valuable non-invasive measurements, there can be fixed biases between the BAUSport™ estimates and radiographic assessments, leading to discrepancies in some cases.

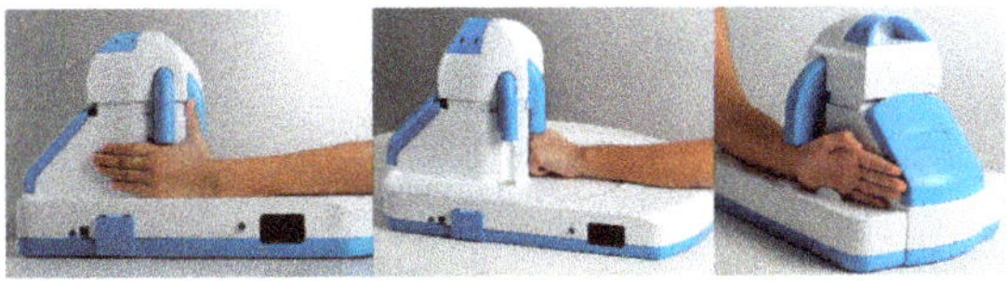

Figure 1. BAUSport™ SonicBone device protocol (1st, 2nd and 3rd hand measurements positions) (source: https://sonicbonemedical.com/product/, accessed on 30 January 2025).

A comparison of the methodological approach of the three maturation assessment methods is presented in Table 1.

Descriptive anthropometric parameters were calculated in the first step to provide an overview of the data, including means, standard deviations, and ranges (minimum and maximum values) for all variables, as well as the percentages of late, on-time, and early maturers (%). Secondly, using a paired samples *t*-test, the difference between the chronological and skeletal age of the participants was calculated. The third step contained correlation analyses with the aim of assessing the relationships between the skeletal age estimates of the three methods (Moore, Mirwald, and BAUSport™), as well as their relationships with other measured parameters. In the fourth step, the authors investigated the results of repeated-measures analysis of variance (ANOVA) with a post hoc Tukey test to compare the skeletal age estimates from each pair of methods (Moore vs. Mirwald, Moore vs. BAUSport™, Mirwald vs. BAUSport™) to test for significant differences. Regression analysis was conducted to explore the predictive power of BAUSport™ method and to assess the influence of other variables like height, weight, sitting height, leg length, hand grip strength, wrist, and hand joint diameter. Normality of data was assessed using the Shapiro–Wilk test prior to parametric statistical procedures. All analyses were conducted using Statistica 14.1, TIBCO Software Inc., Santa Clara, CA, USA.

Table 1. Main methodological differences between the Moore, Mirwald, and BAUSport^{TM} methods.

Characteristic	Moore Method	Mirwald Method	BAUSport^{TM} Method
Type of measurement	Anthropometric	Anthropometric	Ultrasound
Main variables	Height, sitting height, leg length, weight, sex	Same as Moore	Ultrasound-based bone density, speed of sound through bone
Accuracy	Moderate	Moderate	High for skeletal age
Ease of application	Very simple	Very simple	Requires specialized equipment
Application in sports	Monitoring growth and predicting PHV	Same as Moore	Direct assessment of skeletal age and bone development
Main advantage	Improved accuracy compared to Mirwald	Simple and widely used method	Non-invasive alternative to X-ray-based skeletal age assessment
Main limitation	Relies on statistical estimations	May underestimate or overestimate maturation	Calibration issues and still uninvestigated sensitivity to anatomical variations

3. Results

The best Croatian young female volleyball players, members of the Croatian U-17 national team, were tested for this research, and their results are presented in the tables below. Table 2 comprises the results of descriptive statistics for all applied variables (mean values, standard deviations, minimal and maximal results). The assumption of normality was verified using the Shapiro–Wilk test, which yielded non-significant results ($p > 0.05$) across all variables, indicating an approximately normal distribution suitable for parametric testing. The results presented show that the average age of the U-17 players was just under 16 years, while the youngest representative was only 14.7 years old. For comparison, the skeletal age of the participants was numerically higher, averaging 17.29 years. The results of the Moore and Mirwald OFFSET and APHV are basically aligned, and a relatively small standard deviation suggests homogeneity within the sample regarding maturity progression. The range for APHV is minimal, demonstrating a tightly clustered distribution, meaning that the cohort reached PHV around a similar age. However, the upper limits of skeletal age, for both the Moore and the Mirwald method, nearing almost 20 years suggests some advanced maturation cases. The values of skeletal age measured with the use of BAUSport^{TM} device showed somewhat lower values for both mean and minimal and maximal results.

The paired samples *t*-test revealed that the difference between chronological and skeletal age (measured with BAUSport^{TM}) is significant ($t = 5.31$, $p = 0.00$).

For the BAUSport^{TM} measurement method (SonicBone ultrasound device), the maturation time for each individual (i.e., early, on time, late) is defined by the difference between their skeletal age and chronological age at the time of assessment and is presented in years. An individual with a skeletal age greater than their chronological age is considered advanced in maturation for their gender and age. Conversely, an individual with a skeletal age lower than their chronological age is considered delayed in maturation. When skeletal age and chronological age are equal, the individual is considered on time in maturation. In the case of the sample of young female volleyball players, it can be determined from Figure 2 that there were no participants with delayed maturation (those with a chronological age greater than their skeletal age). Only 31% of the participants were in the on-time maturation status, while as many as 69% were in the early stages of maturation.

Table 2. Descriptive statistical parameters of all applied variables (Mean—mean value, SD—standard deviation, Min—minimal result, Max—maximal result).

	Mean	SD	Min	Max
Chronological age	15.89	0.58	14.71	16.49
Height (cm)	180.63	5.06	171.00	187.00
Weight (kg)	67.50	5.83	54.00	75.00
Sitting height (cm)	93.67	3.15	87.30	97.50
Leg length (cm)	86.68	3.12	81.70	92.30
Hand joint diameter (cm)	8.22	0.37	7.70	8.80
Wrist diameter (cm)	5.68	0.27	5.10	6.10
Hand grip strength (kg)	33.67	3.91	26.30	38.70
Moore OFFSET	3.65	0.52	2.83	4.41
Moore APHV	11.24	0.36	10.74	11.98
Moore Skeletal Age	18.55	1.03	16.54	19.76
Mirwald OFFSET	3.54	0.76	2.81	5.95
Mirwald APHV	11.36	1.02	7.91	12.44
Mirwald Skeletal Age	18.43	0.88	16.56	19.82
BAUSport^{TM} Skeletal Age	17.29	0.89	15.46	18.64

Legend: OFFSET—measures the time from the peak height velocity, APHV—age at peak height velocity.

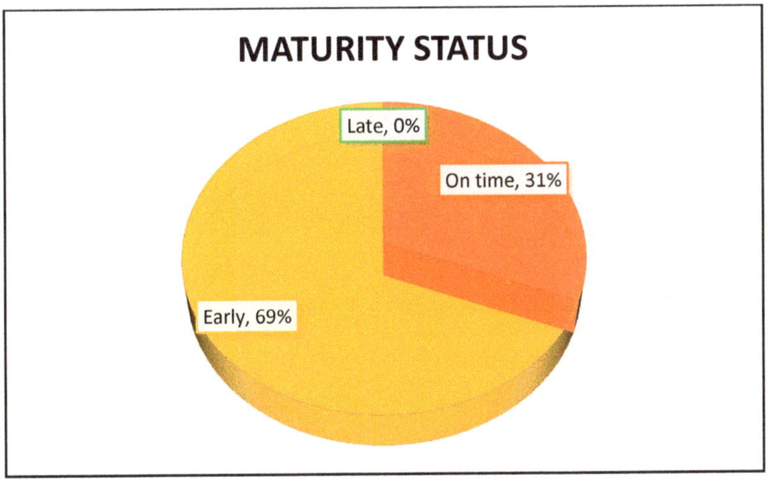

Figure 2. The percentages of late, on-time, and early maturers (%) in Croatian female national volleyball youth team members (U-16) according to the BAUSport^{TM} method.

The correlation analysis drawn from the skeletal age assessment data in Figure 3 brings out several pertinent observations. Between the skeletal age estimates, the Moore and Mirwald methods show a moderate to strong correlation (r = 0.66), which implies an adequate agreement level regarding their assessments of somatic maturity. On the other hand, both methods exhibit weaker correlations with the BAUSport^{TM} skeletal age (Moore: r = 0.18; Mirwald: r = 0.4), indicating that they might be based on different criteria or scales for measuring skeletal development. Height, weight, and other physical measurements show varied degrees of correlation with skeletal age estimates. It is worth highlighting the significant correlation between wrist diameter and BAUSport^{TM} skeletal age (r = 0.74). The strong correlation implies that wrist diameter could be a reliable predictor of skeletal age.

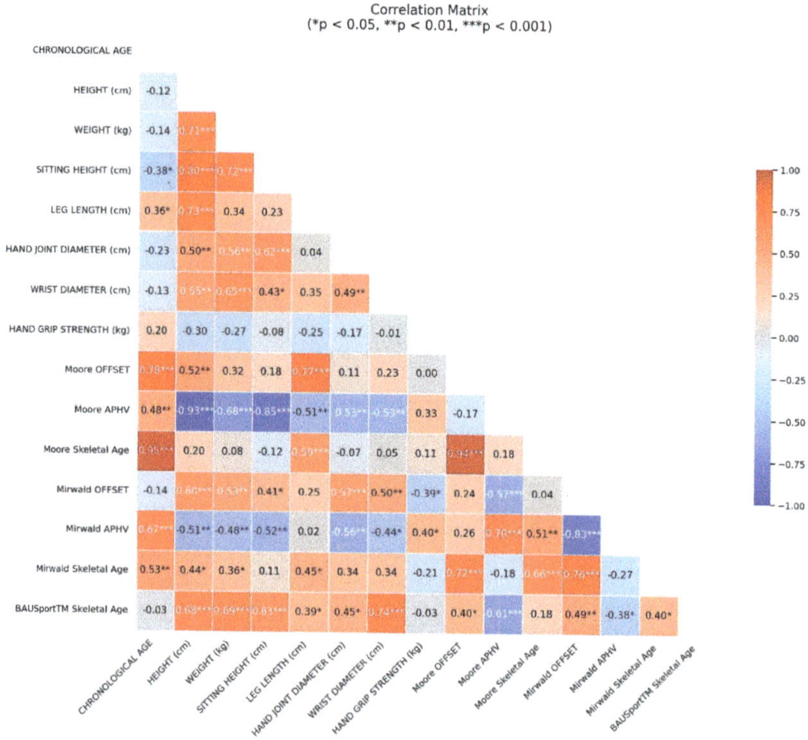

Figure 3. "The Heatmap" of the correlation matrix (the colors indicate the strength of the correlation, where warmer colors (red-orange) denote a positive correlation, and cooler colors (blue) denote a negative correlation). The numerical values inside the squares represent the actual correlation coefficients.

As expected, results of repeated-measures ANOVA with post hoc Tukey test in Table 3 show that there is no significant difference between the Moore and Mirwald methods in estimating skeletal age ($p > 0.05$). Also, the results show that there is a significant difference between the Moore and Mirwald methods and the BAUSport^{TM} method in estimating skeletal age ($p < 0.05$). The Cohen's d values for the post hoc pairwise comparisons showed a small difference between the Moore and Mirwald methods (d = 0.12), while the differences between Moore and BAUSport^{TM} (d = 1.35) and Mirwald and BAUSport^{TM} (d = 1.33) were large, indicating significant differences between these methods.

Table 3. Repeated-measures ANOVA with post hoc Tukey test for skeletal age assessment using the Moore, Mirwald, and BAUSport^{TM} methods.

Comparison	Mean Difference	*p*-Value	Interpretation
Moore vs. Mirwald	0.115	0.935	no significant difference ($p > 0.05$)
Moore vs. BAUSport^{TM}	−1.252	0.001	significant difference ($p < 0.05$)
Mirwald vs. BAUSport^{TM}	−1.136	0.004	significant difference ($p < 0.05$)

The regression analysis from Table 4 shows that height, weight, sitting height, leg length, hand joint, and wrist diameter explained 69% of BAUSport^{TM} skeletal age estimation, with wrist diameter being the only significant predictor.

Table 4. The regression analysis results for the BAUSport™ method.

Method	Coefficients	
BAUSport™	height: 0.035, weight: 0.027, sitting height: 0.017, leg length: 0.019, hand joint diameter: 0.048, wrist diameter: 1.626 *	R = 0.83 R^2 = 0.69 p = 0.04

Legend: * significant predictor ($p < 0.05$).

4. Discussion

The U17 national team serves as a platform for developing the most talented young volleyball players in the country. They are selected from domestic clubs and gather for training camps and tournaments. Since this team precedes the elite senior national team, it is extremely important to monitor the characteristics, abilities, and biological age of the players to ensure the proper development and progress of each individual athlete. This allows for an individualized approach to training, load adjustment, and injury prevention, thereby maximizing the potential for success at the highest level of competition. Proper monitoring of biological age helps in understanding the physical development of the players, ensuring that they develop optimally and are prepared for the challenges of senior competition. In this context, the data contribute to understanding skeletal development patterns within a highly homogenous group of young athletes, with implications for training, talent identification, and health monitoring. A sample of elite Croatian young volleyball players demonstrated a tendency toward early maturation, as the difference between chronological and biological age proved to be significant. One of the key aspects of this finding is the accelerated biological development of volleyball players, which indicates early pubertal maturity. This phenomenon has already been documented in previous research showing that female athletes, especially in sports requiring high levels of physical performance, may exhibit advancements in biological development compared to their less physically active peers [22]. Similar findings were reported on male volleyball players, identifying how young volleyball players classified as "early" seemed to show anthropometric characteristics linked to better performance at the tournament (higher height, upper arm and calf muscle area, fat mass percentage, and total fat-free mass) [23]. To further explore the risk of bias in talent selection, it is essential to recognize that the preference for early-maturing athletes may inadvertently disadvantage those with later maturation, who may not yet exhibit their full potential. To mitigate this bias, talent identification systems should consider not only physical attributes but also the long-term development trajectory, incorporating measures to assess athletes' maturity stages and accounting for the potential advantages of late-maturing athletes in the future.

Accelerated biological development may be associated with various factors, including genetic predisposition, physical activity levels, nutrition, and exposure to stress. It is important to note that although advanced skeletal age may enable earlier participation in competitions with older athletes, it also carries certain risks, such as an increased likelihood of injuries due to earlier closure of growth plates (epiphyseal plates) and excessive stress on joints and bones [24]. Further, early-maturing athletes may be more susceptible to overtraining due to their accelerated physical development and increased training intensity. This can lead to an imbalance between training load and recovery, increasing the risk of injury and hindering long-term performance progress. Overtraining can increase the risk of early burnout due to higher physical demands and psychological pressure. Athletes who mature early may push themselves too hard, leading to fatigue, and potential long-term performance declines. The difference between chronological and skeletal age can significantly impact how these athletes are trained and monitored. Since biologically older athletes may exhibit higher levels of physical strength, endurance, and explosiveness, coaches might be inclined to increase training intensity to capitalize on these advantages. However, it is crucial to ensure that training is tailored to the individual needs of athletes, considering not only their physical capacities but also their long-term health.

This is particularly important during the period of accelerated growth ("peak height velocity"), which occurs at different times for each individual. Unsurprisingly, the risk of injury significantly increases during the most intense growth periods [25,26], so coaches must exercise particular caution in planning the training process for growing athletes. One strategy may include regular monitoring of skeletal age and other indicators of biological development to optimize training methods and reduce injury risk. For instance, research suggests that high-intensity training should be adjusted according to the degree of biological maturity, allowing for long-term sports development without compromising health [27]. In addition to monitoring immediate performance, it is crucial to adopt a long-term perspective when evaluating an athlete's career. While short-term success may be a motivating factor, focusing solely on immediate results can overlook the potential for long-term development and sustainability in an athlete's career. Taking a more holistic approach to training and progression, considering maturation status and long-term goals, helps ensure that athletes reach their full potential without prematurely peaking or risking injury.

These results show an interesting and potentially significant pattern in the maturation of young female volleyball players. First, the fact that 69% of the participants were classified in the early maturation stage may indicate that physical development plays a key role in the selection of young volleyball players at the national level. Early maturation is associated with faster physical development, including growth in height, strength, and muscle mass, all of which are key components of success in volleyball. These results may reflect the tendency of selectors to choose players who are physically superior to their peers, which can have a direct impact on their on-court performance.

Second, only 31% of the participants were in the on-time maturation status, which is a relatively low percentage compared to expected distributions in the general population. These data may suggest that girls who mature in accordance with their chronological age may be at a disadvantage in the selection process for top-level sports teams, as their peers with earlier maturation may physically outmatch them. In the context of long-term development, it is important to consider the implications of these results. Early maturation may bring certain advantages in the earlier stages of a career but can also lead to early burnout, increased injury risk, and a decline in long-term sports performance [28]. These risks are especially pronounced in athletes who specialize early in one sport, which can result in overtraining and increased stress on the young body. Therefore, it would be useful to investigate how early maturation affects the careers of these athletes and whether there is a need to adjust selection criteria to ensure the long-term sustainability and success of young volleyball players.

The Moore and Mirwald methods for assessing skeletal age and somatic maturation are based on the concept of predicting APHV. These methods utilize growth curves that plot the anthropometric measurements over time. However, like all predictive methods, they have limitations. The accuracy of these methods can be influenced by genetic, nutritional, and environmental factors, and they may not be as precise as other more technologically advanced methods [3]. Also, the authors [2] caution against using maturity offset as a continuous measure, and instead recommend considering it as a categorical variable.

On the other hand, the BAUSport™ method is a more recent and technologically advanced approach for assessing skeletal age and somatic maturation, particularly in youth athletes. Unlike the Moore and Mirwald methods, which rely on anthropometric measurements and growth curves, the BAUSport™ method utilizes ultrasound technology to evaluate skeletal age. The technology measures parameters such as the speed of sound (SOS) and the distance attenuation factor (ATN) through the bone. Studies have shown that the BAUSport™ method has high repeatability and reliability [12,15,21]. It has been found to be comparable to traditional X-ray-based methods in terms of accuracy in skeletal age assessments. While the BAUSport™ method is still under development and refinement, particularly in terms of reproducibility and eliminating confounding factors, it shows promising results, as many recent studies used this technology for assessing skeletal maturity purposes [11,21]. Despite these promising aspects, further validation of the measurement process is necessary. In our study, we encountered a case where the BAUSport™ device was unable to obtain a measurement for one participant, displaying an error message instead. Our hypothesis is that this issue may have been caused by the participant's unusually large hand size, specifically the diameter of the wrist joint or the thickness of the metacarpal region, which may have exceeded the device's measurement range. Due to the inability to obtain a valid measurement, this participant had to be excluded from the study. This limitation underscores the need for manufacturers to investigate the root causes of such errors and assess whether the device's sensor calibration and measurement algorithm can be optimized to accommodate a wider range of hand dimensions. Future research should explore whether adjustments in sensor sensitivity, hardware design, or software parameters could improve measurement reliability across diverse anatomical variations. A potential extension of the BAUSport™ device could involve adapting its calibration and measurement algorithms to assess skeletal age beyond the limits of youth athletes, enabling its application in longevity research and aging studies by evaluating bone health, density, and structural changes over time in adult and elderly populations.

The strongly positive correlation between the Moore and Mirwald methods indicates a high agreement level in their estimates of skeletal age, which was expected. Both methods rely on somatic indicators and growth variables, which is probably the reason for their correlation. This confirms their suitability for assessing skeletal maturity, especially in homogeneous groups. However, methodological similarities between the two may hinder their ability to detect subtleties of skeletal development, particularly when maturation is advanced or delayed. The weaker correlations between the skeletal age indicated by BAUSport™ and those derived from the Moore and Mirwald methods imply that BAUSport™ utilizes different standards or scales in evaluating skeletal maturity. This difference lies in the fact that BAUSport™ relies on a direct assessment of bone using ultrasound technology. Although this raises questions about the comparability of methods, it also opens up the possibility that BAUSport™ may provide valuable information on skeletal development that somatic methods may fail to detect. The lower skeletal age values observed in the BAUSport™ method may reflect the device's sensitivity to structural bone maturity rather than somatic development proxies, underscoring its potential for the early detection of maturation-related injury risk.

There is a remarkable correlation between the diameter of the wrist and BAUSport™ skeletal age. This makes for a good indicator of skeletal maturity since, probably, it is involved with an area of bone development that correlates directly with both chronological and biological aging. The regression analysis further confirmed the influence of the measured variables on the skeletal age measured by BAUSport™, but with wrist diameter standing out as the only significant predictor. This may be because measures such as height, weight, leg length, and sitting height are more influenced by somatic growth, while the BAUSport™ method accurately tracks the development of specific bone structures. These observations point to an idea that could be related to the results from a study using a novel method of wrist skeletal maturity [29]. The authors used epiphyseal–metaphyseal ratios of the first and third metacarpals, combined with chronological age and sex, and showed improved accuracy in estimating skeletal maturity compared to the Greulich and Pyle technique, especially in preadolescents. Such results could also lead to the fact that the wrist is an important factor in maturity estimations. These results further confirm that skeletal age assessment requires a focus on specific anatomical indicators rather than general somatic measurements. Therefore, this supports the idea that making measurements of wrist diameter is a workable, non-invasive approach to estimating skeletal age in situations where there may not be access to sophisticated imaging equipment like BAUSport™. On the other hand, BAUSport™ SonicBone technology is a valuable tool for skeletal age assessment in children. Skeletal growth has critical phases and the BAUSport™ technology can determine these phases during periods of growth, like the "dangerous zone", during which injuries are more common. Children are usually at a greater risk of skeletal injuries during a growth spurt, as they develop rapidly [24,30]. The danger of injury in this phase is what makes it the dangerous zone. By accurate assessment of skeletal maturity, the BAUSport™ system allows practitioners to track growth and development, such as when certain individuals are prone to injuries and when specific training can be incorporated. The injuries are a result of overtraining, so early detection of the disabling illness can aid prevention. Skeletal age assessment using BAUSport™ SonicBone technology is extremely precise and safe, but should still be combined with anthropological methods like Moore's and Mirwald's methods. Their distinct methods of determining skeletal maturity and the varying standards they use result in different outcomes, which cause disparity and confusion in practice. The integration of BAUSport™ technology and anthropological methods provides a more nuanced perspective on the growth and development of an athlete, including skeletal age.

Further studies are needed, though, to determine if it is consistent across different populations and age groups. Also, research should explore the impact of genetic, nutritional, and environmental factors on the correlation between anthropometric measures and skeletal age. One of the most effective ways to investigate genetic factors would be through twin studies, which allow for the comparison of monozygotic and dizygotic twins to distinguish genetic influences from environmental effects on the relationship between anthropometric measures and skeletal age. Researchers could incorporate dental age assessment as an additional indicator of biological age, exploring its potential relationship with skeletal age and building on previous studies that have confirmed the connection between these two parameters [31]. Finally, validation of the model on longitudinal samples is necessary to determine its reliability over time.

5. Conclusions

The findings indicate that 69% of the national team members, with zero late participants, were early maturers, suggesting a potential selection bias favoring physically advanced athletes, which may overlook late-maturing individuals with long-term athletic potential. This highlights the need for a more holistic talent identification approach that considers biological maturation alongside skill development and long-term athletic progression to prevent premature exclusion of late bloomers.

In terms of practical implications, coaches, sports doctors, and selectors should continuously monitor biological maturation to ensure that athletes' development is optimized in accordance with their individual growth. This includes being mindful of the risks of early maturation, such as the potential for overtraining, early burnout, and growth-related injuries.

The study confirms that while the Moore and Mirwald methods are traditional and useful tools for estimating the timing of a youth athlete's growth spurt, the BAUSport™ method represents a significant advancement in this field. This method offers a non-invasive, safe, fast, and reliable alternative for effectively monitoring the maturation of young athletes. The observed limitations in applying the BAUSport™ method to a subset of participants emphasize the need for further validation of BAUSport™ ultrasound-based techniques across diverse populations to ensure broader applicability and methodological reliability.

The three methods used in this study have different approaches to maturity assessment:

- The Moore and Mirwald methods are simple, quick, and easy to apply, but they do not directly measure skeletal age, relying instead on statistical models.
- The BAUSport™ method uses direct ultrasound measurements, providing greater precision and reliability, but requires specialized equipment and further validation.

A combination of ultrasound methods and anthropometric measurements could improve the accuracy of skeletal maturation assessment and optimize talent selection. The main contribution of this research is demonstrating that while the Moore and Mirwald methods offer valuable insights for coaches and sports scientists, the BAUSport™ method has the potential to become a new standardized method for monitoring the skeletal development of young athletes. These findings underscore the importance of integrating both anthropometric prediction models and direct skeletal assessment tools into longitudinal talent development strategies. Such a dual-modality approach may provide a more accurate, ethical, and individualized pathway for youth athlete monitoring and selection. Future studies should include comparative analyses with other advanced methods, such as genetic markers, hormonal assessments, or MRI technology, to further enhance the understanding of biological maturation and its role in sports talent development. Additionally, further investigation is needed to optimize the calibration of the BAUSport™ device to accommodate a wider range of anatomical variations and ensure its applicability across diverse populations.

Author Contributions: Conceptualization, D.C.; methodology, D.C. and M.C.; software, M.B., A.K. and M.C.; validation, M.B., A.K. and T.B.; formal analysis, M.B., T.B., A.K. and M.C.; investigation, M.B., A.K. and T.B.; resources, D.C.; data curation, D.C., A.K. and M.C.; writing—original draft preparation, D.C. and A.K.; writing—review and editing, M.B. and M.C.; visualization, M.B. and T.B.; supervision, D.C.; project administration, A.K. All authors have read and agreed to the published version of the manuscript.

Funding: This research was supported by the Croatian Science Foundation under Project Grants No. [IP-2020-02-3366], No. [DOK-2021-02-8613], and No. [IP-2024-05-8340]. All grants contributed to different aspects of the study in compliance with the funding regulations.

Institutional Review Board Statement: This study was conducted according to the Declaration of Helsinki, and the Ethics Committee of the Faculty of Kinesiology University of Split, Croatia, approved the described protocol (permit number: 003-08/20-04/00121818-205-02-05-20-006, date: 26 February 2020).

Informed Consent Statement: Written informed consent to publish this paper was obtained from the participants.

Data Availability Statement: The data presented in this study are available upon request from the corresponding author.

Conflicts of Interest: Author Drazen Cular was the owner of the company Einstein, Startup for Research, Development, Education, Trade and Services. The remaining authors declare that the research was conducted in the absence of any commercial or financial relationships that could be construed as a potential conflict of interest.

References

1. Müller, L.; Müller, E.; Hildebrandt, C.; Kapelari, K.; Raschner, C. The Assessment of Biological Maturation for Talent Selection—Which Method Can Be Used? *Sportverletz. Sportschaden* **2015**, *29*, 56–63.
2. Mirwald, R.L.; Baxter-Jones, A.D.G.; Bailey, D.A.; Beunen, G.P. An Assessment of Maturity from Anthropometric Measurements. *Med. Sci. Sports Exerc.* **2002**, *34*, 689–694.
3. Monasterio, X.; Gil, S.M.; Bidaurrazaga-Letona, I.; Cumming, S.P.; Malina, R.M.; Williams, S.; Larruskain, J. Estimating Maturity Status in Elite Youth Soccer Players: Evaluation of Methods. *Med. Sci. Sports Exerc.* **2024**, *56*, 102–112. [CrossRef] [PubMed]
4. Moore, S.A.; McKay, H.A.; Macdonald, H.; Nettlefold, L.; Baxter-Jones, A.D.; Cameron, N.; Brasher, P.M. Enhancing a Somatic Maturity Prediction Model. *Med. Sci. Sports Exerc.* **2015**, *47*, 1755–1764. [CrossRef]
5. Cossio-Bolaños, M.; Vidal-Espinoza, R.; de Campos, L.F.C.C.; Sulla-Torres, J.; Cossio-Bolaños, W.; Albornoz, C.U.; Gómez-Campos, R. Equations Predicting Maturity Status: Validation in a Cross-Sectional Sample to Assess Physical Growth and Body Adiposity in Chilean Children and Adolescents. *Endocrinol. Diabetes Nutr.* **2021**, *68*, 689–698. [CrossRef] [PubMed]
6. Fransen, J.; Bush, S.; Woodcock, S.; Novak, A.; Deprez, D.; Baxter-Jones, A.D.; Lenoir, M. Improving the Prediction of Maturity from Anthropometric Variables Using a Maturity Ratio. *Pediatr. Exerc. Sci.* **2018**, *30*, 296–307. [CrossRef] [PubMed]
7. Moraes Macêdo, M.; Linhares, R.V.; Filho, J.F. Equations for Determining Bone Age and Sexual Maturation of Children and Adolescents. *Rev. Salud Pública* **2015**, *17*, 267–276.
8. Malina, R.M.; Dompier, T.P.; Powell, J.W.; Barron, M.J.; Moore, M.T. Validation of a Noninvasive Maturity Estimate Relative to Skeletal Age in Youth Football Players. *Clin. J. Sport Med.* **2007**, *17*, 362–368. [CrossRef]
9. Gouvea, M.; Cyrino, E.S.; Ribeiro, A.S.; da Silva, D.R.P.; Ohara, D.; Valente-Dos-Santos, J.; Coelho-E-Silva, M.J.; Ronque, E. Influence of skeletal maturity on size, function and sport-specific technical skills in youth soccer players. *Int. J. Sports Med.* **2016**, *37*, 464–469. [CrossRef]
10. Suh, J.; Heo, J.; Kim, S.J.; Park, S.; Jung, M.K.; Choi, H.S.; Choi, Y.; Oh, J.S.; Lee, H.I.; Lee, M.; et al. Bone Age Estimation and Prediction of Final Adult Height Using Deep Learning. *Yonsei Med. J.* **2023**, *64*, 679–686. [CrossRef]
11. Leyhr, D.; Murr, D.; Basten, L.; Eichler, K.; Hauser, T.; Lüdin, D.; Romann, M.; Sardo, G.; Höner, O. Biological Maturity Status in Elite Youth Soccer Players: A Comparison of Pragmatic Diagnostics with Magnetic Resonance Imaging. *Front. Sports Act. Living* **2020**, *2*, 587861. [CrossRef] [PubMed]
12. Cumming, S.; Pi-Rusiñol, R.; Rodas, G.; Drobnic, F.; Rogol, A.D. The Validity of Automatic Methods for Estimating Skeletal Age in Young Athletes: A Comparison of the BAUSport Ultrasound System and BoneXpert with the Radiographic Method of Fels. *Biol. Sport* **2023**, *41*, 61–67. [CrossRef]
13. Rüeger, E.; Hutmacher, N.; Eichelberger, P.; Löcherbach, C.; Albrecht, S.; Romann, M. Ultrasound Imaging-Based Methods for Assessing Biological Maturity during Adolescence and Possible Application in Youth Sport: A Scoping Review. *Children* **2022**, *9*, 1985. [CrossRef]
14. Towlson, C.; Salter, J.; Ade, J.D.; Enright, K.; Harper, L.D.; Page, R.M.; Malone, J.J. Maturity-Associated Considerations for Training Load, Injury Risk, and Physical Performance in Youth Soccer: One Size Does Not Fit All. *J. Sport Health Sci.* **2021**, *10*, 403–412. [CrossRef] [PubMed]
15. Rachmiel, M.; Naugolni, L.; Mazor-Aronovitch, K.; Koren-Morag, N.; Bistritzer, T. Bone Age Assessments by Quantitative Ultrasound (SonicBone) and Hand X-ray Based Methods Are Comparable. *Isr. Med. Assoc. J.* **2017**, *19*, 533–538. [PubMed]
16. Sasaki, M.; Motegi, E.; Soejima, U.; Nomura, M.; Kaneko, Y.; Shimizu, T.; Takeuchi, F.; Yamaguchi, T.; Yamanaka, S.; Yamaguchi, H. Appraisal of Bone Maturity Age Derived from Broadband Ultrasonic Attenuation in Japanese Children and Adolescents. *Bull. Tokyo Dent. Coll.* **2003**, *44*, 37–42. [CrossRef]
17. Shimura, N.; Koyama, S.; Arisaka, O.; Imataka, M.; Sato, K.; Matsuura, M. Assessment of Measurement of Children's Bone Age Ultrasonically with Sunlight BonAge. *Clin. Pediatr. Endocrinol.* **2005**, *14* (Suppl. 24), S24_17–S24_20. [CrossRef]
18. Olivete Júnior, C.; Rodrigues, E.L.L. Bone Maturity: Estimation by Means of Eklof and Ringertz Method Simplifications. *Radiol. Bras.* **2010**, *43*, 13–18. [CrossRef]
19. Baxter-Jones, A.D.; Thompson, A.M.; Malina, R.M. Growth and Maturation in Elite Young Female Athletes. Sports Med. *Arthrosc. Rev.* **2002**, *10*, 42–49.
20. Cameron, N. The Tanner-Whitehouse II Skeletal Maturity Method: Rationale and Applicability. *Clin. Pediatr. Endocrinol.* **1993**, *2*, 9–18. [CrossRef]
21. Ruf, L.; Cumming, S.; Härtel, S.; Hecksteden, A.; Drust, B.; Meyer, T. Construct Validity of Age at Predicted Adult Height and BAUS Skeletal Age to Assess Biological Maturity in Academy Soccer. *Ann. Hum. Biol.* **2021**, *48*, 101–109. [CrossRef] [PubMed]
22. Malina, R.M.; Bouchard, C.; Bar-Or, O. *Growth, Maturation, and Physical Activity*; Human Kinetics: Champaign, IL, USA, 2004.
23. Grigoletto, A.; Mauro, M.; Toselli, S. Differences in Body Composition and Maturity Status in Young Male Volleyball Players of Different Levels. *J. Funct. Morphol. Kinesiol.* **2023**, *8*, 162. [CrossRef] [PubMed]
24. Caine, D.J.; Goodwin, B.J. Risk Factors for Injury in Pediatric and Adolescent Sports. In *Injury in Pediatric and Adolescent Sports*; Springer: Berlin/Heidelberg, Germany, 2016; pp. 191–203.

25. Rumpf, M.C.; Cronin, J. Injury Incidence, Body Site, and Severity in Soccer Players Aged 6–18 Years: Implications for Injury Prevention. *Strength Cond. J.* **2012**, *34*, 20–31. [CrossRef]
26. van der Sluis, A.; Elferink-Gemser, M.T.; Coelho-e-Silva, M.J.; Nijboer, J.A.; Brink, M.S.; Visscher, C. Sport Injuries Aligned to Peak Height Velocity in Talented Pubertal Soccer Players. *Int. J. Sports Med.* **2014**, *35*, 351–355. [CrossRef]
27. McManus, A.M.; Armstrong, N. Physiology of Elite Young Female Athletes. *Med. Sport Sci.* **2011**, *56*, 23–46.
28. Whitney, K.E.; d'Hemecourt, P.A.; Stracciolini, A. Youth Sport Specialization: Risks, Benefits, and Mental Health Considerations. In *Psychological Considerations in the Young Athlete*; Springer: Berlin/Heidelberg, Germany, 2023; pp. 179–191.
29. Huang, L.F.; Furdock, R.J.; Uli, N.; Liu, R.W. Estimating Skeletal Maturity Using Wrist Radiographs during Preadolescence: The Epiphyseal: Metaphyseal Ratio. *J. Pediatr. Orthop.* **2022**, *42*, e801–e805. [CrossRef] [PubMed]
30. Wik, E.H.; Martínez-Silván, D.; Farooq, A.; Cardinale, M.; Johnson, A.; Bahr, R. Skeletal Maturation and Growth Rates Are Related to Bone and Growth Plate Injuries in Adolescent Athletics. *Scand. J. Med. Sci. Sports* **2020**, *30*, 894–903. [CrossRef]
31. Bala, M.; Pathak, A.; Jain, R.L. Assessment of Skeletal Age Using MP3 and Hand-Wrist Radiographs and Its Correlation with Dental and Chronological Ages in Children. *J. Indian Soc. Pedod. Prev. Dent.* **2010**, *28*, 95–99. [CrossRef]

Journal of
Functional Morphology and Kinesiology

MDPI

Article

Posture Status Differences Between Preschool Boys and Girls

Anida Kapo-Gurda [1,*], Amin Efendić [2], Indira Mahmutović [1], Siniša Kovač [2], Husnija Kajmović [2], Safet Kapo [2] and Jožef Šimenko [3]

1 Faculty of Education, University of Sarajevo, 71000 Sarajevo, Bosnia and Herzegovina; imahmutovic@pf.unsa.ba

2 Faculty of Sports and Physical Education, University of Sarajevo, 71000 Sarajevo, Bosnia and Herzegovina; aminefendic@hotmail.com (A.E.); sinisa.kovac@fasto.unsa.ba (S.K.); husnija.kajmovic@fasto.unsa.ba (H.K.); safet.kapo@fasto.unsa.ba (S.K.)

3 Faculty of Sport, University of Ljubljana, 1000 Ljubljana, Slovenia; jozef.simenko@fsp.uni-lj.si

* Correspondence: akapo@pf.unsa.ba

Abstract: Background/Objectives: The preschool period plays an essential role in shaping a child's overall development, which influences physical, emotional, social, and cognitive growth. At this stage, establishing proper postural habits is essential, as it can have lasting effects on health, well-being, helps to prevent future issues, and supports overall development. Therefore, the present work aims to determine the differences in postural status between boys and girls of preschool age. **Methods**: The sample of participants consisted of 92 children (n = 46 boys and n = 46 girls); the average age for girls was 5.41 ± 0.30 years and for boys it was 5.53 ± 0.31 years. Data were collected using licensed state-of-the-art diagnostic equipment, Contemplas 3D Posture Compact, using 16 variables to assess postural status with a Mann–Whitney U test. **Results**: The results of this study indicate that boys have more pronounced deformities in the following variables: shoulder displacement ($p = 0.047$), pelvic obliquity ($p = 0.000$), sag. distance cervical spine–sacrum ($p = 0.029$), sag. distance thoracic spine–sacrum (SDTS) ($p = 0.016$), and sag. distance lumbar spine–sacrum (SDLS) ($p = 0.005$). **Conclusions**: This study confirmed gender differences in postural characteristics in preschool children. Boys showed a greater tendency towards postural deviations, indicating the necessity for specific interventions and programs to improve their posture. On the basis of the results of this research, it is recommended to carry out cross-cultural research that would enable the comparison of results among children from different environments and cultural contexts in order to determine possible differences and particularities in the development of postural characteristics. Future research should include larger and more diverse samples of participants, including children from rural and urban areas, in order to ensure the representativeness and generalizability of the results. In addition, conducting a longitudinal study that would monitor the postural characteristics of children through different developmental stages is suggested, aiming to identify critical periods for intervention and to determine, more precisely, development trends within the context of gender differences.

Keywords: assessment; body posture; deformities; early childhood; gender differences

1. Introduction

The preschool period encompasses the period from the third year until the beginning of primary schooling (five to seven years) [1]. During that period, a child masters various activities, processes a large amount of different information, develops his speech to such a level that brings him close to the speech of adults, begins to think logically, and develops his senses [2]. The important aspects of child development at this age is the development of postural habits and patterns that can have a lifelong impact on the health and well-being of a child. In recent years, abnormalities related to body posture, often called postural defects, have increasingly been diagnosed in children [3–7]. In writing on this topic, many authors have tried to define what posture, in fact, is. Posture is considered a descriptive term for the relative position of body segments during rest or physical activity. Therefore, good posture implies an optimal relation between the reduction in load on the spinal column and the reduction in muscle work [8], i.e., the musculoskeletal balance that protects from the formation and gradual development of postural disorders in the structures responsible for keeping the body upright or stable, whether during movement or at rest [9,10]. On the other hand, researchers describe the postural status as a quality interrelationship of individual body segments aiming at maintaining the correct and upright body position without disturbing its stable position or falling [11]. Therefore, it is considered that each motor task is performed successfully only if there is no significant disturbance of body stability (equilibrium position).

Correct body posture implies correct relations between all body segments. This should be especially emphasized when it comes to children of preschool age because that period of development is of crucial importance. Given that postural control undergoes significant transitions between ages 4 and 6 [12], this study aims to assess whether early deviations can already be observed in this developmental phase. It starts with establishing control of the head, then the torso, and finally achieving postural stability while standing.

The sensorimotor system, which controls postural stability, undergoes significant changes between the ages of four and six, reaching full maturity between the ages of seven and ten [13]. Because of the sensitivity of a child's body, establishing proper posture is critically important during the preschool years and the early stages of schooling [14]. In this period, children begin to adopt basic patterns of movement and body posture. Correct posture is not only an aesthetic issue but also has important functional and health implications.

Primary health care emphasizes the importance of preventing, diagnosing, and treating postural deformities in children early. Identifying postural disorders is extremely important, especially at the preschool age, as poor posture can indicate serious health issues if left uncorrected. However, these problems are often overlooked [15–18]. The reason lies in the necessity of the early formation of a "pattern of proper body posture", which, if created in early childhood, not only contributes to the proper growth and development of children but also has a positive impact on their health and quality of life [19]. Poor body posture in children of preschool and school age is an indicator of health problems that can become very serious if not corrected in time [20]. Nevertheless, these issues often go unnoticed until it is too late. This highlights the importance of educators' role in supporting healthy growth and development. By consistently monitoring and periodically evaluating children's posture using 3D analysis models, many health problems can be identified early—before they become serious. The Contemplas Templo 3D apparatus [18,21] is one of these methods, which enables the objective and detailed measurement of body position, i.e., postural status, by analysis in three-dimensional space. Contemplas 3D technology is a modern method that combines optical analysis, computer algorithms, and specialized software to obtain precise data on body posture. The method is non-invasive and identifies asymmetries in postural segments (head, shoulders, torso, pelvis) and changes in the curvature of the spine, such as kyphosis, lordosis, or scoliosis, with clear reference values and visual interpretation. Unlike traditional postural assessment methods, 3D motion capture provides quantifiable and objective data, reducing observer bias and improving reproducibility [20].

However, the results of previous research are inconsistent regarding the dynamics of the development of sagittal curves of the spinal column and the identification of critical periods regarding the postural status of children and adolescents [13,18,21–24]. Common to all these studies is their recognition of specific age-related dynamics. Moreover, one key finding is the identification of patterns in upright posture development, showing that as children grow, changes generally lead to reduced spinal mobility and more pronounced deviations from correct posture. Therefore, the present study aimed to test these assumptions with the main aim of discovering the differences in postural status between boys and girls aged 5 to 6 in the Sarajevo Canton area.

2. Materials and Methods

2.1. The Sample of Participants

The present study is of a descriptive cross-sectional design, where the sample of participants consisted of 92 children, of which n = 46 girls (5.41 ± 0.30 years; 115.82 ± 5.61 cm, 21.05 ± 3.37 kg (body mass); 16.74 ± 2.19 kg (fat-free mass; 15.84 ± 2.09 kg muscle mass)) and n = 46 boys (5.53 ± 0.31 years; 119.29 ± 5.39 cm, 23.44 ± 5.12 kg (body mass); 17.58 ± 2.93 kg (fat-free mass); 16.53 ± 2.78 kg (muscle mass)) from the Public Institution "Children of Sarajevo", from the area of Sarajevo Canton. The sample was selected using a simple random sampling method, by which an equal share of boys and girls was achieved. The G*Power software (Version 3.1.9.7, Universität Kiel, Kiel, Germany) was used to determine the number of participants through a priori power analysis for an independent T-test [25]. The results were taken by the following assumptions: two-tailed test, with significance level set at $\alpha = 0.05$, study power set at 0.80, effect size set at 0.3, and distribution ratio set at 1 [21]. The recommended total number of samples was 82 (41 per group). However, we recruited 46 participants per group, resulting in an actual study power of 0.95. Considering that all the children included in the research were minors, their parents/guardians signed a written consent form regarding the participation of their children in the research. The participants had no associated diseases from the spectrum of possible influence on postural disorders/deformations.

All procedures were carried out in accordance with recommendations of the Helsinki Declaration and ethical standards of the Ethics Committee of the University of Sarajevo. Additionally, the study received an ethical committee approval from the Ministry of Education, Science, and Youth of Sarajevo Canton No. 11/03-34-37155/20.

2.2. Testing Protocol

2.2.1. Anthropometry

Body height was measured with an anthropometer, according to Martin. During the measurement, the subjects were barefoot and in underwear (girls in undershirts) and stood in an upright position on a firm horizontal surface. The head of the subject was in the horizontal Frankfort plane. The examiner stood on the left side of the subject and controlled whether the anthropometer device was placed vertically and directly along the back of the body. The horizontal arm of the anthropometer is lowered to the top of the head (vertex point) and supported firmly but without pressure. The measurement result for body height values was read with an accuracy of 0.1 cm. An experienced examiner performed the measurements.

2.2.2. Body Composition

Body composition was measured using a Tanita BC-420MA digital scale (Tanita BC 420 MA Segmental Body Composition Analyzer, Tanita Corp., Tokyo, Japan, 2015) [26].

The scale's measuring platform features four electrode plates on which the subject stands barefoot, wearing only underwear. A low-strength direct current is transmitted through these contact plates to the body, and the scale measures the body's total electrical resistance. Using the integrated software, along with inputted data such as body height, age, and sex, the device calculated the following variables: body mass, muscle mass, and fat-free mass. The results for body composition are displayed with a precision of 0.1 kg and 0.1%.

2.2.3. Posture Assessment

The CONTEMPLAS Templo professional motion analysis software is a reliable and modern tool for diagnosing and detecting postural disorders. Due to its 3D analysis, Contemplas provides much more data on postural status compared to previous methods used to diagnose this problem. The validity of the system was confirmed by the Sports University of Cologne with an acceptable to excellent ICC (ICC > 0.70) agreement to the 3D Vicon system [27], while the inter and intra-rater reliability was also reported with acceptable to excellent consistency (ICC > 0.75) [22]. Such a method provides a wide range of possibilities for many specific needs and may be adapted to many hardware solutions with the help of which various analyses of movement and passive positions can be performed, resulting in conclusions about needs, knowledge, or cause of deformity [10]. 3D Compact is an analysis for assessing the postural status and status of the legs using three cameras that synchronically record and analyze the body from three perspectives simultaneously. In the 3D analysis, retroreflective marker balls were used (Figure 1b) [28].

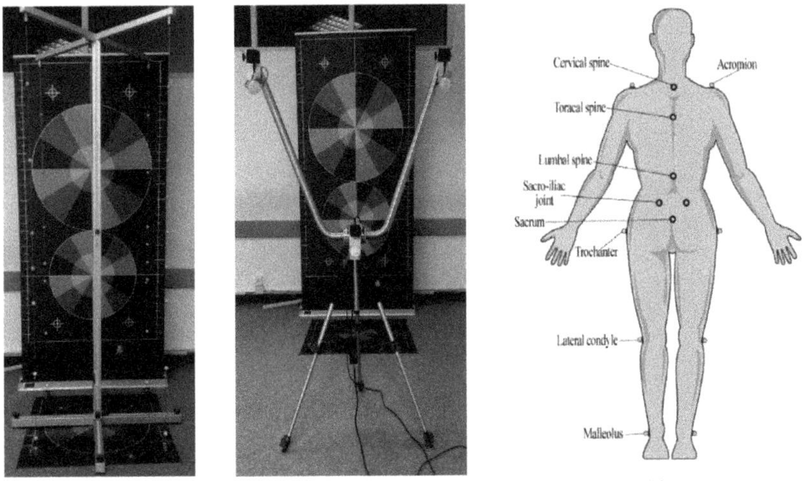

(a) (b) (c)

Figure 1. The Contemplas Templo apparatus ((**a**)—calibration frame; (**b**)—V camera frame; (**c**)—marker positions for 3D posture Compact protocol) [29].

Postural status in a standing position involves testing postural characteristics using 3D Contemplas software [27]. The procedure involves projecting the position of specific points on the body, after which a 3D kinematic model is created in the frontal plane. The CONTEMPLAS GmbH TEMPLO photometric apparatus, which includes a camera system and software analysis, by which it determines the position of marked points in space according to the 3D protocol, detects the positions of body segments in the calibrated space [22]. The measuring plate is placed on a flat surface. After determining the flat surface and placing the measuring plate, the Contemplas 3D posture compact mode is placed and fixed on the surface to avoid movement during the children's positioning and additional space calibration. The 3D calibrator is placed on the surface with fluorescent markers. The 3D calibrator must be placed exactly in the center of the measuring plate, and the upper and lower beams, together with the vertical beam, must be ideally aligned and leveled (Figure 1a). The next step is to install a "V" frame that supports three cameras, which enable 3D analysis (Figure 1c). The camera's distance from the center of the measuring plate must be at least 2 m and 15 cm. The images captured by the camera must be sharpened in software programming to begin space calibration. After completing the calibration, the 3D calibrator is packed away and testing can begin. An experienced examiner placed 16 reflective markers on the body of the subject, who only wore underwear, on specific points on the body: acromion (left and right), cervical spine, thoracic spine (kyphosis), lumbar spine (lordosis), crista iliaca posterior superior (left and right), sacrum, trochanter major (left and right), condylus laterallis (left and right), and malleolus laterallis (left and right), after which the subject is positioned in calibrated space with the back turned to the cameras, feet parallel and hip-width apart, and in the frontal plane. After that, the subject is instructed to take an upright position, look straight ahead, and relax his arms next to his body. Then, the projection takes place for 12 s, and photos are taken using three cameras (Basler acA645-100 gm/gc). The subject is photographed in an upright position. After the test, the markers were removed from the subject and placed on the next subject to be tested. The procedure of assembling and testing the instrument calibration is repeated whenever the place of testing is changed [22].

Four experienced certified measurers from the Sports Institute, with more than 7 years of experience working with the equipment, carried out the complete measurement procedure. To avoid subjectivity, only one person placed reflective markers in accordance with the protocol.

1. Sample of variables

The variables used for the purpose of this research provide basic information about the status of body position. The sample variables consist of 16 parameters. Shoulder displacement (SD) measures lateral shoulder asymmetry in the frontal plane; pelvic obliquity (PO) displays the elevated/lowered left/right pelvic side in the frontal plane; shoulder rotation (SR) measures left/right (L/R) shoulder displacement in the transversal plane; pelvic rotation (PO) measures L/R rotation in the transversal plane; trochanter rotation (TR) measures L/R trochanter rotation in the transversal plane; condylus rotation (CR) measures knee rotation in the transversal plane; the sagittal distance of the cervical spine (SDCS) indicates the distance of the most protruded cervical (neck) vertebra in regards to the vertical line projection of the sacrum in the sagittal plane; the sag. sacrum/sagittal distance of the thoracic spine (SDTS) indicates the distance of the thoracic spine in regard to the vertical line projections of the sacrum in the sagittal plane; sag. distance lumbar spine–sacrum/sagittal distance of the lumbar spine (SDLS) indicates the distance of the lumbar (lower) spine in regards to the vertical line projection of the sacrum in the sagittal plane; varus/valgus left/X/O left leg (VVL) indicates the varus/valgus alignment angle of the left leg (medial/lateral) at the knee joint; varus/valgus right/X/O right leg (VVR) indicates the varus/valgus alignment angle of the right leg (medial/lateral) at the knee joint; flexion/extension left leg hyperextension/flexion of the left leg (FEL) indicates the hyperextension and flexion of the left leg at the knee joint in the sagittal plane; flexion/extension right leg (FER) indicates the hyperextension and flexion of the right leg at the knee joint in the sagittal plane; the frontal cervical spine (CS) indicates the distance of the cervical spine in the frontal plane in relation to the vertical line projection of the sacrum; the frontal thoracic spine (TS) indicates the distance of the thoracic spine in the frontal plane in relation to the vertical line projection of the sacrum; frontal lumbar spine (LS) indicates the distance of the lumbar spine in the frontal plane in relation to the vertical line projection of sacrum. These measurements were obtained using the 3D posture compact test protocol of the Contemplas measuring instrument [20]. The obtained parameters indicate possible deviations from zero (normal) values of the posture status in all three planes. A higher deviation value (positive or negative) implies a higher level of deformity [29].

2.3. Statistical Analysis

The data analysis was performed using SPSS V30.0 statistical software for the social sciences (SPSS Inc., Chicago, IL, USA). The following descriptive statistics were used to determine the basic descriptive parameters: mean and standard deviation (SD). The normality of data assessment was determined using the Kolmogorov–Smirnov test. As variables were not normally distributed, the nonparametric Mann–Whitney U test was used to determine differences between boys and girls in postural status. Cohen's r value was used to measure the effect size with $r \geq 0.10$ indicating a small effect; $r \geq 0.30$ indicating a medium effect; and $r \geq 0.50$ indicating a large effect [30–32]. There were no missing data. The Alfa level was set at $p \leq 0.05$.

3. Results

Perusing the results of the posture recording by the Contemplas–Templo 3D compact mode procedure (Table 1), large variations in the angular parameters that describe the status of the legs in boys and girls are noticeable. A wide range of variable values show the status of the participants' legs: varus/valgus left (VVL)/X/O left leg, varus/valgus right (VVR)/X/O right leg, flexion/extension left leg (FEL)/hyperextension/flexion left leg, and flexion/extension right leg (FER)/hyperextension/flexion right leg for relations between extension and flexion of the knee joint in the sagittal plane (legs are either slightly flexed or in hyperextension—sagittal plane, which shifts the angle ratios, e.g., from −163.18° to +174.27°). A similar finding is also present in the varus/valgus relation of the upper leg in relation to the lower leg (frontal plane of −173.97° to +179.55°). Such a range of results is determined by the position of the active, spherical benchmarks that represent the center of the hip joint—laterally, the center of the knee joint—laterally and the lateral malleolus, both for the right and for the left side (the position relationship of the mentioned benchmarks valorizes the angular values of the above variables). The position of the spherical markers is adjusted by software for 3D analysis and, in this case, reflects the state of a potential slight flexion or hyperextension in the knees in the upright position, recorded from the back for the variables flexion/extension left leg (FEL)/hyperextension/flexion of the left leg and flexion/extension right leg (FER)/hyperextension/flexion of the right leg. In addition, the spherical rappers reflect the potentially accentuated "Q" angle position of the center of the hip joint and the center of the knee joint, ending at the corresponding malleolus, in relation to the status of the subjects' legs.

Table 1. A table containing descriptive statistics of selected variables, Mann–Whitney U test between genders, and effect size.

Variables	Girls		Boys		Mean Rank		U	Z	*p*	*r*
	Mean	SDv	Mean	SDv	Girls	Boys				
SD (cm)	−4.7	8.78	−9.94	12.32	52.0	41.0	803.5	−1.99	0.047 *	0.21
PO (cm)	0.49	11.8	−3.62	3.06	56.4	36.6	603.5	−3.55	0.000 *	0.37
SR (°)	26.95	42.4	32.6	46.13	44.8	48.2	981.5	−0.60	0.550	0.06
PR (°)	26.6	45.9	27.26	53.88	47.9	45.1	994.5	−0.50	0.620	0.05
TR (°)	40.87	47.4	49.99	58.69	43.9	49.1	937.5	−0.94	0.347	0.10
CR (°)	4.17	55.1	16.84	73.68	42.1	50.9	857.5	−1.57	0.117	0.16
SDCS (cm)	−18.1	45.7	−44.7	48.98	52.6	40.4	778.5	−2.18	0.029 *	0.23
SDTS (cm)	−8.67	37.9	−27.8	33.21	53.2	39.8	748.5	−2.42	0.016 *	0.25

Table 1. *Cont.*

Variables	Girls		Boys		Mean Rank		U	Z	p	r
	Mean	SDv	Mean	SDv	Girls	Boys				
SDLS (cm)	1.75	20.7	−4.85	9.69	54.3	38.7	699.5	−2.80	0.005 *	0.29
VVL (°)	10.47	84.3	22.12	62.22	41.9	51.1	848.5	−1.64	0.102	0.17
VVR (°)	−27.1	92	−46.1	101.1	48.0	45.0	987.5	−0.55	0.582	0.06
FEL (°)	5.34	74.1	−14.8	68.07	47.5	45.5	1014	−0.35	0.728	0.04
FER (°)	−36.1	91.7	−53.3	101.2	49.3	43.7	927.5	−1.02	0.308	0.11
CS (cm)	0.22	14.7	−0.59	27.89	48.3	44.7	975.5	−0.64	0.519	0.07
TS (cm)	2.4	11.5	2.99	26.67	49.5	43.5	921.5	−1.07	0.286	0.11
LS (cm)	1.14	7.13	2.97	14.62	49.4	43.6	926	−1.03	0.303	0.11

Legend: All values are presented as mean ± standard deviation. Variables marked with * indicate statistically significant differences ($p \leq 0.05$); standard deviation (SDv); shoulder displacement (SD); pelvic obliquity (PO); shoulder rotation (SR); pelvic rotation (PO); trochanter rotation (TR); condylus rotation/knee rotation (CR); sag. distance cervical spine–sacrum/sagittal distance of the neck part of the spinal column (SDCS); sag. distance thoracic spine–sacrum/sagittal distance of the thoracic spine (SDTS); sag. distance lumbar spine–sacrum/sagittal distance of the lumbar part of the spinal column (SDLS); varus/valgus left/X/O left leg (VVL); varus/valgus right/X/O right leg (VVR); flexion/extension left leg hyperextension/flexion of the left leg (FEL); flexion/extension right leg/hyperextension of the right leg (FER); frontal cervical spine/frontal distance of the neck part of the spinal column (CS); frontal thoracic spine/frontal distance of the chest part of the spinal column (TS); frontal lumbar spine/frontal distance of the lumbar part of the spinal column (LS); Mann–Whitney U test statistic (U); Mann–Whitney U test Z statistic (Z); Mann–Whitney U test significance value (p); Cohen's r effect size (r).

By analyzing the results in Table 1, it was observed that boys exhibited significantly greater deviation from 0 in the following variables: shoulder displacement (SD) with $p = 0.047$; $r = 0.21$ (small effect), pelvic obliquity (PO) with $p = 0.000$; $r = 0.37$ (medium effect), sag. distance cervical spine–sacrum (SDCS) with $p = 0.029$; $r = 0.23$ (small effect), sag. distance thoracic spine–sacrum (SDTS) with $p = 0.016$; $r = 0.25$ (small effect), and sag. distance of the lumbar part of the spinal column (SDLS) with $p = 0.005$; $r = 0.29$ (small effect) (Figure 2). The effect size in the SD, SDCS, SDTS, and SDLS variables was small, while in the PO variable, it was considered medium, suggesting that the observed differences may have small to moderate clinical relevance.

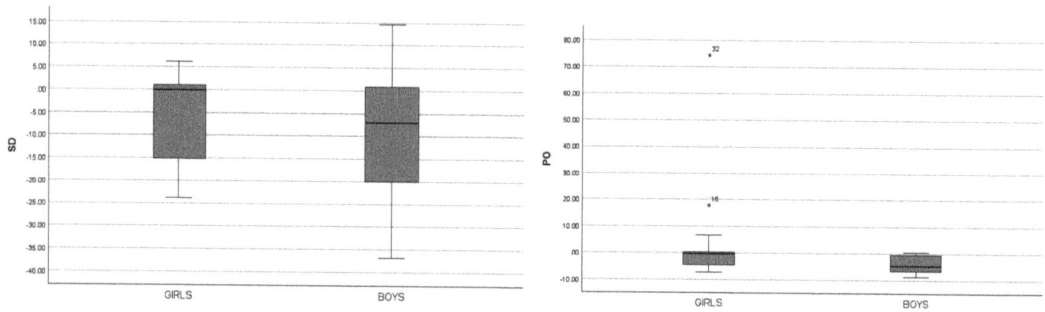

Figure 2. *Cont.*

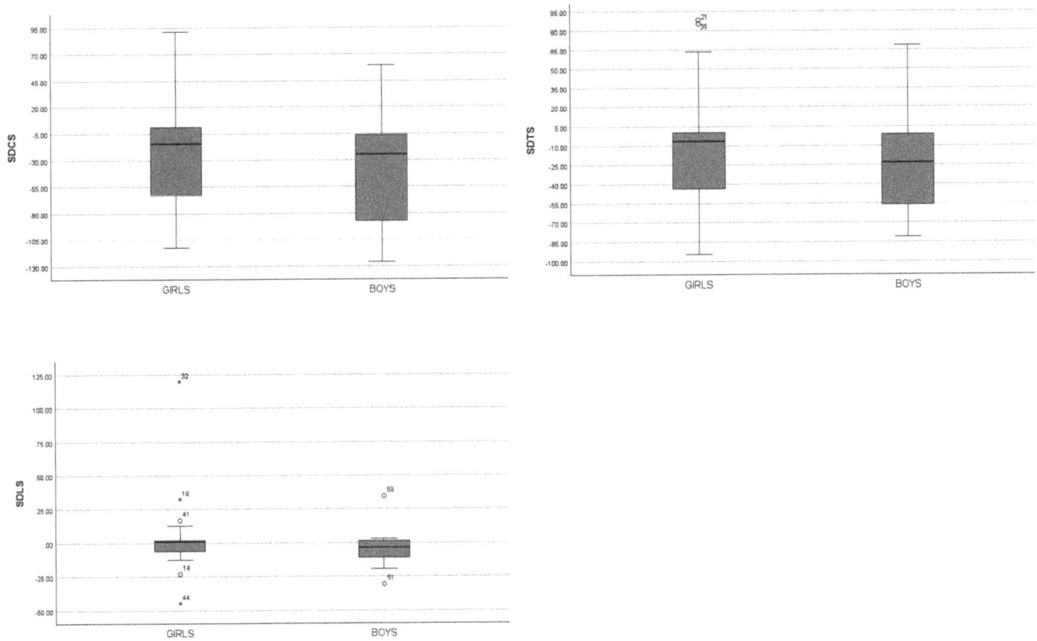

Figure 2. Boxplots of significant variables for boys and girls. * = an outlier.

4. Discussion

The present study highlighted gender differences in postural characteristics in preschool children. Boys showed a greater tendency towards postural deviations, indicating the necessity for specific interventions and programs to improve their posture.

Postural deformities in children have been the subject of numerous studies for many years [3,7,20–22,33–35]. The prevalence of postural deformities (30–80%) is very broad, reflecting variations in assessment methodologies, age groups, and definitions of postural abnormalities. A study on children aged 5–18 years [33] reported particularly high rates of scoliosis (86.5%), kyphosis (71%), and lordosis (28.4%), with 70% of participants exhibiting multiple disorders simultaneously. Similarly, research on school-aged children (6–15 years) [36] identified high frequencies of forward head posture (53.5%), shoulder elevation (74.3%), and winged scapulae (66.3%), highlighting the widespread nature of postural deviations. Beyond structural misalignments, the impact of body composition on postural health was evident in a study on children aged 3–18 years [37], where overweight and obese participants demonstrated significantly higher rates of postural errors (69.2% and 78.6%, respectively).

Large-scale research involving 595,057 Chinese children [34] found an overall prevalence of incorrect posture at 65.3%, with higher rates among older students and girls, suggesting a need for targeted intervention strategies. In younger populations, research on children aged 8–10 years [3] revealed that hyperlordosis (24.1%) was the most frequent condition, while scoliotic posture (33.3%) and flat back syndrome (18.4%) were also prevalent. Additionally, a study on children aged 6–8 years [35] reported fallen foot arches in 65% of participants, with 30% showing spinal curvature deformities in the sagittal plane and 13% in the frontal plane, reinforcing the interconnection between foot structure and spinal posture.

Further emphasizing the severity of postural issues, research conducted in Bulgaria on children aged 6–11 years [24] found incorrect posture in 58.85% of cases, while 23.67% were diagnosed with spinal deformities, raising concerns about the long-term implications of these conditions. Lastly, a study focusing on children aged 7–12 years [38] reported postural disorders in the sagittal plane in 83.9% of participants, indicating that spinal misalignments are highly prevalent in this age group.

Primary health care considers the prevention, early diagnosis, and treatment of postural deformities in children to be of great importance [16]. Poor posture in preschool children is an indicator of health problems that can become very serious if not corrected in time [15]. However, these issues are often not detected early enough. This underscores the important role educators play in supporting healthy growth and development. Through the regular monitoring and assessment of children's posture, many health problems could be identified and addressed before they progress and leave serious consequences.

This research indicates that boys have more pronounced deformities in the following variables (Figure 2): shoulder displacement (SD) $p = 0.047$; $r = 0.21$, pelvic obliquity (PO) $p = 0.000$; $r = 0.37$, sag. distance cervical spine–sacrum (SDCS) $p = 0.029$; $r = 0.23$, sag. distance thoracic spine–sacrum (SDTS) $p = 0.016$; $r = 0.016$, and in sag. distance of the lumbar part of the spinal column (SDLS) $p = 0.005$; $r = 0.29$ compared to girls. The obtained values indicate statistically significant differences between the groups with small to medium effect sizes. Based on the results obtained, it is evident that boys have a more pronounced shoulder deviation (SD) from the ideal position compared to girls. Such a finding may indicate that growth patterns or developmental characteristics can influence the resulting postural differences, or that boys have certain differences in postural habits compared to girls. In Slovenian studies, postural deformities were found in one-half of the children. Deviations in the position of the shoulders and shoulder blades were recorded in more than 80% of children, while flat feet affected 65% of them [35].

Gender differences in this age group have been explored by researchers from several perspectives. The literature suggests that boys generally develop muscle strength later than girls, which potentially leads to weaker postural muscles during early childhood. This delayed development has been connected to higher incidences of postural issues, such as a winged scapula and shoulder imbalances, among boys, suggesting that scapular fixation occurs in boys at a later age [39]. The additional factor mentioned is flexibility. It has been highlighted that boys with limited flexibility exhibit more pronounced knee asymmetry and anteroposterior body tilt compared to girls, which contributes to a higher prevalence of postural deformities [40]. Research on posture gender differences, that is not directly connected to the same age group, suggests that the aforementioned increased flexibility in females might contribute to greater spinal, pelvic, and sacral mobility, which influences better posture maintenance in females [41]. Moreover, it has been noted that girls had better postural stability than boys, which may enable girls to maintain proper posture more effectively [42]. Additionally, the research has reported significant differences in male and female morphological characteristics in biceps and thigh and calf skin folds [12]. Furthermore, the results show that boys have a greater deformity compared to girls when it comes to the pelvic tilt (PO) variable. In Slovenian studies, pelvic anteversion and head protrusions were recorded in almost half of the cases [24]. Pelvic obliquity (PO) represents an abnormal position of the pelvic bone. This condition can lead to abnormal posture due to the compression and misalignment of the spine in order to compensate for any misalignment. The most common causes of an asymmetrical pelvis are uneven leg length, scoliosis of the spine, and muscle imbalances or contractures. These problems often occur in combination. Excessive sitting, poor posture, and muscle weakness can increase the risk of developing a tilted pelvis. A tilted or asymmetrical pelvis can also be the result of functional or structural problems. The most common structural causes are scoliosis and uneven leg lengths. Additionally, the variations in knee alignment observed in the sample (e.g., 'X' and 'O' leg positioning valorized by variables varus/valgus left (VVL)/X/O left leg, and varus/valgus right (VVR)/X/O right leg) suggest differing biomechanical adaptations in boys and girls. Knowing these facts, it can be determined that the results with large numerical ranges in the mentioned variables are adequate for clarifying the part of the investigation related to the postural qualities of the treated population in terms of potential leg deformities.

The obtained results for the variables sag. distance cervical spine–sacrum (SDCS) and sag. distance thoracic spine (SDTS) indicate a deviation from the physiological curves of the spinal column in the sagittal plane. The upright position of a person is conditioned by the continuous maintenance of balance between the paravertebral musculature and centripetal forces, as well as gravity. In the formation of the upright position in humans during evolution, physiological formations were created: lordosis in the cervical and lumbar parts of the spinal column, and kyphosis in the thoracic part. Within physiological limits, these curves are a normal phenomenon, while their increase or decrease is considered pathological [43]. It has been reported that deformities in the sagittal plane in first-year elementary school students can show up in 73.9% of the sample [44]. Excessive pelvic anteversion can lead to compensatory lumbar hyperlordosis, increasing mechanical stress on the lower back and predisposing individuals to chronic musculoskeletal discomfort. It was found that boys have a more frequent problem with cervical lordosis and with kyphotic body posture, as well as with a problem of holding the head, which should stand in the extension of the body, but which is actually moved forward. Boys' posture is more often characterized by hyperkyphosis and lumbar hyperlordosis compared to girls [45,46]. The results of longitudinal studies indicate a higher incidence and increased tendency of boys toward the deformity of thoracic kyphosis [47]. Increased kyphosis and scoliosis, as compensatory postural outliers, in children have been connected to the increasingly greater lack of physical activity among children at this age [45].

The present study confirms the existence of significant sex differences in postural characteristics among preschool-aged children, with boys exhibiting more pronounced postural deviations in the sagittal plane, particularly in the shoulder and spinal regions. These findings indicate early manifestations of postural irregularities, underscoring the need for preventive interventions at this developmental stage. Similar patterns of postural deviations have been reported in previous studies that utilized the Contemplas 3D Posture Compact as a measurement instrument for posture analysis across the same and different age groups. Previous research has identified significant variations in postural parameters among children of different ages. One study [48] analyzed children aged 11–12 years and reported substantial differences in shoulder and pelvic rotation, sagittal distances, and flexion/extension parameters, confirming the high sensitivity of 3D posture assessment methods and highlighting the importance of early detection and prevention. Another study [21] investigated postural status in children aged 5–11 years and found considerable deviations in both the frontal and sagittal planes, including valgus knee alignment, knee joint hyperextension, and spinal asymmetry. Similarly, research on postural differences across multiple age groups (5–8, 9–11, and 12–14 years) [22] revealed a negative trend of increasing postural deformities associated with higher BMI, with boys showing a higher prevalence of thoracic kyphosis and shoulder girdle asymmetry, whereas girls exhibited a greater tendency toward lumbar lordosis and pelvic rotation. Additionally, a study focusing on children aged 6–9 years [29] confirmed sex-based differences in postural alignment, with girls being more prone to lumbar lordosis and valgus knee positioning, while boys demonstrated a higher occurrence of thoracic kyphosis. Moreover, an analysis of postural status in children aged 4–13 years [49] found that nearly 30% of participants had poor posture, with lower limb deformities being more common among girls. Overall, a vast majority of studies report significant differences between boys and girls while only one study we found reported no differences [44].

In line with these findings, the present study further supports the consistency of postural deviations identified using the Contemplas 3D Posture Compact system. By comparing our results with previous research, we observed that preschool-aged children exhibit similar patterns of postural misalignments to those reported in older age groups. This suggests that the early detection of postural irregularities using this technology can facilitate timely interventions aimed at mitigating progressive postural deformities. The reliability and precision of the Contemplas 3D Posture Compact system, as demonstrated across various studies, reinforce its applicability as a standardized tool for posture assessment in both research and clinical settings.

When assessing posture, X-ray imaging is still the gold standard [50]. However, optical analysis systems like the Contemplas Templo 3D have their own strengths and weaknesses. X-ray imaging has been reported to be particularly useful for detailed clinical skeletal assessments, making it an essential tool for diagnosing spinal conditions like scoliosis [51]. However, a major drawback is its exposure to ionizing radiation, which raises concerns about repeated use, particularly in children and young patients [52–54]. A noticeable limitation of X-rays is that they provide a static snapshot of posture, which may not fully capture functional posture changes/adaptations during movement [55,56]. Additionally, it is not readily accessible to clinicians [57]. On the other hand, photogrammetry with Contemplas Templo 3D and similar optical analysis systems are considered low-cost and completely non-invasive, making them safer for repeated use, especially for posture monitoring over time [58–60]. However, one drawback is their sensitivity to environmental conditions, such as lighting and camera positioning [61]. Additionally, having experienced raters can directly impact the reliability of clinical measures [62]. Therefore, having experienced raters with clinical and academic experience in postural assessment and human anatomy with regular online training is advisable to ensure high-quality posture analysis [57,62]. The literature suggests a two-level approach, where firstly, the photogrammetric methods should be used to assess the posture and, in case of concern, the second-level approach with radiography would be used, minimizing the need for repeated X-rays [63].

The modern way of life, especially hypokinesia, represents a real threat to maintaining a normal upright posture. It is assumed that the background of this body posture problem is most likely a result of daily irregular sitting for several hours [14], which leads to changes in the position of the head, thoracic and lumbar spine [64], inactivity of muscles responsible for maintaining correct upright posture, i.e., inactivity of the so-called antigravity musculature, which can lead to impaired bad postures, both at the level of the spinal column and lower extremities, incorrect position of the spine, weakness of the abdominal muscles, as well as the use of modern electronic devices [65,66]. Early intervention strategies, such as incorporating targeted postural exercises into school curricula, may help mitigate the impact of prolonged sitting and electronic device usage on musculoskeletal health [67–69]. Additionally, early screening is essential, and the usage of a photometric apparatus, like in the present study, which offers a fast and objective assessment of body posture, is advisable.

Postural deformities in children have become increasingly common. Rather than being primarily hereditary, these issues seem to nowadays largely stem from modern lifestyle changes affecting both children and their parents. Factors such as excessive screen time, sedentary behavior, and insufficient physical activity play a major role. Additionally, poor nutrition, a fast-paced and stressful lifestyle with limited time for healthy habits, and the rising prevalence of non-communicable diseases like obesity all contribute to the growing concern [70]. Additionally, the obtained results also emphasize the importance of education and intervention measures for correcting poor posture in children, considering that this can cause incurable damage to the musculoskeletal system, and neurological and pathological damage in the future. Moreover, studies have suggested that specially programmed and planned exercises and physical activity can substantially contribute to the development of good posture and should be viewed as an important factor of prevention in the daily activities of children [23,24,71,72]. Research has shown that specific structured movement activities in youth ages may be more effective in the development of motor coordination, agility, and speed of movement in children [73].

Based on the results of this research, several directions for future research and application in practice are recommended:

- Conducting cross-cultural research that would enable the comparison of results among children from different environments and cultural contexts, which would determine eventual differences and specificities in the development of postural characteristics.
- Including a larger sample of participants in future research, including children from rural and urban areas, to ensure the representativeness and generalizability of the results.
- Conducting a longitudinal study that would follow the postural characteristics of children through different developmental stages, aiming to identify critical periods for intervention and determine more precisely developmental trends in the context of gender differences.
- Conducting research with a larger sample of subjects and by including factors of physical activity and lateralization could provide more detailed insights into these variables and enable a better understanding of their influence on postural status.

The present study needs to acknowledge some limitations. One of them is that the participants' physical activity was not monitored, nor was their body lateralization analyzed. The lack of the analysis of these factors can limit the interpretation of the results because physical activities and lateralization can significantly impact postural status and the development of deformities. Therefore, further studies should account for that. The obtained results indicate statistically significant differences between the recorded groups. Although the effect size of these differences is small, these asymmetries should be interpreted carefully. Additionally, a larger sample size would be beneficial.

5. Conclusions

The results of this research indicate that boys have more pronounced postural deformities than girls, which is especially evident in variables such as shoulder displacement, pelvic tilt, and sagittal deviations in the cervical and thoracic spine. For early screening, the usage of a photometric apparatus that offers a fast and objective assessment of body posture is advisable. The differences in posture between boys and girls may be influenced by their physical development, daily habits, and the types of activities they engage in. Factors such as growth patterns, muscle balance, and lifestyle choices—like prolonged sitting and the frequent use of electronic devices—could contribute to postural issues. To better understand these differences, future research should explore how posture develops over time in larger, more diverse groups. Cross-cultural studies and investigations into physical activity and movement preferences could provide valuable insights into the key factors shaping children's posture.

Author Contributions: Conceptualization, A.K.-G., S.K. (Safet Kapo) and I.M.; methodology, A.K.-G., A.E., I.M. and S.K. (Siniša Kovač); software, H.K. and J.Š.; validation, A.K.-G., H.K. and S.K. (Siniša Kovač); formal analysis, H.K.; investigation, A.K.-G., S.K. (Safet Kapo), A.E., and I.M.; resources, S.K. (Safet Kapo); data curation, A.K.-G. and A.E.; writing—original draft preparation, A.K.-G., J.Š.; writing—review and editing, I.M., S.K. (Safet Kapo), S.K. (Siniša Kovač) and J.Š.; visualization, H.K. and J.Š.; supervision, S.K. (Safet Kapo); project administration, S.K. (Safet Kapo); funding acquisition, J.Š. All authors have read and agreed to the published version of the manuscript.

Funding: This research received no external funding.

Institutional Review Board Statement: This study was conducted following the Declaration of Helsinki and approved by the Ministry of Education, Science, and Youth of Sarajevo Canton by the Institutional Review Board (No. 11/03-34-37155/20, approved on 2 October 2020).

Informed Consent Statement: Informed consent was obtained from all subject's parents/guardians involved in the study.

Data Availability Statement: The data supporting this study's findings are available from the corresponding author upon reasonable request.

Acknowledgments: We express our sincere gratitude to the Ministry of Education, Science, and Youth of the Sarajevo Canton for approving the implementation of this research, thus enabling its successful execution and contribution to scientific advancement. Special thanks also go to the Faculty of Sport and Physical Education for their exceptional support in providing top-quality equipment, whose precision and accuracy were crucial for the validity and reliability of all measurements. We would also like to thank the measurers, whose professional engagement was invaluable in maintaining high standards in the data collection process, significantly contributing to the quality and credibility of the research.

Conflicts of Interest: The authors declare no conflicts of interest.

References

1. Skender, N.; Pistotnik, B.; Čolakhodžić, E. *Basics of Movement in Sports*; Faculty of Education, University of Bihać: Bihać, Bosnia and Herzegovina, 2007.
2. Augustinović, I.; Halilović, S.; Dobrić, A.; Kovačevska, K.; Kovač, L.; Matošević, L.; Kosterović, A.; Begović, H.; Jakešević, B.; Barnjak, T.; et al. *The Present Development Program of Pre-School Upbringing and Education of the Canton of Central Bosnia*; Ministry of Education, Science, Culture and Sports of the Canton of Central Bosnia/Central Bosnia Canton: Travnik, Bosnia and Herzegovina, 2018.
3. Baranowska, A.; Sierakowska, M.; Owczarczuk, A.; Olejnik, B.J.; Lankau, A.; Baranowski, P. An Analysis of the Risk Factors for Postural Defects among Early School-Aged Children. *J. Clin. Med.* **2023**, *12*, 4621. [CrossRef] [PubMed]
4. Plandowska, M.; Lichota, M.; Górniak, K. Postural Stability of 5-Year-Old Girls and Boys with Different Body Heights. *PLoS ONE* **2019**, *14*, e0227119. [CrossRef] [PubMed]
5. Peulić, J.; Katanić, B.; Jovanović, N.; Bjelica, D. Exploring Anthropometric Characteristics, Weight Status, and Posture among Preschool Children in Serbia. *J. Anthr. Sport Phys. Educ.* **2024**, *8*, 19–24. [CrossRef]
6. Rusek, W.; Baran, J.; Leszczak, J.; Adamczyk, M.; Baran, R.; Weres, A.; Inglot, G.; Czenczek-Lewandowska, E.; Pop, T. Changes in Children's Body Composition and Posture during Puberty Growth. *Children* **2021**, *8*, 288. [CrossRef] [PubMed]
7. Vranešić-Hadžimehmedović, D.; Arapović, M.; Hodžić, A.; Mašić, S. Prevalence of spinal deformities in preschool children living in the SOS children's village in Sarajevo. *Homosporticus* **2024**, *26*, 18–24. [CrossRef]
8. Standring, S. *Gray's Anatomy: The Anatomical Basis of Clinical Practice*; Elsevier Churchill Livingstone: Edinburgh, UK, 2005.
9. Protić-Gava, B.; Šćepanović, T. *Basis of Kinesitherapy and Applied Corrective Gymnastic*; University of Novi Sad: Novi Sad, Serbia, 2016.
10. Madić, D. (Ed.) *Improving Testing Abilities on Postural and Spinal Column Status—Spinelab*; Faculty of Sport and Physical Education: Novi Sad, Serbia, 2014.
11. Čović, N. The Influence of a Program of Motor Control Exercises on Body Composition, Postural Status and Motor Skills in Young Athletes. Ph.D. Thesis, Faculty of Sports and Physical Education, University of Sarajevo, Sarajevo, Bosnia and Herzegovina, 2020.
12. McEvoy, M.P.; Grimmer, K. Reliability of Upright Posture Measurements in Primary School Children. *Biomed. Cent. Ser. Musculoskelet. Disord.* **2005**, *29*, 6. [CrossRef]
13. Romanov, R.; Stupar, D.; Međedović, B.; Brkin, D. Postural Status of Preschool Children in the Territory of Novi Sad. *TIMS Acta* **2014**, *8*, 129–135. [CrossRef]
14. Sabo, E. Postural Status of Preschool Children in Novi Sad. *Pedagog. Real.* **2006**, *54*, 108–113.
15. Džibrić, D. Differences in Morphological and Postural Status between Primary School Males and Females. *Sport Sci. Pract. Asp.* **2019**, *16*, 2. [CrossRef]
16. Kolybacz, A.; Niewiem, M.; Buczyńska, A.; Woś, H. Health Status of 9-Year-Old Children in Katowice. *Stand. With. Pediatr.* **2019**, *16*, 537–554. (In Polish)

17. Glavaš, J.; Rumboldt, M.; Karin, Ž.; Matković, R.; Kresina, S.; Dragaš-Zubalj, N.; Aljinović, J. The role of school medicine in the early detection and management of adolescent idiopathic scoliosis. *Wien. Klin. Wochenschr.* **2023**, *135*, 273–281. [CrossRef]
18. Rusnák, R.; Kolarová, M.; Aštaryová, I.; Kutiš, P. Screening and Early Identification of Spinal Deformities and Posture in 311 Children: Results from 16 Districts in Slovakia. *Rehabil. Res. Pract.* **2019**, *2019*, 4758386. [CrossRef] [PubMed]
19. Miletić, A.; Protić-Gava, B. Gender-Related Morphological Characteristics in Preschool Children of Kolubara District. *EQOL J.* **2017**, *9*, 31–36. [CrossRef]
20. Kovač, S.; Kajmović, H.; Rađo, I.; Manić, G. Trend Projections of Body Deformities Occurrence between the Ages of 5 and 12, Metrically Objectified and Estimated by 3D Postural Status Screening. *Sporty Man* **2014**, *16*, 5–13.
21. Kovač, S.; Kapo, S.; Alić, H.; Manić, G. Postural Quality Analyses for Children Recorded in Kinematic 2D and 3D Contemplas Method. *Tech. Technol. Educ. Manag.* **2015**, *10*, 99–106K.
22. Kapo, S.; Rađo, I.; Smajlović, N.; Kovač, R.; Talović, M.; Doder, L.; Čović, N. Increasing Postural Deformity Trends and Body Mass Index Analysis in School-Age Children. *Slov. J. Public Health* **2018**, *57*, 1. [CrossRef]
23. Bićanin, P.; Milenković, S.; Radovanović, D.; Gajević, A.; Ivanović, J. Postural Disorders in Preschool Children in Relation to Gender. *Univ. Facts Phys. Educ. Sports* **2017**, *15*, 1–10. [CrossRef]
24. Mitova, S. Frequency and Prevalence of the Postural Disorders and Spinal Deformities in the Children of Primary School Age. *Res. Kinesiol.* **2015**, *43*, 21–24.
25. Faul, F.; Erdfelder, E.; Buchner, A.; Lang, A.G. Statistical Power Analyses Using G*Power 3.1: Tests for Correlation and Regression Analyses. *Behav. Res. Methods* **2009**, *41*, 1149–1160. [CrossRef]
26. Tanita Corporation. *BC-420MA Body Composition Monitor Instruction Manual*; Tanita Corporation: Tokyo, Japan, 2015.
27. Widhalm, K.; Durstberger, S.; Greisberger, A.; Wolf, B.; Putz, P. Validity of Assessing Level Walking with the 2D Motion Analysis Software TEMPLO and Reliability of 3D Marker Application. *Sci. Rep.* **2024**, *14*, 1427. [CrossRef]
28. Šćepanović, T.; Marinković, D.; Korovljev, D.; Madić, D. The Status of the Spine in the Sagittal Plane in Girls. In Proceedings of the 5th International Scientific Conference Contemporary Kinesiology, Split, Croatia, 28–30 August 2015.
29. Ćirić, A.; Čaušević, D.; Bejdić, A. Differences in Postural Status between Boys and Girls Aged 6–9 Years Recorded with the 3D Method of the Measuring Instrument Contemplas. *Homosporticus* **2015**, *17*, 15–20.
30. Cohen, J. *Statistical Power Analysis for the Behavioral Sciences*, 2nd ed.; Routledge: New York, NY, USA, 1988.
31. Fritz, C.O.; Morris, P.E.; Richler, J.J. Effect Size Estimates: Current Use, Calculations, and Interpretation. *J. Exp. Psychol. Gen.* **2012**, *141*, 2–18. [CrossRef] [PubMed]
32. Tomczak, M.; Tomczak, E. The Need to Report Effect Size Estimates Revisited: An Overview of Some Recommended Measures of Effect Size. *Trends Sport Sci.* **2014**, *21*, 19–25.
33. Dop, D.; Pădureanu, V.; Pădureanu, R.; Niculescu, S.-A.; Drăgoescu, A.N.; Moroșanu, A.; Mateescu, D.; Niculescu, C.E.; Marcu, I.R. Risk Factors Involved in Postural Disorders in Children and Adolescents. *Life* **2024**, *14*, 1463. [CrossRef]
34. Yang, L.; Lu, X.; Yan, B.; Huang, Y. Prevalence of Incorrect Posture among Children and Adolescents: Finding from a Large Population-Based Study in China. *iScience.* **2020**, *23*, 101043. [CrossRef]
35. Kolarova, M.; Kutiš, P.; Rusnak, R.; Hrčková, Z.; Hudáková, Z.; Lysá, L.; Luliak, M.; Babel'a, R. Analysis of Body Segments and Postural State in School Children. *Neuroendocrinol. Lett.* **2019**, *40* (Suppl. S1), 17–23.
36. Batistão, M.V.; Carreira Moreira, R.F.; Cote Gil Coury, H.J.; Salasar, L.E.B.; Sato, T.O. Prevalence of Postural Deviations and Associated Factors in Children and Adolescents: A Cross-Sectional Study. *Fisioter. Mov.* **2016**, *29*, 777–785. [CrossRef]
37. Maciałczyk-Paprocka, K.; Stawińska-Witoszyńska, B.; Kotwicki, T.; Sowińska, A.; Krzyżaniak, A.; Walkowiak, J.; Krzywińska-Wiewiorowska, M. Prevalence of incorrect body posture in children and adolescents with overweight and obesity. *Eur. J. Pediatr.* **2017**, *176*, 563–572. [CrossRef]
38. Bubanj, S.; Đorđević, S.; Milenković, S.; Stanković, R.; Vidojević, M.; Đokić, M. Postural Disorders and Muscle Power in Primary School Children. *Acta Fac. Medicae Naissensis* **2021**, *38*, 270–278.
39. Penha, P.J.; Casarotto, R.A.; Sacco, I.C.N.; Marques, A.P.; João, S.M.A. Qualitative Postural Analysis among Boys and Girls of Seven to Ten Years of Age. *Braz. J. Phys. Ther.* **2008**, *12*, 386–391. [CrossRef]
40. Coelho, J.J.; Graciosa, M.D.; de Medeiros, D.L.; Pacheco, S.C.; da Costa, L.M.; Ries, L.G. Influence of Flexibility and Gender on the Posture of School Children. *Rev. Paul. Pediatr.* **2014**, *32*, 223–228. [CrossRef] [PubMed]
41. Mizukoshi, R.; Yagi, M.; Yamada, Y.; Yokoyama, Y.; Yamada, M.; Watanabe, K.; Nakamura, M.; Nagura, T.; Jinzaki, M. Gender Differences in Spinal Mobility During Postural Changes: A Detailed Analysis Using Upright CT. *Sci. Rep.* **2024**, *14*, 9154. [CrossRef]
42. Smith, A.W.; Ulmer, F.F.; Wong, D.P. Gender Differences in Postural Stability Among Children. *J. Hum. Kinet.* **2012**, *33*, 25–32. [CrossRef] [PubMed]
43. Gajić, D. Kinesitherapy in Spine Deformity. Master's Thesis, Faculty of Health Care, Pan-European University, Banja Luka, Bosnia and Herzegovina, 2009.

44. Đorđević, S.; Jorgić, B.; Milenković, S.; Stanković, R.; Stanković, M. The Incidence of Spinal Postural Disorders in First-Grade Elementary-School Students. *Facta Univ. Ser. Phys. Educ. Sport* **2020**, *18*, 381–390. [CrossRef]
45. Đorđević, S.; Stanković, M.; Jorgić, B.; Milenković, S.; Smailović, S.; Katanić, B.; Jelaska, I.; Pezelj, L. The Association of Sagittal Spinal Posture among Elementary School Pupils with Sex and Grade. *Children* **2024**, *11*, 446. [CrossRef]
46. Wojtków, M.; Szkoda-Poliszuk, K.; Szotek, S. Influence of Body Posture on Foot Load Distribution in Young School-Age Children. *Acta Bioeng. Biomech.* **2018**, *20*, 101–107.
47. Poussa, M.S.; Heliövaara, M.M.; Seitsamo, J.T.; Könönen, M.H.; Hurmerinta, K.A.; Nissinen, M.J. Development of Spinal Posture in a Cohort of Children from the Age of 11 to 22 Years. *Eur. Spine J.* **2005**, *14*, 738–742. [CrossRef]
48. Skender, N.; Kurtović, N.; Kovač, S.; Šabić, E.; Ćeleš, N. Two Different Methods Comparative Analysis of Determining Body Deformities in Students. *Homosporticus* **2023**, *25*, 18–23. [CrossRef]
49. Kojić, M. Differences in Indicators of Postural Status between Boys and Girls from Srem. *Exerc. Qual. Life* **2014**, *6*, 17–22.
50. Ng, S.Y.; Bettany-Saltikov, J. Imaging in the Diagnosis and Monitoring of Children with Idiopathic Scoliosis. *Open Orthop. J.* **2017**, *11*, 1500–1520. [CrossRef]
51. Weigel, S.; Dullien, S.; Grifka, J.; Jansen, P. Comparison Between Rasterstereographic Scan and Orthopedic Examination for Posture Assessment: An Observational Study. *Front. Surg.* **2024**, *11*, 1461569. [CrossRef] [PubMed]
52. Kutanzi, K.R.; Lumen, A.; Koturbash, I.; Miousse, I.R. Pediatric Exposures to Ionizing Radiation: Carcinogenic Considerations. *Int. J. Environ. Res. Public Health* **2016**, *13*, 1057. [CrossRef] [PubMed]
53. Inoue, D.; Shigematsu, H.; Nakagawa, Y.; Takeshima, T.; Tanaka, Y. The Influence of Posture on Instability Evaluation Using Flexion-Extension X-Ray Imaging in Lumbar Spondylolisthesis. *Asian Spine J.* **2021**, *15*, 308–316. [CrossRef]
54. Dorfman, A.L.; Fazel, R.; Einstein, A.J.; Applegate, K.E.; Krumholz, H.M.; Wang, Y.; Christodoulou, E.; Chen, J.; Sanchez, R.; Nallamothu, B.K. Use of Medical Imaging Procedures with Ionizing Radiation in Children: A Population-Based Study. *Arch. Pediatr. Adolesc. Med.* **2011**, *165*, 458–464. [CrossRef]
55. Dreischarf, M.; Pries, E.; Bashkuev, M.; Putzier, M.; Schmidt, H. Differences Between Clinical "Snap-Shot" and "Real-Life" Assessments of Lumbar Spine Alignment and Motion—What Is the "Real" Lumbar Lordosis of a Human Being? *J. Biomech.* **2016**, *49*, 638–644. [CrossRef]
56. Sugiyama, N.; Kai, Y.; Koda, H.; Morihara, T.; Kida, N. Agreement in the Postural Assessment of Older Adults by Physical Therapists Using Clinical and Imaging Methods. *Geriatrics* **2024**, *9*, 40. [CrossRef] [PubMed]
57. Hébert-Losier, K.; Abd Rahman, F. Reliability of postural measures in elite badminton players using Posture Pro 8. *Physiother. Theory Pract.* **2017**, *34*, 483–494. [CrossRef]
58. Van Niekerk, S.M.; Louw, Q.; Vaughan, C.; Grimmer-Somers, K.; Schreve, K. Photographic measurement of upper-body sitting posture of high school students: A reliability and validity study. *BMC Musculoskelet. Disord.* **2008**, *9*, 113. [CrossRef]
59. Ruivo, R.M.; Pezarat-Correia, P.; Carita, A.I.; Vaz, J.R. Reliability and validity of angular measures through the software for postural assessment. *Rehabilitación* **2013**, *47*, 223–228.
60. Ha, S.M.; Kwon, O.Y.; Weon, J.H.; Kim, M.H.; Kim, S.J. Reliability and validity of goniometric and photographic measurements of clavicular tilt angle. *Man. Ther.* **2013**, *18*, 367–371. [CrossRef]
61. Alam, M.F.; Zaki, S.; Sharma, S.; Nuhmani, S. Establishing the Reliability of the GaitON Motion Analysis System: A Foundational Study for Gait and Posture Analysis in a Healthy Population. *Sensors* **2024**, *24*, 6884. [CrossRef] [PubMed]
62. Ludwig, O.; Hammes, A.; Kelm, J.; Schmitt, E. Assessment of the posture of adolescents in everyday clinical practice: Intra-rater and inter-rater reliability and validity of a posture index. *J. Bodyw. Mov. Ther.* **2016**, *20*, 761–766. [CrossRef] [PubMed]
63. Roggio, F.; Ravalli, S.; Maugeri, G.; Bianco, A.; Palma, A.; Di Rosa, M.; Musumeci, G. Technological Advancements in the Analysis of Human Motion and Posture Management through Digital Devices. *World J. Orthop.* **2021**, *12*, 467–484. [CrossRef]
64. Claus, A.P.; Hides, J.A.; Moseley, G.L.; Hodges, P.W. Thoracic and Lumbar Posture Behaviour in Sitting Tasks and Standing: Progressing the Biomechanics from Observations to Measurements. *Appl. Ergon.* **2016**, *53*, 161–168. [CrossRef] [PubMed]
65. Straker, L.M.; O'Sullivan, P.B.; Smith, A.; Perry, M. Computer Use and Habitual Spinal Posture in Australian Adolescents. *Public Health Rep.* **2007**, *122*, 634. [CrossRef]
66. Simov, S.B.; Milinić, S.M.; Stojanović, D.O. Frequency of Poor Posture and Flat Feet in Preschool Children. *Apollo Med. Aescul.* **2011**, *9*, 5–8.
67. Slováková, M.; Mandzáková, M. Effect of an exercise program on the body posture of young school-aged pupils. *J. Phys. Educ. Sport* **2024**, *24*, 747–753. [CrossRef]
68. Araújo, C.L.; Moreira, A.; Carvalho, G.S. Postural Education Programmes with School Children: A Scoping Review. *Sustainability* **2023**, *15*, 10422. [CrossRef]
69. Miñana-Signes, V.; Monfort-Pañego, M.; Valiente, J. Teaching Back Health in the School Setting: A Systematic Review of Randomized Controlled Trials. *Int. J. Environ. Res. Public Health* **2021**, *18*, 979. [CrossRef]

70. Lopez-Fuenzalida, A.; Rodriguez Channels, C.; Reyes Ponce, A.; Contreras Molina, A.; Quezada, J.F.; Polanco, C.A. Association between Nutritional Status and Flat Foot Prevalence in Chilean Children from 6 to 10 Years of Age. *Nutr. Hosp.* **2016**, *33*, 249–254. [CrossRef]
71. Fernandes, O.; Basilio, F. Tai Chi training can improve postural control. *Rev. Artes Marciales Asiáticas* **2016**, *11*, 42–43. [CrossRef]
72. Wąsik, J.; Motow-Czyz, M. Comparative analysis of body posture in child and adolescent taekwon-do practitioners and non-practitioners. *Ido Mov. Cult. J. Martial Arts Anthropol.* **2015**, *15*, 35–40. [CrossRef]
73. Kojić, F.; Arsenijević, R.; Grujić, G.; Toskić, L.; Šimenko, J. Effects of Structured Physical Activity on Motor Fitness in Preschool Children. *Children* **2024**, *11*, 433. [CrossRef] [PubMed]

Journal of
Functional Morphology and Kinesiology

MDPI

Article

Body Asymmetry and Sports Specialization: An Exploratory Anthropometric Comparison of Adolescent Canoeists and Kayakers

Carlos Abraham Herrera-Amante [1,2,3], William Carvajal-Veitía [3,4], Rodrigo Yáñez-Sepúlveda [5], Fernando Alacid [3,6], Juan Gavala-González [3,7], José Francisco López-Gil [8,*], Jorge Olivares-Arancibia [9] and César Octavio Ramos-García [1,2,3,*]

1 Nutritional Assessment and Nutritional Care Laboratory (LECEN), Division of Health Sciences, Tonalá University Center, University of Guadalajara (UdeG), Tonalá 45425, Mexico; carlos.amante@academicos.udg.mx
2 Research Division, Ibero-American Institute of Sports Sciences and Human Movement (IICDEM), Guadalajara 44510, Mexico
3 Ibero-American Network of Researchers in Applied Anthropometry (RIBA^2), 04120 Almería, Spain; wiliamcarvajal790@gmail.com (W.C.-V.); falacid@ual.es (F.A.); jgavala@us.es (J.G.-G.)
4 Institute of Sports Medicine (IMD), Havana 10800, Cuba
5 Faculty of Education and Social Sciences, Universidad Andres Bello, Viña del Mar 2520000, Chile; rodrigo.yanez.s@unab.cl
6 Department of Education, Health Research Centre, University of Almería (UAL), 04120 Almeria, Spain
7 Department of Physical Education and Sports, Universidad de Sevilla (US), 41013 Sevilla, Spain
8 One Health Research Group, Universidad de Las Américas, Quito 170124, Ecuador
9 Grupo AFySE, Investigación en Actividad Física y Salud Escolar, Escuela de Pedagogía en Educación Física, Facultad de Educación, Universidad de las Américas, Santiago 7500975, Chile; jolivares@udla.cl
* Correspondence: josefranciscolopezgil@gmail.com (J.F.L.-G.); cesar.ramos@cutonala.udg.mx (C.O.R.-G.)

Abstract: Background/Objectives: The evaluation of body asymmetry between the right and left sides of the body is crucial in the context of sports because of its potential impact on performance and injury prevention. This study analyzed the anthropometric differences between the right and left sides of the body in adolescent (13.0 [2.0] years) male canoeists and kayakers from Spain. This study aimed to explore the asymmetries associated with these disciplines. **Methods**: Anthropometric assessments were carried out on 27 male adolescents (13 canoeists and 14 kayakers). A total of 58 anthropometric variables were assessed, including 23 pairs of bilateral variables and 12 unilateral measurements. The evaluations included basic measures, skinfolds, girths, lengths, and breadths. Relative differences between sides were calculated via the bilateral asymmetry index (%BAI). Nonparametric tests, such as the Wilcoxon signed-rank test for within-group comparisons and the Mann–Whitney U test for between-group comparisons, were applied. **Results**: Comparisons between body sides within the groups revealed significant asymmetries in the subscapular skinfold (p = 0.010) in canoeists and in the mid-thigh girth (p = 0.041) in kayakers. Among the groups, differences were found in the subscapular skinfold (p = 0.010) and the bicep skinfold (p = 0.038) on the right side. **Conclusions**: Although significant differences were found in some variables, no distinctive profile of the differences between canoeists and kayakers was established in the categories analyzed. These results suggest that, in general, body asymmetries do not significantly distinguish between these disciplines within the sample studied. Further research is needed to better understand the implications of these differences for performance and injury prevention in specific adolescent sports contexts.

Keywords: water sports; anthropometry; body composition; athletic injuries; muscle imbalances; bilateral deficits

1. Introduction

Symmetry is defined as the quality of an object that exhibits exact equality in size, shape, and structure across its two halves when divided along an axis [1]. In contrast, asymmetry refers to variations in characteristics between one side of the body and the other and can be influenced by genetic and environmental factors and the differential use of body segments [2,3]. In the sports domain, the study of body symmetry is highly relevant, as it provides valuable information for performance evaluations, development and maturation, injury prevention, training optimization, and equipment innovation [1,4–6]. Sports with unilateral gestures, such as tennis, rowing, and fencing, often induce notable asymmetries due to their repetitive motion patterns, which differ significantly from those in bilateral or symmetrical sports such as swimming or gymnastics [1,7]. These differences underscore the need to understand asymmetries across sports disciplines.

Canoeing and kayaking are two paddle sports that, despite their similarities, differ significantly in terms of their biomechanics and muscular demands. Canoeing (specifically Canadian canoeing) requires paddlers to kneel on one knee while executing unilateral strokes on one side of the body using a single-bladed paddle. This repetitive motion predominantly engages the upper body, particularly the deltoids, latissimus dorsi, and trapezius, while also placing asymmetric demands on the core and lower limbs due to the required kneeling position [8]. Recent research has demonstrated the importance of assessing asymmetries in athletes, particularly in sports like canoeing, where repetitive movements may lead to imbalances [9].

In contrast, kayaking is performed in a seated position with the legs extended forward and the use of a double-bladed paddle, which allows for alternating strokes on both sides of the body. This movement pattern generally distributes muscular effort more symmetrically across the upper body, engaging the pectorals, latissimus dorsi, and obliques, while still requiring strong lower-body stabilization [8]. However, despite its more balanced movement mechanics, kayaking can still lead to asymmetries, particularly in terms of muscle activation patterns and joint loading over time.

Among the various methodologies for evaluating body symmetry, anthropometry stands out as a valuable and highly reproducible technique. Its ability to enable bilateral and segmental analyses allows for meaningful insights into the impact of body asymmetries [10]. However, anthropometric assessments have traditionally focused on the right side of the body, as international standards, such as those established by the International Society for the Advancement of Kinanthropometry (ISAK), prioritize right-side measurements in their protocols [11]. This approach may overlook the importance of systematically monitoring asymmetries to better understand their practical implications during training and in competitive settings. Furthermore, recent advancements have highlighted the potential of three-dimensional scanning technologies and advanced imaging tools in capturing more detailed asymmetry metrics, which could complement traditional anthropometry [12]. While these tools offer precision, they remain underutilized in youth sports research.

Sports such as canoeing and kayaking are particularly intriguing in the context of asymmetry, as their movements impose specific biomechanical demands that influence muscular and skeletal development [8]. Studies have shown that paddlers often develop more significant asymmetries than other athletes do in bilateral sports, such as swimming, primarily because of the repetitive and unidirectional forces exerted during paddling [13,14]. These sport-specific asymmetries may affect performance and play an important role in injury susceptibility, especially during adolescent growth spurts, as athletes experience rapid physical changes [15].

Examining asymmetries could help elucidate their potential impact on the development of different sports modalities. Research has shown significant differences in anthropometric dimensions, such as body girths, breadths, and lengths, favoring the dominant side of the body [16,17]. Furthermore, the effects of differences in body composition symmetry on performance have been documented [18]. However, some studies have not found an association between training volume and the magnitude of asymmetry [7], while other studies identify training as having a strong influence [5,19,20]. The divergence in these findings suggests that the relationship between asymmetry and performance is complex and multifactorial and could be influenced by age, training intensity, and duration [18,21]. These varying results underscore the need for more research to understand how training impacts symmetry and how these effects may differ across sports modalities. Currently, there is limited information on what levels of asymmetry might be beneficial or detrimental to athletic performance [16]. A recent study on kettlebell athletes found notable symmetry in their body composition and strength, particularly in the upper body, suggesting that training characteristics may play a key role in mitigating asymmetries. These findings contribute to the ongoing debate on the relationship between asymmetry and performance, highlighting the need for further research that considers factors such as age, level of competition, and type and amount of training [3].

Beyond their implications for performance, addressing asymmetries in adolescent athletes may provide broader benefits regarding their long-term health and injury prevention. During adolescence, physical growth occurs at an accelerated pace, creating a critical window where muscle imbalances and joint stresses may lead to chronic issues if left unaddressed [22]. Early intervention programs incorporating an asymmetry assessment could guide coaches and healthcare professionals in designing tailored training regimens. By identifying and mitigating excessive asymmetries, such initiatives could enhance biomechanical efficiency while reducing the likelihood of the overuse injuries common in water sports [15,20].

One less explored aspect of applied anthropometry is the difference in the symmetry of anthropometric profiles across sports modalities. Studies in Sports Science have focused primarily on physical capacity and strength rather than body composition. A study by Stagi and associates [3] examined the relationship between symmetry, body composition, and physical performance, contributing to the broader discussion on how asymmetry may influence athletic outcomes. However, several questions remain unanswered: do differences in the symmetry of anthropometric profiles correspond to the motor performance demands of each sport modality? What trends are characteristic of each sport? In which body region is asymmetry more pronounced, the upper or lower body? Previous findings indicate that asymmetries may evolve over time due to sport-specific demands, with younger athletes often showing less pronounced differences than seasoned competitors do [23,24]. The dynamic nature of asymmetries underscores the importance of longitudinal studies within sports research.

The literature suggests that most sports exhibit pronounced functional asymmetries due to their specific demands [13,21]. However, the classification of sports as symmetrical or asymmetrical is based mainly on their motor gestures. In practice, all athletes exhibit asymmetries within a generally permissible range. For example, Canadian canoeing and kayaking have been classified as asymmetrical [13,23] and symmetrical [13], respectively. However, there is insufficient evidence to determine whether the asymmetrical or symmetric load in each sport is reflected in a specific anthropometric profile or if it varies by competitive age group [23]. Other studies have also suggested that certain degrees of asymmetry, particularly in strength and muscle mass, may confer competitive advantages in asymmetric sports, while excessive imbalances increase the risk of injury [25,26]. In this context, Krzykała and associates [13] noted that during the biological development of young canoe athletes, specialized training could cause asymmetries, leading to differences in muscle mass between sides. However, prolonged training appears to reduce lower-limb muscle mass asymmetry among older competitors [23].

Furthermore, a few studies have highlighted the differences between kayak paddlers and canoe paddlers, which have been attributed to the continuous physical development that kayakers require to stay competitive and the need for young canoeists to focus more intensively on refining their technical skills [8,15]. Given the limited literature on adolescent paddlers, this study fills a critical gap by exploring these dynamics at a developmental stage, where interventions may yield significant long-term benefits [20,27]. This research addresses these discrepancies by comparing the differences in the symmetry of the anthropometric profiles of Spanish adolescents practicing Canadian canoeing and those practicing kayaking. The findings of this study could enhance the application of anthropometry as, despite its high reproducibility, its use in profiling symmetry differences has been limited. This underutilization stems from the small number of anthropometric indicators typically employed [1,5,19,20,28] or the reliance on alternative methods such as dual-energy X-ray absorptiometry (DXA), bioelectrical impedance analysis (BIA), or other body composition studies [2,3,13,18,21]. Our hypothesis posits that asymmetries in young practitioners are more pronounced in sports disciplines such as Canadian canoeing compared to kayaking.. By addressing this hypothesis, this paper aims to inform coaches, sports scientists, and medical professionals about the complex relationship between training regimens and asymmetry, ultimately contributing to optimized performance and injury prevention strategies [29].

2. Materials and Methods

2.1. Study Design

A cross-sectional descriptive study was designed. The participants reported to the testing area where data collection was conducted only once. The Strengthening the Reporting of Observational Studies in Epidemiology (STROBE) criteria for cross-sectional research were followed in the design of this study [30,31].

2.2. Setting

This study was conducted in Seville, Spain. Individual information (demographics, descriptive data, sports discipline, and experience) was collected after a brief questionnaire was completed. Signed parental consent was obtained from the parents or legal guardians of all participants. Parents and study participants were fully and appropriately informed about the participation requirements and the purpose, risks, and benefits of the study. All measurements were taken in the presence of other athletes and their coaches to ensure a comfortable and familiar environment for the participants. This study was carried out in accordance with the ethical principles for medical research outlined in the international guidelines for good clinical practice and the Declaration of Helsinki [32]. The Institutional Ethical Committee of the University of Murcia approved this study (SKMBT-C25211110314021).

2.3. Participants

Anthropometric data from 27 adolescent Spanish males (13 canoeists and 14 kayakers), with a median age of 13.0 years (IQR: 2.0) for both groups, were analyzed. The anthropometric data were collected during the preseason period and include data from athletes from three Sevillian clubs who met the necessary standards to compete in their category in national championships. Athletes who met the following inclusion criteria were invited to participate in this study: (i) those who attended at least 90% of training sessions and (ii) had a performance level that allowed them to compete in national championships in their category. The exclusion criteria included (i) not providing written consent (parental consent) for the procedures to be conducted or data to be disclosed for research purposes at the time of the evaluations.

Arriving at the assessment area without appropriate clothing was considered a removal criterion.

2.4. Variables

A total of 58 anthropometric variables were assessed, including 23 pairs of bilateral variables and 12 unilateral measurements, following the guidelines established by the International Society for the Advancement of Kinanthropometry (ISAK) [11]. All anthropometric measurements were taken two or three times (with a third measurement taken if the difference between the first two measurements exceeded 5% for skinfolds and 1% in the remaining measures), and the mean or the median value was used for data analysis, respectively. The technical error of measurement (TEM) was calculated according to Pederson and Gore [33].

2.5. Measurements

Anthropometric measurements were performed based on the international standards established by the ISAK [11]. These protocols are specifically designed for the evaluation of the right side of the body; however, they were adapted in this study to assess both sides. Measurements were carried out by two certified anthropometrists: one level 3 (ISAK L3) and one level 2 (ISAK L2) anthropometrist. The level 3 anthropometrist, as they were the most experienced, conducted all the measurements, while the level 2 anthropometrist assisted and recorded the data.

Body mass (kg) was determined using a digital scale with a precision of 50 g (SECA® 874, Hamburg, Germany). Stature (cm) and sitting height (cm) were assessed with a 1 mm precision stadiometer (SECA® 217, Hamburg, Germany). Skinfold thickness (mm) was measured with a skinfold caliper with a precision of 0.2 mm (Harpenden, British Indicators, Crymych, UK). Girths (cm) were measured with a flexible, nonstretchable metal tape with a precision of 1 mm (SmartMet Kinanthropometric Assessment®, Jalisco, Mexico). Lengths (cm) and breadths (cm) were measured using a segmentometer and large bone caliper with a precision of 1 mm (SmartMet Kinanthropometric Assessment®, Jalisco, Mexico). All instruments were calibrated before the evaluations to minimize measurement errors.

2.6. Statistical Methods

Statistical analysis was performed in R Studio version 4.4.1, and we evaluated 46 anthropometric variables measured on both sides of the body (23 variables on the right side and 23 on the left side). Relative differences between measurements on the left and right sides of the body were calculated using the bilateral asymmetry index (%BAI) proposed by Impellizzeri and associates [34] and modified in this study based on previous research [9], as shown in the following equation:

$$\%\text{BAI} = \left(\frac{\text{Dominant side} - \text{Non dominant side}}{\text{Dominant side}}\right) \times 100$$

The Shapiro–Wilk test was applied to assess the normality of these relative differences. Variables whose differences did not follow a normal distribution were analyzed via nonparametric tests. The Wilcoxon test was used to compare the medians of the differences within groups, and the Mann–Whitney *U* test was used to compare the medians between groups, considering a significance level (alpha) of 0.05.

3. Results

Table 1 presents the descriptive statistics of the training time and unilateral variables assessed in the canoeist and kayaker groups. Both groups have an average of two years of sports practice and spend 10.7 h per week training.

Table 1. General characteristics of the study sample and their unilateral anthropometric variables, including the differences between canoeists and kayakers.

	Variable	Canoe (n = 13)	Kayak (n = 14)	p *
		Median (IQR)	Median (IQR)	
General information	Age (years)	13.0 (2.0)	13.0 (2.0)	1.000
	Length of practice (years)	1.0 (5.0)	1.75 (6.5)	0.023 *
	Weekly training (h)	10.0 (9.5)	12.0 (21.5)	0.058
Basics	Body mass (kg)	56.2 (32.2)	50.8 (46.8)	0.452
	Stature (cm)	160.4 (25.1)	165.2 (18.8)	0.182
	Sitting height (cm)	79.6 (16.5)	81.3 (15.5)	0.234
	Arm span (cm)	163.6 (28.8)	166.4 (20.7)	0.512
Girths and Breadths	Chest girth (cm)	83.0 (30.5)	79.8 (32.9)	0.716
	Waist girth (cm)	70.2 (26.0)	68.5 (26.9)	0.846
	Hip girth (cm)	85.1 (17.5)	84.0 (31.1)	0.482
	Iliospinale length (cm)	101.5 (16.8)	101.3 (76.8)	0.680
	Biacromial breadth (cm)	29.0 (9.2)	29.5 (15.4)	0.450
	Biiliocristal breadth (cm)	19.8 (11.6)	21.0 (12.6)	0.295
	Transverse chest breadth (cm)	23.2 (7.9)	22.4 (12.5)	1.000
	Antero-posterior chest breadth (cm)	17.0 (8.0)	16.3 (13.0)	0.367

Values are presented as medians and interquartile ranges; * statistical significance according to Mann–Whitney U test ($p < 0.05$).

Table 2 presents the side-by-side analysis of the canoeing and kayaking groups. The hemibodies of the paddlers in both groups showed homogeneity in variables describing their skeletal size in terms of length and breadth ($p > 0.05$). The only significant asymmetries were found in the subscapular skinfold ($p = 0.010$) and the flexed and tensed arm girth in canoeists ($p = 0.055$). For the kayakers, differences between hemibodies were observed only in the mid-thigh girth ($p = 0.041$).

In evaluating the differences between the groups (canoeing versus kayaking) on a single side, it was found that both disciplines were homogeneous in most anthropometric variables when comparing each side ($p > 0.05$). However, the kayakers differed from the canoeists in the subscapular skinfold ($p = 0.010$) and the bicep skinfold ($p = 0.038$) on the right side, whereas on the left side, the differences were limited to the subscapular skinfold ($p = 0.010$).

Figure 1 shows the direction of the asymmetries found via the %BAI. The specific observations for each type of measurement are as follows.

Graph (a): Asymmetry trends in skinfolds. Compared with kayakers, canoeists presented marked rightward asymmetry in their subscapular and bicep skinfolds. Kayakers had a higher %BAI in their triceps skinfold. The average values for the rest of the variables were relatively similar. Canoeists had higher extreme values for their subscapular, bicep, and abdominal skinfolds, whereas kayakers had lower extreme values for their tricep and suprailiac skinfolds.

Graph (b): Asymmetry trends in girths. Compared with kayakers, canoe paddlers presented with an average rightward asymmetry in their relaxed arm, flexed and tensed arm, and ankle girths. Kayak paddlers had a higher index for their maximum and middle-thigh girths. The average values for the remaining indices were relatively similar. Canoeists presented higher extreme values for their maximum thigh, whereas kayakers presented higher extreme values for their arm, forearm, and thigh girths.

Graph (c): Asymmetry trends in lengths. Kayak paddlers presented greater heterogeneity in their acromial-radial, radial-stylion, and trochanterion tibial lateral lengths. Canoe paddlers presented higher extreme values for their thigh length (1 cm gluteal), whereas kayakers presented higher extreme values for their acromial-radial, radial-stylion, and trochanterion tibial lateral lengths.

Graph (d): Asymmetry trends in breadths. The average index values are similar for all three breadths in both groups. The asymmetry index for the humerus breadth is more heterogeneous among canoeists, whereas for kayakers, the asymmetry indices of the styloid and femur breadths are more heterogeneous.

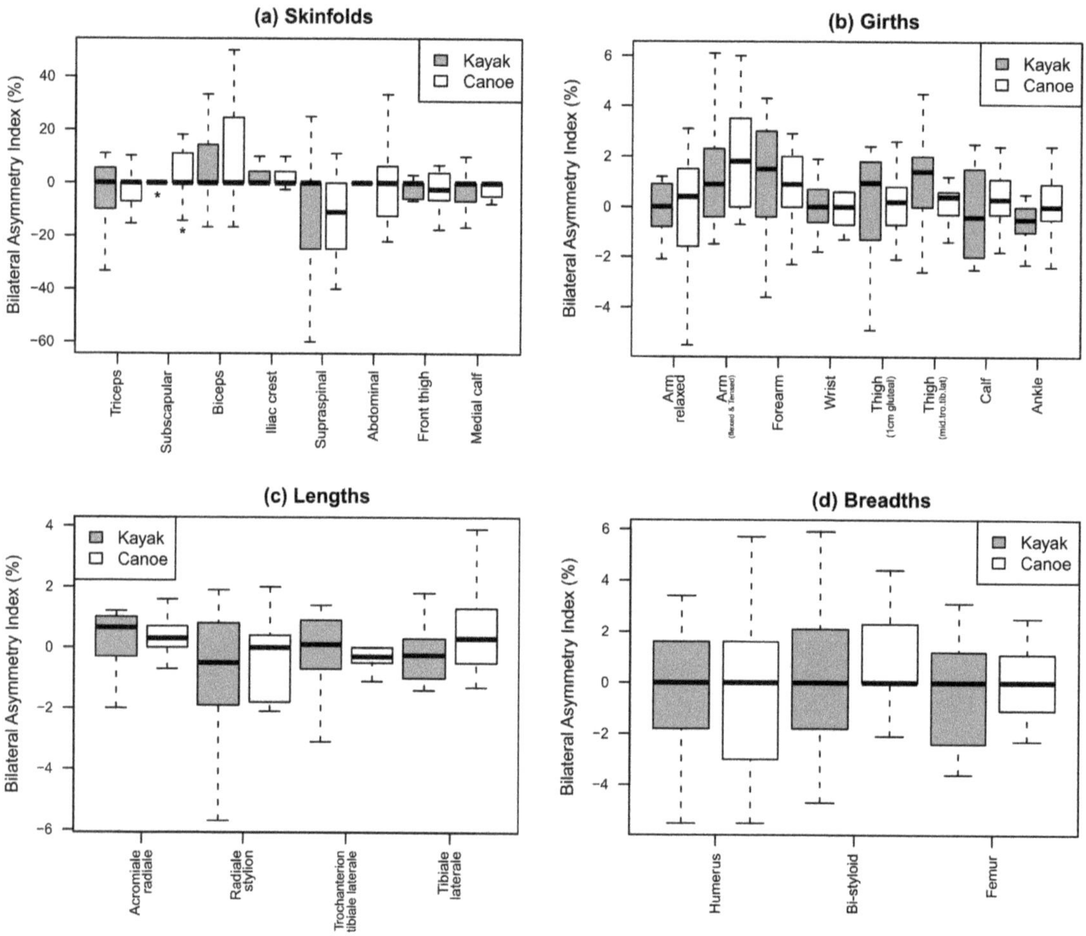

Figure 1. Bilateral asymmetry indices (%BAI) for skinfolds, girths, lengths, and breadths. Box plots marked with an asterisk (*) indicate statistically significant differences, considering a significance level of $\alpha = 0.05$.

Table 2. Comparison of anthropometric asymmetries between canoe and kayak athletes: intra- and intergroup analysis.

Variable		Canoe (n = 13)			Kayak (n = 14)			All (n = 27)	
		Right Side	Left Side	p	Right Side	Left Side	p	p (Right Side)	p (Left Side)
Skinfolds (mm)	Triceps	13.0 (7.6)	13.0 (7.0)	0.784	9.2 (4.2)	9.0 (4.2)	0.951	0.113	0.084
	Subscapular	9.0 (6.0)	9.0 (6.0)	0.275	6.5 (2.0)	7.0 (2.0)	1.000	0.010 **	0.010 **
	Biceps	7.0 (6.0)	5.0 (6.0)	0.143	4.0 (3.0)	4.0 (2.0)	0.203	0.038 *	0.117
	Iliac crest	12.0 (11.0)	12.0 (8.0)	0.240	10.0 (5.8)	9.5 (6.0)	0.203	0.188	0.158
	Supraspinal	9.0 (9.0)	10.0 (8.0)	0.017 *	6.5 (7.5)	7.8 (6.2)	0.256	0.224	0.144
	Abdominal	11.0 (8.0)	10.0 (6.0)	0.916	8.5 (4.1)	8.5 (4.9)	0.766	0.273	0.263
	Front thigh	16.0 (6.0)	17.0 (7.0)	0.120	14.5 (4.2)	15.0 (4.8)	0.457	0.067	0.061
	Medial calf	14.0 (8.0)	14.0 (8.0)	0.386	12.5 (4.8)	12.5 (5.8)	0.390	0.196	0.188
Girths (cm)	Arm (relaxed)	26.0 (4.1)	25.9 (4.7)	0.674	24.0 (1.5)	24.4 (0.9)	1.000	0.382	0.356
	Arm (flexed and tensed)	27.9 (4.6)	27.9 (5.9)	0.055 *	26.8 (2.0)	26.8 (1.7)	0.131	0.716	0.716
	Forearm	23.5 (2.5)	23.0 (1.8)	0.655	23.4 (1.4)	23.6 (1.1)	0.123	0.734	0.644
	Wrist	15.6 (1.0)	15.5 (0.9)	0.412	15.6 (0.8)	15.4 (0.9)	0.535	0.752	0.808
	Thigh (1 cm gluteal)	52.2 (9.0)	52.1 (7.8)	0.834	50.0 (2.9)	49.2 (1.2)	0.379	0.275	0.344
	Mid-thigh	48.0 (7.7)	47.4 (8.3)	0.272	44.2 (2.6)	44.1 (2.8)	0.041 *	0.452	0.409
	Calf	34.0 (5.9)	34.6 (4.9)	0.125	33.0 (2.3)	33.3 (3.0)	0.777	0.734	0.846
	Ankle	21.0 (3.0)	22.3 (3.0)	1.000	21.8 (0.9)	22.0 (1.8)	0.148	0.808	0.884
Lengths (cm)	Acromiale-radiale	30.3 (3.4)	30.5 (3.1)	0.326	30.4 (1.6)	30.4 (1.8)	0.344	0.903	1.000
	Radiale-stylion	24.5 (3.1)	24.0 (2.6)	0.258	25.4 (3.8)	25.3 (4.3)	0.419	0.308	0.331
	Trochanterion tib.lat	37.4 (7.6)	38.0 (6.4)	0.448	44.1 (5.2)	43.8 (5.1)	0.889	0.094	0.120
	Tibiale laterale	37.3 (2.5)	37.5 (2.2)	0.600	37.2 (2.9)	37.3 (2.4)	0.220	1.000	1.000
Breadths (cm)	Humerus	5.9 (0.7)	5.9 (0.7)	0.124	5.8 (0.6)	5.8 (0.6)	0.811	0.769	1.000
	Bistyloid	4.6 (0.6)	4.6 (0.6)	0.573	4.8 (0.4)	4.8 (0.4)	0.719	0.306	0.329
	Femur	8.7 (0.6)	8.7 (0.6)	0.723	8.5 (0.6)	8.6 (0.3)	0.608	0.465	0.480

Values are presented as medians and interquartile ranges; * statistical significance according to Wilcoxon signed-rank test or Mann–Whitney U test ($p < 0.05$); ** statistical significance according to Mann–Whitney U test ($p < 0.01$).

Interestingly, the observed asymmetries were not uniform across all participants, by reflected individual variations. These differences suggest that future research could explore how intrinsic factors (e.g., genetics, growth) and extrinsic factors (e.g., training intensity) contribute to these asymmetries.

Additional Analyses

In addition to the main comparisons, further analyses were conducted to explore possible interactions and differences within demographic subgroups. The role of age in the observed differences was investigated, and no significant interactions were found ($p = 0.761$). The technical error of measurement (TEM) was 5.36% for skinfolds and 2.07% for the remaining variables, indicating a high level of reliability in the anthropometric assessments. A sensitivity analysis was also performed to assess the robustness of the results against variations in measurement methods and participant inclusion criteria. The findings remained consistent ($p = 0.121$), supporting the internal validity of the observed differences between the right and left sides in the anthropometric measurements.

4. Discussion

4.1. Key Findings

The main objective of this study was to assess the differences in asymmetry between canoeists and kayakers using anthropometric measurements, focusing on the possible differences derived from the asymmetrical nature of canoeing and the more symmetrical demands of kayaking. On the right side of the body, kayakers showed significant differences from canoeists, specifically in the subscapular skinfold and bicep skinfold. In contrast, on the left side, differences were limited to the subscapular skinfold. These asymmetries may reflect sport-specific adaptations. These findings are consistent with previous studies indicating that repetitive movement patterns in asymmetrical sports often lead to localized imbalances in muscle and soft tissue development [35]. Additionally, the absence of significant skeletal asymmetry highlights that these adaptations are more likely to occur in soft tissues than in bone structure during adolescence [36]. Some studies have reported similarities and differences with the present research, although many were conducted with athletes from other sports [24,27,37]. For instance, one study found that football players exhibited significant asymmetries in their knee extensors and flexors, especially in the U13 and U15 categories, but these asymmetries diminished in older categories (U17). These findings suggest that long-term adaptations and balanced training can mitigate initial asymmetries [37]. In our research, although the motor gestures in canoeing are asymmetrical, young paddlers did not show significant differences in their anthropometric profiles, which may be due to early adaptations and balanced training programs. This finding supports the idea that asymmetries in young athletes are often transient and respond to well-structured training interventions [38].

Our initial hypothesis predicted a greater number of asymmetries due to the biomechanical nature of each discipline. Existing studies suggest that canoeing is an asymmetrical sport due to the unilateral movements performed during paddling, while kayaking is considered more symmetrical, as it uses both sides of the body more evenly [13,29]. However, contrary to this expectation, no other significant differences were observed between canoeists and kayakers beyond those that have already been mentioned.

Another key finding was that canoeists had less experience practicing the sport (1.4 years) and fewer weekly training hours (9 h) compared to kayakers (2.6 years and 12.2 h). However, despite this significant difference in experience, no characteristic pattern emerged when comparing the right and left sides of the body between the two modalities. These results align with a study conducted by Saal and associates, which suggests that athletes may exhibit limited asymmetries during their developmental phase, highlighting the importance of balanced training programs to address emerging imbalances [38].

Overall, no significant differences were found between canoeists and kayakers in terms of symmetry; this study highlights the potential of anthropometry, as a highly reproducible tool, for monitoring possible asymmetries over time. The findings suggest that regular anthropometric evaluations could be invaluable for assessing the effectiveness of training programs and preventing the development of asymmetries that could increase the risk of long-term injuries. Additionally, these results support the idea that asymmetries in young athletes may not be significant in the early stages of training but could develop over time as they continue their practice and specialize in their sport.

4.2. Limitations

The main limitation of this study is the sample size, as no prior calculation was made, which may have affected its statistical power in detecting significant differences. Future studies with larger samples could provide more robust and generalizable results.

Another relevant limitation is the significant difference in sports experience between the groups analyzed. On average, the kayaking group had 1.2 more years of experience than the canoeing group, which may have influenced the results and prevented an equitable comparison. Evaluating adolescents with a broader spectrum of experience in both sports could help us understand how muscular and structural adaptations affect asymmetry and to what extent balanced training programs and technical experience can reduce these differences [13,16,29]. A recent study revealed that young athletes practicing asymmetrical sports, such as tennis and volleyball, showed greater asymmetries compared to those participating in symmetrical sports like triathlon and gymnastics. However, these differences were more pronounced in athletes with more years of experience [24]. Although it may seem contradictory, the lack of significant asymmetries in athletes with few years of experience suggests that asymmetries develop over time and with sport specialization. Therefore, early monitoring and corrective measures play a key role in minimizing excessive asymmetries [25].

Additionally, another important factor is the possible influence of growth and puberty on anthropometric characteristics, as adolescents undergo significant physical changes during their maturation. Factors such as the onset of puberty and growth spurts could have affected the observed asymmetries. Since this study did not assess these developmental factors, it is not possible to fully determine their influence on the results. Ramos-García and colleagues [39] highlight the importance of considering growth in anthropometric studies of adolescents, suggesting that future research that includes growth assessments could provide a more comprehensive understanding of how maturation affects the development of asymmetry in young athletes.

It is important to note that biological maturation does not always follow a linear pattern [22,39], so its impact on the development of asymmetries may vary among individuals. Future studies could benefit from the inclusion of biological age assessments or maturation analyses to better understand these effects and distinguish between changes attributable to natural development and those derived from specific training.

In relation to the points previously discussed, Krzykała and colleagues [13] noted that young canoe athletes, due to specialized training, may develop sport-specific asymmetries during their biological development. However, in our study population, this asymmetry did not manifest, possibly due to the reduced sample size (previously mentioned as the main limitation of this study) and the relatively shorter sports experience of the canoeing group compared to the kayaking group.

An important limitation of this study is the lack of a longitudinal analysis, which would have allowed for observing the evolution of asymmetries over time, and another is the absence of asymmetry assessments using advanced complementary tools, such as three-dimensional imaging and biomechanical analyses. These tools could have provided a more detailed view of how technique and movement mechanics influence the development of asymmetries. A longitudinal approach would allow for evaluating how asymmetries evolve across different stages of sports development and establish whether there are critical moments at which an intervention could be more effective. Additionally, future studies could focus on key variables such as training load progression and sports specialization to determine their impact on the emergence and magnitude of asymmetries.

In line with this perspective, a longitudinal study has suggested that asymmetries may evolve depending on the training load and competitive level, with more experienced athletes typically exhibiting more refined biomechanical adaptations [21]. This highlights the importance of structured intervention programs, which could not only correct imbalances but also improve performance by leveraging controlled asymmetries. Specifically, in water sports, integrating biomechanical analyses into training protocols can provide valuable information to optimize stroke mechanics and minimize the risk of injury [27], which is essential and allows coaches to design more specific and effective interventions [29]. In this context, the use of advanced tools (such as those mentioned above) would have increased our accuracy in detecting asymmetries in our study population.

4.3. Interpretation

Several studies suggest that asymmetrical differences between young and adult athletes can be attributed to factors such as muscle development, technique, long-term adaptations, and injury history, reflecting how the body adapts to the specific demands of a sport over time [13,16,25,26]. The absence of significant asymmetries in this study may be due to the athletes still being in an early stage of their development, where training adaptations have not yet fully consolidated. At this phase, any imbalances may be transient, as the body is still in the process of adjusting and maturing.

Longitudinal studies have suggested that asymmetries may evolve based on the training load and competitive level of the athlete, with more experienced athletes often exhibiting more refined biomechanical adaptations [21]. This finding emphasizes the potential of structured intervention programs, not only to address imbalances but also to enhance performance by leveraging controlled asymmetries. Specifically, in water sports, integrating biomechanical analyses into training protocols could provide useful insights to optimize stroke mechanics and minimize the risk of injury [27].

Although both groups engage in intense physical activity, the specific demands of each discipline may induce slightly different adaptations in certain body regions. The few significant differences observed in the subscapular and bicep skinfolds on the right side, as well as in the subscapular skinfold on the left side, suggest that although no significant asymmetries were found overall, each sport may generate subtle morphological variations (as previously mentioned). These differences could be related to rowing technique, muscle activation patterns, and the load distribution in each discipline.

Recent studies highlight the importance of considering the specific mechanical demands of each sport when interpreting asymmetry data. It has been proposed that sports with unilateral gestures, such as canoeing, may induce compensatory adaptations that do not necessarily negatively affect performance [23]. In fact, these adaptations could optimize athlete efficiency, even in the presence of asymmetries, especially during the early stages of sports development. However, further research is still needed to determine whether these adaptations offer performance advantages or whether they pose long-term risks [24].

4.4. Generalization

The generalization of these results should be made with caution. Although our findings align with previous studies in suggesting that asymmetries may not be significant in the early stages of training [24,27], other works have reported marked asymmetries in young athletes [13,23,37]. This indicates that the emergence of asymmetries could depend on various factors, such as the specific demands of the sport, the duration and intensity of training, and the individual characteristics of the athletes. These factors could significantly influence the manifestation of asymmetries. To better understand these aspects, it would be helpful to conduct studies with larger and more diverse samples. Furthermore, additional variables such as training techniques, genetic predisposition, and injury history should be considered when evaluating the generalization of the results, as they may play an important role in the observed differences.

4.5. Relevance and Practical Applications

The continuous assessment of asymmetries using more reproducible tools, such as anthropometry, could directly impact the development of balanced training programs aimed at mitigating potential asymmetries, thereby improving the performance and health of athletes.

In summary, these findings provide a valuable foundation for understanding the asymmetries in young canoeists and kayakers. However, further studies are needed to generalize these results to other populations and sporting contexts. Understanding at what point asymmetries cease to be functional and become risk factors could be key to designing specific interventions [14]. As mentioned in the Introduction, most studies in this field have used a limited number of anthropometric indicators to assess asymmetry differences [1,5,19,20,28]. In this regard, the present study is one of the first to address these differences using a broad and detailed anthropometric profile, making it an important reference for future research in this field.

5. Conclusions

This study identified significant differences in the anthropometric asymmetry profiles of adolescent canoeists and kayakers, but these differences were observed in only 3 of the 58 variables analyzed. These findings suggest that the differences were more limited and less pronounced than initially expected. These results highlight the need for further research to examine the relationship between sports experience and asymmetries, with the aim of identifying adaptive patterns in aquatic sport disciplines. A better understanding of these patterns could help optimize the design of balanced training plans aimed at supporting performance and reducing injury risk, with an emphasis on the importance of monitoring asymmetries from an early stage.

Furthermore, this study reinforces the value of anthropometry as a highly reproducible tool for monitoring asymmetries in young athletes. Future research integrating this with long-term monitoring and/or biomechanical analyses could provide deeper insights into these adaptations, offering practical applications for coaches and sports professionals.

Author Contributions: Conceptualization, C.A.H.-A., W.C.-V., J.G.-G., F.A. and C.O.R.-G.; methodology, C.A.H.-A., W.C.-V., R.Y.-S., F.A. and J.O.-A.; software, C.A.H.-A., R.Y.-S. and J.F.L.-G.; validation, J.G.-G., F.A., J.F.L.-G., J.O.-A. and C.O.R.-G.; formal analysis, C.A.H.-A., W.C.-V. and R.Y.-S.; investigation, C.A.H.-A., J.G.-G., F.A. and C.O.R.-G.; resources, R.Y.-S., F.A., J.F.L.-G. and J.O.-A.; data curation, C.A.H.-A., W.C.-V., J.G.-G. and F.A.; writing—original draft preparation, C.A.H.-A., W.C.-V., R.Y.-S., J.F.L.-G. and C.O.R.-G.; writing—review and editing, J.G.-G., F.A. and J.O.-A.; visualization, F.A., J.F.L.-G. and C.O.R.-G.; supervision, F.A. and C.O.R.-G.; project administration, J.G.-G., F.A. and

C.O.R.-G.; funding acquisition, R.Y.-S., J.O.-A. and J.F.L.-G. All authors have read and agreed to the published version of the manuscript.

Funding: This research received no external funding.

Institutional Review Board Statement: This study was conducted in accordance with the Declaration of Helsinki and approved by the Institutional Ethical Committee of the University of Murcia (SKMBT-C25211110314021 on 24 October 2024).

Informed Consent Statement: Signed parental consent was obtained from the guardians of all participants under 18 years of age. Parents and study participants were fully and appropriately informed about the participation requirements and the purpose, risks, and benefits of the study.

Data Availability Statement: The original contributions presented in this study are included in the article. Further inquiries can be directed to the corresponding author(s).

Acknowledgments: The authors would like to express their gratitude to the paddlers, coaches, and parents for their invaluable assistance and commitment, which made this study possible. We also thank the participating clubs for welcoming us and granting us access to their most valuable asset, their young athletes.

Conflicts of Interest: The authors declare that they have no conflicts of interest.

Abbreviations

The following abbreviations are used in this manuscript:

DXA	Dual-energy X-ray absorptiometry
BIA	Body Electrical Impedance Analysis
STROBE	Strengthening the Reporting of Observational Studies in Epidemiology
TEM	Technical error of measurement
%BAI	Bilateral asymmetry index

References

1. Maloney, S.J. The relationship between asymmetry and athletic performance: A critical review. *J. Strength Cond. Res.* **2019**, *33*, 2579–2593. [PubMed]
2. Burdukiewicz, A.; Pietraszewska, J.; Andrzejewska, J.; Chromik, K.; Stachoń, A. Asymmetry of musculature and hand grip strength in bodybuilders and martial artists. *Int. J. Environ. Res. Public Health* **2020**, *17*, 4695. [CrossRef]
3. Stagi, S.; Mulliri, G.; Doneddu, A.; Ghiani, G.; Marini, E. Body Composition and Strength Symmetry of Kettlebell Sport Athletes. *Biology* **2023**, *12*, 440. [CrossRef] [PubMed]
4. Chapelle, L.; Bishop, C.; D'Hondt, J.; Rommers, N.; D'Hondt, E.; Clarys, P. Development of Upper and Lower Extremity Functional Asymmetries in Male and Female Elite Youth Tennis Players: A Longitudinal Study. *J. Sports Med. Phys. Fit.* **2023**, *63*, 1269–1284.
5. Badau, A.; Badau, D. Identifying the Differences in Symmetry of the Anthropometric Parameters of the Upper Limbs in Relation to Manual Laterality between Athletes Who Practice Sports with and without a Ball. *Symmetry* **2024**, *16*, 558. [CrossRef]
6. Sebastian, K.; Hassan, M. Inter Limb Symmetry of Selected Foot Anthropometric Measures and Its Relationship to Vertical Jumping Ability. *Int. J. Health Sci.* **2022**, *6* (Suppl. S2), 2571–2578.
7. Chapelle, L.; D'Hondt, E.; Rommers, N.; Clarys, P. Development of Upper-Extremity Morphological Asymmetries in Male and Female Elite Youth Tennis Players: A Longitudinal Study. *Pediatr. Exerc. Sci.* **2023**, *63*, 91–97.
8. Alacid, F.; Marfell-Jones, M.; Muyor, J.M.; López-Miñarro, P.A.; Martínez, I. Kinanthropometric Comparison Between Young Elite Kayakers and Canoeists. *Coll. Antropol.* **2015**, *39*, 119–124. [PubMed]
9. Yáñez-Sepúlveda, R.; Herrera-Amante, C.A.; Clemente-Suárez, V.J.; Vasquez-Bonilla, A.; Alacid, F.; Tuesta, M.; Báez-San Martín, E.; Giakoni-Ramírez, F.; Cristi-Montero, C. Anthropometry, Body Composition, Somatotype, and Asymmetry of Canoe Sprint World Champion: A Case Study. *Nutr. Health* **2025**. *ahead of print*. [CrossRef]
10. Jones, W. Significance of Anthropometry and Its Practical Applications in Different Fields. *J. Ergon.* **2023**, *13*, 1.
11. Esparza Ros, F.; Vaquero-Cristóbal, R.; Marfell-Jones, M. *International Standards for Anthropometric Assessment*; International Society for the Advancement of Kinanthropometry (ISAK): Murcia, Spain, 2019; 114p.
12. Smith, B.; McCarthy, C.; Dechenaud, M.E.; Wong, M.C.; Shepherd, J.; Heymsfield, S.B. Anthropometric Evaluation of a 3D Scanning Mobile Application. *Obesity* **2022**, *30*, 1181–1188.

13. Krzykała, M.; Karpowicz, K.; Karpowicz, M.; Bartkowiak, S.; Demuth, A.; Czerniak, U.; Janowski, J. Somatic Characteristics, Morphological Asymmetry, and Postural Stability of Youth Male Canoeists Compared to Control: A Cross-Sectional Study. *PLoS ONE* **2023**, *18*, e0285997.
14. Ackland, T.R.; Ong, K.B.; Kerr, D.A.; Ridge, B. Morphological Characteristics of Olympic Sprint Canoe and Kayak Paddlers. *J. Sci. Med. Sport* **2003**, *6*, 285–294.
15. López-Plaza, D.; Alacid, F.; Muyor, J.M.; López-Miñarro, P. Differences in Anthropometry, Biological Age and Physical Fitness Between Young Elite Kayakers and Canoeists. *J. Hum. Kinet.* **2017**, *57*, 181–190.
16. Afonso, J.; Peña, J.; Sá, M.; Virgile, A.; García-de-Alcaraz, A.; Bishop, C. Why Sports Should Embrace Bilateral Asymmetry: A Narrative Review. *Symmetry* **2022**, *14*, 1993. [CrossRef]
17. Chapelle, L.; Rommers, N.; Clarys, P.; D'Hondt, E. Whole-Body Morphological Asymmetries in High-Level Female Tennis Players: A Cross-Sectional Study. *J. Sports Sci.* **2021**, *39*, 777–782. [PubMed]
18. Ojeda-Aravena, A.; Warnier-Medina, A.; Brand, C.; Morales-Zúñiga, J.; Orellana-Lepe, G.; Zapata-Bastias, J.; Tuesta, M. Relationship Between Body Composition Asymmetry and Specific Performance in Taekwondo Athletes: A Cross-Sectional Study. *Symmetry* **2023**, *15*, 2087. [CrossRef]
19. Santos, K.B.; Bento, P.C.; Payton, C.; Rodacki, A.L. Symmetry in the Front Crawl Stroke of Different Skill Levels of Able-Bodied and Disabled Swimmers. *PLoS ONE* **2020**, *15*, e0229918.
20. Bartolomeu, R.F.; Rodrigues, P.; Santos, C.C.; Costa, M.J.; Barbosa, T.M. Is there any effect of symmetry on velocity of the four swimming strokes. *Symmetry* **2021**, *14*, 12. [CrossRef]
21. Warmenhoven, J.; Smith, R.; Draper, C.; Harrison, A.; Bargary, N.; Cobley, S. Force coordination strategies in on-water single sculling: Are asymmetries related to better rowing performance? *Scand. J. Med. Sci. Sports* **2018**, *28*, 1379–1388.
22. Malina, R.M.; Bouchard, C.; Bar-Or, O. *Growth, Maturation, and Physical Activity*; Human Kinetics: Champaign, IL, USA, 2004; 712p.
23. Rynkiewicz, M.; Baumgarten, M.; Mroczkowski, A. Does Paddling in Canoes Cause Asymmetry of Muscle Mass and Fat Mass in High-Level Athletes? *Rocz. Lubus.* **2018**, *44*, 355–366.
24. Kalata, M.; Maly, T.; Hank, M.; Michalek, J.; Bujnovsky, D.; Kunzmann, E.; Zahalka, F. Unilateral and Bilateral Strength Asymmetry Among Young Elite Athletes of Various Sports. *Medicina* **2020**, *56*, 683. [CrossRef] [PubMed]
25. Parpa, K.; Michaelides, M. Anterior-Posterior and Inter-Limb Lower Body Strength Asymmetry in Soccer, Basketball, Futsal, and Volleyball Players. *Medicina* **2022**, *58*, 1080. [CrossRef] [PubMed]
26. Šarabon, N.; Smajla, D.; Maffiuletti, N.A.; Bishop, C. Strength, Jumping and Change of Direction Speed Asymmetries in Soccer, Basketball and Tennis Players. *Symmetry* **2020**, *12*, 1664. [CrossRef]
27. Gao, Z. The Effect of Application of Asymmetry Evaluation in Competitive Sports: A Systematic Review. *J. Phys. Act. Health* **2022**, *6*, 257–272.
28. Badau, D.; Badau, A.; Joksimović, M.; Manescu, C.O.; Manescu, D.C.; Dinciu, C.C.; Margarit, I.R.; Tudor, V.; Mujea, A.M.; Neofit, A.; et al. Identifying the Level of Symmetrization of Reaction Time According to Manual Lateralization Between Team Sports Athletes, Individual Sports Athletes, and Non-Athletes. *Symmetry* **2023**, *16*, 28. [CrossRef]
29. Álvarez-Yates, T.; Iglesias-Caamaño, M.; Cuba-Dorado, A.; Serrano-Gómez, V.; Ferreira-Lima, V.; Nakamura, F.Y.; García-García, O. Explanatory Model for Elite Canoeists' Performance Using a Functional Electromechanical Dynamometer Based on Detected Lateral Asymmetry. *Symmetry* **2024**, *16*, 347. [CrossRef]
30. Elm, E.V.; Altman, D.G.; Egger, M.; Pocock, S.J.; Gøtzsche, P.C.; Vandenbroucke, J.P. Strengthening the Reporting of Observational Studies in Epidemiology (STROBE) Statement: Guidelines for Reporting Observational Studies. *BMJ* **2007**, *335*, 806–808.
31. Vandenbroucke, J.P.; von Elm, E.; Altman, D.G.; Gøtzsche, P.C.; Mulrow, C.D.; Pocock, S.J.; Schlesselman, J.J.; Egger, M. Strengthening the Reporting of Observational Studies in Epidemiology (STROBE): Explanation and Elaboration. *Int. J. Surg.* **2014**, *12*, 1500–1524.
32. World Medical Association. World Medical Association Declaration of Helsinki: Ethical Principles for Medical Research Involving Human Subjects. *JAMA* **2013**, *310*, 2191–2194. [CrossRef] [PubMed]
33. Pederson, D.; Gore, C. Anthropometry Measurement Error. In *Anthropometrica*; Norton, K., Olds, T., Eds.; UNSW Press: Sydney, Australia, 1996; pp. 77–96.
34. Impellizzeri, F.M.; Rampinini, E.; Maffiuletti, N.; Marcora, S.M. A Vertical Jump Force Test for Assessing Bilateral Strength Asymmetry in Athletes. *Med. Sci. Sports Exerc.* **2007**, *39*, 2044. [CrossRef] [PubMed]
35. Abelleira-Lamela, T.; Vaquero-Cristóbal, R.; Esparza-Ros, F.; Marcos-Pardo, P.J. Biomechanical Adaptations in Kayakers of Different Competitive Levels and the Relationship with the Kayak Elements. *Appl. Sci.* **2020**, *10*, 8389. [CrossRef]
36. Manna, T.; Adhikari, S. A Comparative Study of Anthropometric and Physical Profiles of Male Junior Rowers, Kayakers and Canoers. *Med. Sport. J. Rom. Sports Med. Soc.* **2018**, *14*, 3028–3036.
37. Kalata, M.; Hank, M.; Bujnovsky, D.; Michalek, J.; Varjan, M.; Kunzmann, E.; Zahalka, F.; Maly, T. Bilateral Strength Asymmetry in Elite Youth Soccer Players: Differences Between Age Categories. *Symmetry* **2021**, *13*, 1982. [CrossRef]

38. Saal, C.; Chaabene, H.; Helm, N.; Warnke, T.; Prieske, O. Network Analysis of Associations Between Anthropometry, Physical Fitness, and Sport-Specific Performance in Young Canoe Sprint Athletes: The Role of Age and Sex. *Front. Sports Act. Living* **2022**, 4, 1038350. [CrossRef]
39. Ramos-García, C.O.; López-Plaza, D.; Herrera-Amante, C.A.; Alacid, F. Evaluation of Maturation Among Adolescent Athletes. In *Human Growth and Nutrition in Latin American and Caribbean Countries*; Datta Banik, S., Ed.; Springer International Publishing: Cham, Switzerland, 2023; pp. 199–212.

Journal of
Functional Morphology and Kinesiology

MDPI

Article

Relationship Between Body Mass Index and Fat Mass Percentage with Proprioception in Children

Eduardo Guzmán-Muñoz [1,2,*], Yeny Concha-Cisternas [3,4], Guillermo Méndez-Rebolledo [1], Manuel Vásquez-Muñoz [5,6], Pablo Valdés-Badilla [7,8], Tomás Herrera-Valenzuela [9], Cristian Núñez-Espinosa [10,11] and Jordan Hernández-Martínez [12,13]

1 Escuela de Kinesiología, Facultad de Salud, Universidad Santo Tomás, Talca 3460000, Chile; guillermomendezre@santotomas.cl
2 Pedagogía en Educación Física, Facultad de Educación, Universidad Autónoma de Chile, Talca 3460000, Chile
3 Vicerrectoría de Investigación e Innovación, Universidad Arturo Prat, Iquique 1100000, Chile; yenyconchaci@santotomas.cl
4 Department of Physical Education, Faculty of Education Sciences, University ofCádiz, 11519 Puerto Real, Spain
5 Centro de Observación y Análisis de Datos en Salud, Facultad de Medicina y Ciencias de la Salud, Universidad Mayor, Santiago 8580745, Chile; manuel.vasquez@umayor.cl
6 Escuela de Medicina, Facultad de Medicina y Ciencias de la Salud, Universidad Mayor, Santiago 8580745, Chile
7 Departament of Physical Activity Sciences, Faculty of Education Sciences, Universidad Católica del Maule, Talca 3460000, Chile; pvaldes@ucm.cl
8 Sports Coach Career, School of Education, Universidad Viña del Mar, Viña del Mar 2200055, Chile
9 Department of Physical Activity, Sports and Health Sciences, Faculty of Medical Sciences, Universidad de Santiago de Chile (USACH), Santiago 8370003, Chile; tomas.herrera@usach.cl
10 Escuela de Medicina, Universidad de Magallanes, Punta Arenas 6200000, Chile; cristian.nunez@umag.cl
11 Centro Asistencial Docente e Investigación, Universidad de Magallanes, Punta Arenas 6200000, Chile
12 Department of Physical Activity Sciences, Universidad de Los Lagos, Osorno 5290000, Chile; jordan.hernandez@ulagos.cl
13 Programa de Investigación en Deporte, Sociedad y Buen Vivir, Universidad de los Lagos, Osorno 5290000, Chile
* Correspondence: eguzmanm@santotomas.cl; Tel.: +56-977666128

Abstract: Background/Objectives: Childhood obesity is linked to motor and sensorimotor impairments, including proprioceptive deficits. While research has predominantly focused on lower limb proprioception, less is known about the impact on upper limbs. This study investigated the relationship between body mass index, body fat percentage, and proprioception of children aged 11–12 years. Methods: A quantitative, correlational, observational design was employed. BMI was calculated from weight and height measurements, body fat percentage was assessed via bioelectrical impedance analysis, and proprioception was measured using an active repositioning test with inertial sensors in 44 children. Results: Significant correlations were found between BMI and positional errors in the shoulder (r = 0.64, $p < 0.001$), elbow (r = 0.36, $p = 0.007$), and knee (r = 0.42, $p = 0.002$). Regarding body fat percentage, significant correlations were observed with positional errors in the shoulder (r = 0.28, $p = 0.031$), elbow (r = 0.46, $p < 0.001$), and knee (r = 0.29, $p = 0.030$). Regression analysis showed that BMI and body fat percentage significantly predicted positional errors in the shoulder, elbow, and knee. In the shoulder joint, girls demonstrated lower positional errors compared to boys, influenced by both BMI ($\beta = -1.36$, $p = 0.015$) and body fat percentage ($\beta = -3.00$, $p < 0.001$). Conclusions: Higher BMI and body fat percentage are associated with shoulder, elbow, and knee joint proprioceptive deficits. Interventions targeting weight reduction and proprioceptive training may mitigate these deficits and promote sensorimotor function in children.

Keywords: childhood obesity; proprioception; body mass index; body fat percentage; sensorimotor function; upper limbs; lower limbs

1. Introduction

In the 21st century, overweight and obesity are reaching epidemic levels in many developed and developing countries [1,2]. Overweight and obesity are defined as an abnormal and excessive accumulation of fat that can harm health, manifesting in increased body weight and volume [3]. According to the World Atlas of Obesity, it is projected that by 2035, 39% of children will be obese [4].

Extensive documentation supports the link between childhood obesity and an elevated risk of developing chronic diseases, particularly cardiovascular disease, type 2 diabetes, and specific types of cancer later in life [5–7]. Functionally, childhood obesity can negatively affect various activities of daily life, including balance, stability, and locomotion, increasing predisposition to injury and fall risk [8–10]. Obesity also appears to contribute to reduced efficiency in performing motor action in a bipedal posture, likely due to the limitations in postural control imposed by excess weight [11–13]. Several studies have shown deficits in fine and gross motor skills development among obese individuals, encompassing walking, running, and jumping [14–16].

One of the causes proposed to explain the reported changes in motor skills and motor control in obese children corresponds to the alteration of the sensorimotor system [14,17,18]. The system includes all the afferent, efferent, and central integration components essential for maintaining the functional stability of joints [19]. It processes external and perceives internal sensory information and generates adequate motor responses [19]. Guzmán-Muñoz et al. (2024) reported a longer reaction time in the quadriceps muscle among obese individual compared to their normal-weight peers [20]. This delay highlights the detrimental impact of obesity on motor control, specifically in the integration and execution of neuromuscular responses. Chronic accumulation of adipose tissue and intramuscular fat infiltration is postulated to increase levels of circulating pro-inflammatory cytokines, which may impair muscle function and performance by promoting protein breakdown in muscles [21,22]. In addition, the accumulation of fat mass is associated with a slowdown in motor nerve conduction velocity [23,24], which could contribute to the neuromuscular changes observed in these children.

At the sensory level, studies have shown that children with obesity have a proprioception deficit compared to their normal-weight peers [25,26]. Proprioception is a continuous and unconscious sensory flow from muscles, tendons, joints, and skin, allowing muscle tension, balance, and movement control [27,28]. Two components are essential for an effective motor stabilization strategy: the ability to sense the position of a joint and the ability to perceive the body's movement and its parts [27]. The proprioceptive information comes primarily from muscle spindles, Golgi tendon organs, cutaneous receptors, and capsular mechanoreceptors [28]. In children, the literature has focused specifically on the knee [25,26] and ankle [25] joints, finding a significant proprioceptive deficit in the knee. Despite these results, the relationship between obesity and proprioceptive deficits has only been explored in lower limb joints, and the changes that could be generated in upper limb joints are unknown. Likewise, in these studies, children were classified using BMI [25,26]. However, some authors point out that the body fat percentage would be a more detailed and accurate measurement of body composition to determine the negative effects of obesity [29,30].

This study, therefore, aimed to examine the relationship between BMI and body fat percentage and proprioception in the shoulder, elbow, hip, and knee joints in children. It is hypothesized that children with a higher BMI and body fat percentage will demonstrate a greater proprioceptive deficit in the joints evaluated.

2. Materials and Methods

2.1. Study Design

In this study, we employed a quantitative approach, correlational type, and cross-sectional design. The participants were assessed during a 15-min session in a room set at 21 °C, accompanied by their parents and/or guardians. During the tests, participants wore shorts and were barefoot. Measurements included BMI, body fat percentage, and proprioception.

2.2. Participants

The sampling method used in this research was non-probabilistic, based on convenience. The inclusion criteria included: (i) schoolchildren from a public school located in Talca (Chile); (ii) participants were aged between 11 and 12 years. The exclusion criteria included: (i) individuals with neurological disorders, (ii) those with musculoskeletal injuries in the upper and lower limbs such as fractures, sprains, dislocations, or muscle tears within six months before the assessments, (iii) the presence of any inflammatory or painful conditions affecting the upper and lower limbs at the time of the evaluations, and (iv) reliance on assistive devices for walking.

This study involved a total of 44 schoolchildren, consisting of 24 girls and 20 boys. The girls had an average age of 11.58 ± 0.44 years, a body mass of 48.05 ± 11.27 kg, and an average height of 1.51 ± 0.05 m. The boys had an average age of 11.58 ± 0.41 years, a body mass of 44.92 ± 8.39 kg, and a height of 1.42 ± 0.05 m. In accordance with the principles outlined in the Declaration of Helsinki, informed consent was obtained from both the participants and their parents through signed consent forms. This study was approved by the local Ethics Committee of Universidad Santo Tomás, Chile, under registration number 13320. The sample size was determined using GPower software (Version 3.1.9.6, Franz Faul, Universiät Kiel, Germany) with the multiple linear regression statistical model. For this calculation, an alpha error of 0.05, a power of 0.9, and a number of 2 predictors are considered. The minimum sample size obtained for this study was 36 participants.

2.3. Body Mass Index (BMI)

During the assessments, participants were instructed to wear light clothing (shorts, a light t-shirt, and no shoes) to accurately measure their body weight and standing height. Body weight was recorded using a digital scale Omron Karada HBF-375 (Omron Corporation, Kyoto, Japan; accuracy of 0.1 kg), and height was measured with a stadiometer Seca model 220 (Seca, Hamburg, Germany; accuracy of 0.1 cm). BMI was then calculated by dividing the body weight in kilograms by the square of the height in meters (kg/m^2).

2.4. Body Fat Percentage

Body fat percentage was measured using bioelectrical impedance analysis with the Omron HBF-375 body fat analyzer (Omron HBF-375 Karada Scan; Omron, Kyoto, Japan). This method was chosen due to its validated effectiveness and its frequent use in epidemiological studies for assessing body fat percentage in children [31]. For the measurement, the subject stood upright on the device base, holding the integrated handheld electrodes. The current passed between the foot and hand electrodes, enabling a full-body analysis that estimated body fat percentage. Measurements were conducted under standardized conditions, including 6–8 h of fasting and abstaining from fluid intake or intense physical activity before the assessment [32]. Body fat percentages were classified according to percentiles for sex and age, as defined by McCarthy et al. (2006): normal (2nd–85th percentile), overfat (>85th–95th percentile), and obese (>95th percentile) [33].

2.5. Proprioception

To assess proprioception, it was evaluated using the active repositioning test [25,26,34]. The relative positional error (°) was measured in the shoulder, elbow, hip, and knee joints through an isoinertial measurement unit (IMU). The IMU used for this evaluation was the Trigno Research+ system (Delsys, Boston, USA), with data obtained using the EMGWorks 4.9 software (Delsys, Boston, MA, USA). For each joint, the procedures described below were followed. The subject was placed in a comfortable position to minimize movement and with eyes closed to eliminate visual feedback. An assessor passively moved the joint to a "target" (reference) position and held it for 5 s [25,26,34]. The participant was asked to remember the position of the evaluated segment. The target positions for the joints included 70° of shoulder flexion, 80° of elbow flexion, 70° of hip flexion, and 50° of knee flexion [34]. According to the literature, selecting mid-range joint positions as targets is recommended, as these positions are reported to be more challenging to replicate compared to extremes of the range of motion [35]. Subsequently, the joint was returned passively to a neutral position, and the participant was asked to actively replicate the target position. When the participants indicated that they were in the reference position, the joint angle was recorded and compared to the exposed angle in the target position. The difference between these two values corresponded to the "relative positional error". The joint evaluations were performed three times, and the average of the three measurements was considered [25,26,34].

2.6. Statistical Analysis

Data were analyzed using GraphPad Prism 9.0 statistical software (GraphPad Software, La Jolla, CA, USA). Descriptive statistics, including the mean and standard deviation, were calculated to summarize the sample's characteristics: age, weight, height, BMI, fat mass, and relative positional error. The Shapiro-Wilk test was performed to assess data distribution. Since the data followed a normal distribution, Pearson's correlation test was applied to examine the relationships between BMI and relative positional error, as well as between body fat percentage and relative positional error. A correlation coefficient of from 0 to 0.4 was considered weak, from 0.4 to 0.7 was moderate, and from 0.7 to 1.0 was strong. To determine the influence of gender on the results, multiple linear regression models (95% confidence interval) were performed, where the dependent variable was the relative positional error, and the independent variables were the BMI and body fat percentage adjusted for sex. For this analysis, nutritional status (normal weight, overweight, and obese), body fat percentage (normal, overfat, and obese), and sex (boys/girls) variables were categorized. The goodness of fit was determined using the R2 coefficient. A collinearity diagnosis was made for each variable in the regression models obtained, where variables with values less than 0.10 tolerance and values above 10.0 variance inflation factor (VIF) were eliminated. Statistical significance was set at $p < 0.05$ for all analyses.

3. Results

The sample evaluated obtained an average BMI of 21.93 ± 3.95 kg/m^2, while the mean body fat percentage was 25.18 ± 11.58. Based on BMI, 40.9% of the participants were classified as normal weight, 27.3% as overweight, and 31.8% as obese. On the other hand, based on the body fat percentage, 54.5% were classified as normal, 9.1% as overfat, and

36.4% as obese. Table 1 shows the relative positional error results in the shoulder, elbow, hip, and knee joints.

Table 1. Relative positional error in the shoulder, elbow, hip, and knee joints.

Joint	Mean ± SD	Min-Max
Shoulder (degrees)	4.34 ± 2.33	0.67–9.68
Elbow (degrees)	4.66 ± 2.88	1.00–12.78
Hip (degrees)	3.73 ± 1.66	1.47–8.03
Knee (degrees)	3.88 ± 2.21	0.43–9.97

SD: standard deviation.

Figure 1 shows the results of the correlations between BMI and positional error of the shoulder, elbow, hip, and knee joints. A significant correlation can be observed between BMI and positional error of the shoulder ($p < 0.001$; r = 0.64), elbow ($p = 0.007$; r = 0.36), and knee ($p = 0.002$; r = 0.42). The hip joint had no significant relationship ($p = 0.226$; r = 0.11).

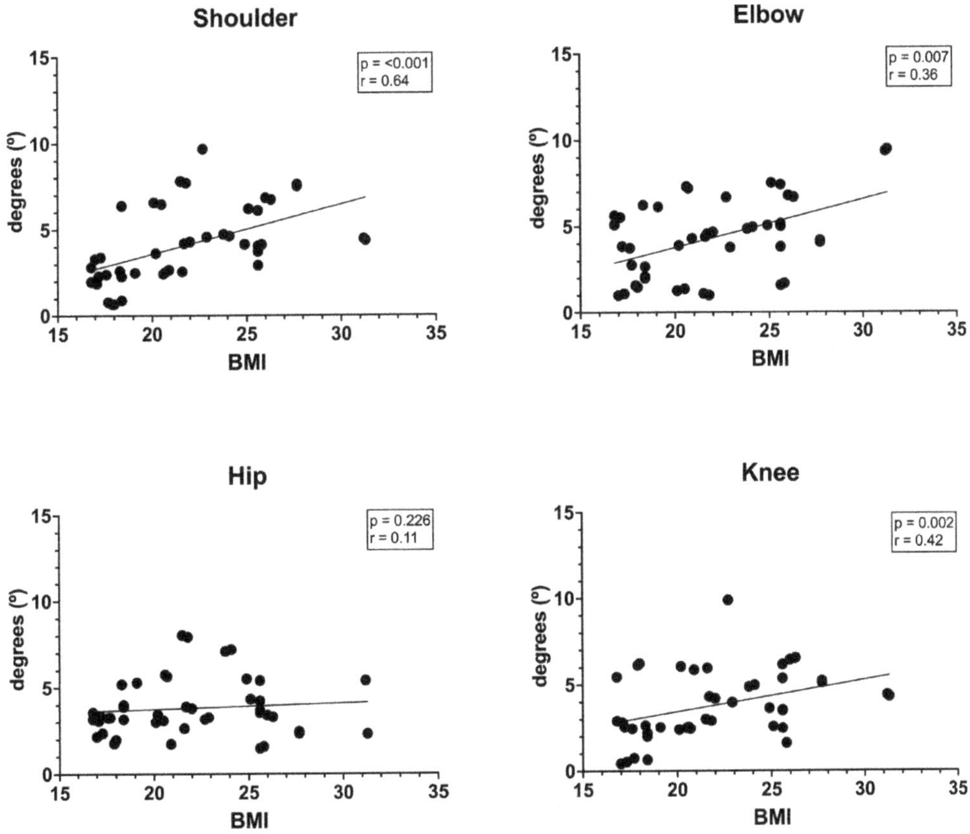

Figure 1. Correlation between BMI and relative positional error.

Regarding body fat percentage, a significant correlation was observed with positional error of shoulder ($p = 0.031$; r = 0.28), elbow ($p < 0.001$; r = 0.46), and knee ($p = 0.030$; r = 0.29) (Figure 2). The hip joint had no significant relationship ($p = 0.449$; r = −0.02).

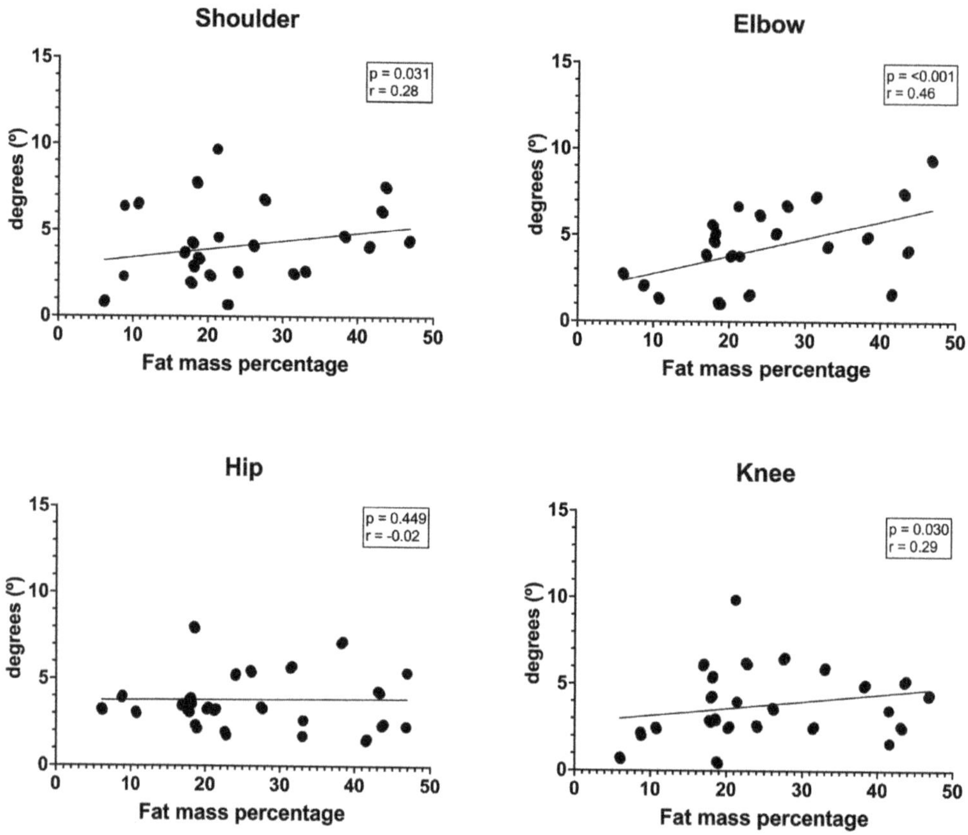

Figure 2. Correlation between body fat percentage and relative positional error.

Table 2 shows the multiple linear regression models obtained for relative positional error time based on BMI. Significant results were found for the shoulder ($R^2 = 0.37$; $p < 0.001$), elbow ($R^2 = 0.19$; $p = 0.013$), and knee ($R^2 = 0.14$; $p = 0.045$) joints. The results indicate that BMI positively correlates with increased relative positional error time in these joints. Particularly, sex significantly influenced results only in the shoulder joint, with girls showing reduced error times compared to boys ($\beta = -1.36$; $p = 0.015$).

Table 2. Multiple linear regression models obtained for relative positional error time according to BMI.

Joint	R^2	Coefficient β	*p value*	95% CI
Shoulder	0.37		<0.001	
Intercept		−1.87	<0.001	−4.97 to −1.22
BMI		0.30	<0.001	0.16 to 0.44
Girls (ref. boys)		−1.36	0.015	−2.46 to −0.27

Table 2. *Cont.*

Joint	R^2	Coefficient β	*p value*	95% CI
Elbow	0.19		0.013	
Intercept		−1.84	ns	−6.01 to 2.32
BMI		0.27	0.004	0.09 to 0.46
Girls (ref. boys)		0.34	ns	−1.13 to 1.81
Hip	0.03		ns	
Intercept		3.14	0.033	0.25 to 6.04
BMI		0.04	ns	−0.08 to 0.173
Girls (ref. boys)		−0.49	ns	−1.51 to 0.52
Knee	0.14		0.045	
Intercept		−0.17	ns	−3.43 to 3.07
BMI		0.18	0.013	0.04 to 0.33
Girls (ref. boys)		−0.25	ns	−1.40 to 0.89

95% CI: 95% confidence interval; ns: no significant.

Table 3 presents the multiple linear regression models for relative positional error time based on body fat percentage. Significant findings were observed for the shoulder ($R^2 = 0.35$; $p < 0.001$), elbow ($R^2 = 0.24$; $p = 0.003$), and knee ($R^2 = 0.11$; $p = 0.047$) joints. The results indicate that increased body fat percentage correlates with higher relative positional error time in these joints. In the shoulder joint, girls again showed reduced error times compared to boys ($\beta = -3.00$; $p < 0.001$). No significant association was found between fat mass percentage and positional error time in the hip joint ($R^2 = 0.03$; p = ns).

Table 3. Multiple linear regression models obtained for relative positional error time according to body fat percentage.

Joint	R^2	Coefficient β	*p value*	95% CI
Shoulder	0.35		<0.001	
Intercept		2.47	<0,001	1.13 to 3.82
Fat percentage		0.13	<0,001	0.06 to 0.19
Girls (ref. boys)		−3.00	<0,001	−4.43 to −1.56
Elbow	0.24		0.003	
Intercept		1.75	0.047	0.02 to 3.48
Fat percentage		0.13	0.001	0,05 to 0.21
Girls (ref. boys)		−1.42	ns	−3.26 to 0.42
Hip	0.03		ns	
Intercept		3.62	<0.001	2.30 to 4.85
Fat percentage		0.02	ns	−0.03 to 0.08
Girls (ref. boys)		−0.83	ns	−2.14 to 0.47
Knee	0.11		0.047	
Intercept		2.57	<0.001	1.15 to 3.98
Fat percentage		0.07	0.028	0.01 to 0.13
Girls (ref. boys)		−1.16	ns	−2.67 to 0.34

95%. CI: 95% confidence interval; ns: no significant.

4. Discussion

The findings of this research indicate that BMI and body fat percentage are correlated with performance in the proprioceptive positional error test in children. Specifically, the results show that higher BMI and a greater body fat percentage are associated with higher positional error values in the shoulder, elbow, and knee joints, indicating lower proprioceptive accuracy. When adjusting these results by sex through multiple linear regression, it was determined that only in the shoulder joint did sex influence the positional error results for both BMI and body fat percentage, with girls demonstrating better proprioception than boys. Previous studies have also reported similar findings, noting poorer proprioception in the knee joint in children with obesity [25,26]. In obese adults, a proprioceptive deficit has also been observed, particularly in the trunk joint [36]. However, unlike our study, these investigations did not include analyses of upper limb joints or consider the sex factor.

One possible explanation for the diminished proprioception observed in children with obesity is the increased body mass, which places additional load on the joints, particularly the knees and ankles [25]. This excess weight can compromise proprioceptive feedback, as evidenced by studies demonstrating a significant decline in knee joint proprioception among obese children, characterized by greater active repositioning errors compared to their normal-weight counterparts [25,26]. Excessive joint loading may exceed the functional capacity of proprioceptors to effectively convey information about joint position and movement sense, potentially exacerbating proprioceptive deficits. However, this hypothesis does not fully account for proprioceptive impairments observed in the upper limbs, such as the shoulder and elbow, which are not subject to direct weight-bearing loads.

An alternative and increasingly supported theory attributes these deficits to systemic inflammation associated with excessive adipose tissue accumulation, as seen in obesity. Adipose tissue, now recognized as a dynamic endocrine and immunological organ, secretes pro-inflammatory cytokines such as tumor necrosis factor-alpha (TNF-α) and interleukin-6 (IL-6) [17]. These mediators disrupt homeostasis by influencing both the central and peripheral nervous systems. Chronic exposure to these cytokines induces oxidative stress and activates inflammatory pathways, including nuclear factor-kappa B (NF-κB) signaling. Such molecular disruptions can impair the function of mechanoreceptors and compromise the structural and functional integrity of nerve fibers that transmit proprioceptive signals [17]. The resulting deficits in sensory feedback from the upper limbs may thus stem from these systemic inflammatory processes rather than direct mechanical overload, highlighting the multifactorial impact of obesity on proprioceptive function.

In this context, another plausible explanation is that the accumulation of excess fat within the muscles and around the joints could disrupt the standard mechanisms of sensory and motor responses. Studies have shown that obese people have a decrease in nerve conduction velocity [23,24,37]. The main cause attributed to this finding is based on nerve compression that causes fat accumulation [38]. Specifically, it has been reported that there is a deceleration of the nerve conduction velocity of the motor nerves and a reduction in the nerve action potential of sensory nerves in obese people compared to healthy subjects [38], which could impact neuromuscular control and proprioception, respectively. Also, the skin stretching from excess adiposity might increase the distance between cutaneous mechanoreceptors, potentially lowering the somatosensory discrimination threshold [39].

Another notable finding of this research was that girls exhibited better proprioception compared to boys, which contrasts with the findings of Das-Yadav et al. (2020), who reported poorer proprioception in adult females compared to adult males. However, in our study, the maturation process in children could be a key factor explaining these results [40]. Sex hormones, particularly estrogen, play a crucial role in the maturation of sensory and neuromuscular systems. Estrogen, the predominant hormone in girls, influences various aspects of the central nervous system by enhancing neuronal plasticity, improving neuromuscular communication efficiency, and optimizing the function of mechanoreceptors, which are responsible for perceiving joint position and movement sense [41,42]. Mosavi-Ghomi et al. (2021) reported that higher levels of estrogen could affect brain areas related to proprioceptive integration and cognition, which in turn may help reduce positional error [43]. On the other hand, it is hypothesized that estrogen has a significant impact on the neurotransmitter gamma-aminobutyric acid (GABA) in the hippocampus, enhancing its balanced inhibitory effect on brain signals. Proper inhibition mediated by GABA is crucial for neuronal modulation, which could improve the accuracy with which the brain processes and integrates proprioceptive signals [44].

On the other hand, our findings showed the absence of a significant correlation between BMI/body fat percentage and hip proprioception. This result may be due to the biomechanical characteristics of this joint. Unlike the knee, which is more exposed to weight-bearing forces and dynamic instability, the hip has a deeper socket and greater structural stability, which may make it less susceptible to proprioceptive impairments related to obesity [45].

The findings of this study emphasize the importance of considering proprioception as a key factor in managing childhood obesity. Given the relationship between increased BMI/body fat percentage and impaired proprioceptive performance, interventions targeting proprioceptive training could play a crucial role in mitigating these deficits. Specifically, balance exercises, joint stabilization activities, and neuromuscular coordination training should be incorporated into physical activity programs aimed at children with obesity to improve proprioceptive accuracy and reduce injury risk. Furthermore, integrating anti-inflammatory dietary strategies and promoting physical activity may help attenuate the systemic inflammation associated with adiposity, which could further support proprioceptive function. Finally, future longitudinal studies focusing on diverse pediatric populations are needed to confirm causal relationships and refine intervention strategies based on these findings.

Among the limitations of this study is its cross-sectional design, which prevents establishing causal relationships between obesity and the observed proprioceptive deficits. Additionally, the sample used, although representative of a specific group, was small and conveniently selected, limiting the generalization of the findings to other pediatric populations. Similarly, complementary biomechanical analyses, such as the assessment of muscle strength or motor control, were not included, nor were additional contextual factors, such as physical activity level, diet, or family history, considered, all of which could have influenced the results. Studies have revealed that these factors can cause alterations in body sensation [46]. For example, it has been seen that individuals with a higher level of physical activity have better body perception [46].

The main strength of this study lies in the use of validated tools such as bioimpedance analysis for estimating body fat percentage and an inertial measurement system (IMU) to assess proprioception, ensuring accuracy and reliability in the data. The focus on a specific population of children aged from 11 to 12 years, a critical stage for motor and proprioceptive development, allows for the generation of relevant findings for early interventions. Additionally, the multivariable analysis adjusted for sex provides the opportunity to identify significant differences in proprioception between boys and girls, considering biological and hormonal factors that may influence the results. Finally, this study's findings have high practical application potential, as they can guide the design of clinical interventions and educational programs aimed at improving proprioception and mitigating the negative effects of obesity in the pediatric population.

5. Conclusions

BMI and body fat percentage are significantly correlated with proprioceptive performance in children, as higher values are associated with greater positional errors in the shoulder, elbow, and knee joints, indicating reduced proprioceptive accuracy. Sex-adjusted analyses revealed that sex only influenced proprioceptive performance in the shoulder joint, where girls demonstrated better accuracy than boys. Based on these results, interventions for children with overweight and obese should extend beyond weight reduction to include training and rehabilitation programs that enhance proprioceptive accuracy and promote motor development. Programs tailored to improve body composition by decreasing body fat percentage and addressing systemic inflammation may help preserve sensorimotor function and reduce the impact of obesity on joint performance. Such integrative approaches could mitigate proprioception deficits, safeguard motor competence, and support the overall physical development of pediatric populations.

Author Contributions: Conceptualization, E.G.-M. and Y.C.-C.; methodology, E.G.-M. and Y.C.-C.; software, E.G.-M.; validation, Y.C.-C. and G.M.-R.; investigation, E.G.-M., G.M.-R. and Y.C.-C.; resources, E.G.-M.; data curation, E.G.-M., G.M.-R. and Y.C.-C.; writing—original draft preparation, E.G.-M., G.M.-R., Y.C.-C., P.V.-B. and C.N.-E.; writing—review and editing, E.G.-M., G.M.-R., Y.C.-C., P.V.-B., M.V.-M., T.H.-V., J.H.-M. and C.N.-E.; visualization, E.G.-M., G.M.-R., Y.C.-C., P.V.-B., M.V.-M., T.H.-V., J.H.-M. and C.N.-E. All authors have read and agreed to the published version of the manuscript.

Funding: This research received no external funding.

Institutional Review Board Statement: The study was conducted in accordance with the Declaration of Helsinki. Ethics approval for this study was obtained from the Ethics Committee of the Universidad Santo Tomás, Chile (No. 133-20).

Informed Consent Statement: Informed consent was obtained from all subjects involved in the study.

Data Availability Statement: The datasets used, and the data analyzed in this study will be made available upon reasonable request to the corresponding author (E.G.-M.).

Conflicts of Interest: The authors declare no conflicts of interest.

References

1. Piché, M.-E.; Poirier, P.; Lemieux, I.; Després, J.-P. Overview of Epidemiology and Contribution of Obesity and Body Fat Distribution to Cardiovascular Disease: An Update. *Prog. Cardiovasc. Dis.* **2018**, *61*, 103–113. [CrossRef]
2. Engin, A. The Definition and Prevalence of Obesity and Metabolic Syndrome. In *Obesity and Lipotoxicity*; Engin, A.B., Engin, A., Eds.; Advances in Experimental Medicine and Biology; Springer International Publishing: Cham, Switzerland, 2017; Volume 960, ISBN 978-3-319-48380-1.
3. Ng, M.; Fleming, T.; Robinson, M.; Thomson, B.; Graetz, N.; Margono, C.; Mullany, E.C.; Biryukov, S.; Abbafati, C.; Abera, S.F.; et al. Global, Regional, and National Prevalence of Overweight and Obesity in Children and Adults during 1980–2013: A Systematic Analysis for the Global Burden of Disease Study 2013. *Lancet* **2014**, *384*, 766–781. [CrossRef]
4. World Obesity Atlas 2023. Available online: https://www.worldobesity.org/ (accessed on 8 October 2024).
5. Biro, F.M.; Wien, M. Childhood Obesity and Adult Morbidities. *Am. J. Clin. Nutr.* **2010**, *91*, 1499S–1505S. [CrossRef]
6. Lopez-Jimenez, F.; Almahmeed, W.; Bays, H.; Cuevas, A.; Di Angelantonio, E.; Le Roux, C.W.; Sattar, N.; Sun, M.C.; Wittert, G.; Pinto, F.J.; et al. Obesity and Cardiovascular Disease: Mechanistic Insights and Management Strategies. A Joint Position Paper by the World Heart Federation and World Obesity Federation. *Eur. J. Prev. Cardiol.* **2022**, *29*, 2218–2237. [CrossRef]
7. Powell-Wiley, T.M.; Poirier, P.; Burke, L.E.; Després, J.-P.; Gordon-Larsen, P.; Lavie, C.J.; Lear, S.A.; Ndumele, C.E.; Neeland, I.J.; Sanders, P.; et al. Obesity and Cardiovascular Disease: A Scientific Statement from the American Heart Association. *Circulation* **2021**, *143*, 54–64. [CrossRef]
8. King, A.C.; Challis, J.H.; Bartok, C.; Costigan, F.A.; Newell, K.M. Obesity, Mechanical and Strength Relationships to Postural Control in Adolescence. *Gait Posture* **2012**, *35*, 261–265. [CrossRef]

9. Guzmán-Muñoz, E.; Mendez-Rebolledo, G.; Núñez-Espinosa, C.; Valdés-Badilla, P.; Monsalves-Álvarez, M.; Delgado-Floody, P.; Herrera-Valenzuela, T. Anthropometric Profile and Physical Activity Level as Predictors of Postural Balance in Overweight and Obese Children. *Behav. Sci.* **2023**, *13*, 73. [CrossRef]
10. Guzmán-Muñoz, E.; Valdés-Badilla, P.; Mendez-Rebolledo, G.; Concha-Cisternas, Y.; Castillo-Retamal, M. Relationship between Anthropometry and Balance of Postural Control in Children 6–9 Years Old. *Nutr. Hosp.* **2018**. [CrossRef]
11. Chivers, P.; Larkin, D.; Rose, E.; Beilin, L.; Hands, B. Low Motor Performance Scores among Overweight Children: Poor Coordination or Morphological Constraints? *Hum. Mov. Sci.* **2013**, *32*, 1127–1137. [CrossRef]
12. Colné, P.; Frelut, M.L.; Pérès, G.; Thoumie, P. Postural Control in Obese Adolescents Assessed by Limits of Stability and Gait Initiation. *Gait Posture* **2008**, *28*, 164–169. [CrossRef]
13. Guzmán-Muñoz, E.; Valdés-Badilla, P.; Castillo-Retamal, M. Postural Control in Children with Overweight and Obesity: A Review of Literature. *Sun* **2021**, *36*, 471–488. [CrossRef]
14. Biino, V.; Pellegrini, B.; Zoppirolli, C.; Lanza, M.; Gilli, F.; Giuriato, M.; Schena, F. Gross Motor Coordination in Relation to Weight Status: A Longitudinal Study in Children and Pre-Adolescents. *Front. Public Health* **2023**, *11*, 1242712. [CrossRef]
15. De Waal, E.; Pienaar, A.E. Influences of Persistent Overweight on Perceptual-Motor Proficiency of Primary School Children: The North-West CHILD Longitudinal Study: Persistent Overweight and Perceptual-Motor Proficiency in Children. *BMC Pediatr* **2021**, *21*, 245. [CrossRef]
16. Gentier, I.; D'Hondt, E.; Shultz, S.; Deforche, B.; Augustijn, M.; Hoorne, S.; Verlaecke, K.; De Bourdeaudhuij, I.; Lenoir, M. Fine and Gross Motor Skills Differ between Healthy-Weight and Obese Children. *Res. Dev. Disabil.* **2013**, *34*, 4043–4051. [CrossRef]
17. Barros, W.M.A.; Silva, K.G.D.; Silva, R.K.P.; Souza, A.P.D.S.; Silva, A.B.J.D.; Silva, M.R.M.; Fernandes, M.S.D.S.; Souza, S.L.D.; Souza, V.D.O.N. Effects of Overweight/Obesity on Motor Performance in Children: A Systematic Review. *Front. Endocrinol.* **2022**, *12*, 759165. [CrossRef]
18. Gill, S.V.; Hung, Y.-C. Effects of Overweight and Obese Body Mass on Motor Planning and Motor Skills during Obstacle Crossing in Children. *Res. Dev. Disabil.* **2014**, *35*, 46–53. [CrossRef]
19. Riemann, B.L.; Lephart, S.M. The Sensorimotor System, Part I: The Physiologic Basis of Functional Joint Stability. *J. Athl. Train* **2002**, *37*, 71–79.
20. Guzmán-Muñoz, E.; Mendez-Rebolledo, G.; Sazo-Rodriguez, S.; Salazar-Méndez, J.; Valdes-Badilla, P.; Nuñez-Espinosa, C.; Herrera-Valenzuela, T. Quadriceps Muscle Reaction Time in Obese Children. *PeerJ* **2024**, *12*, e17050. [CrossRef]
21. Uranga, R.M.; Keller, J.N. The Complex Interactions Between Obesity, Metabolism and the Brain. *Front. Neurosci.* **2019**, *13*, 513. [CrossRef]
22. Tomlinson, D.J.; Erskine, R.M.; Morse, C.I.; Winwood, K.; Onambélé-Pearson, G. The Impact of Obesity on Skeletal Muscle Strength and Structure through Adolescence to Old Age. *Biogerontology* **2016**, *17*, 467–483. [CrossRef]
23. K, C.; Mahesh, U.; Kumar, S.; L, M.; S, S. Effect of Body Mass Index on Ulnar Nerve Conduction Velocity among Medical Students. *Natl. J. Physiol. Pharm. Pharmacol.* **2023**, *1*, 1–3. [CrossRef]
24. Kumar, A.; Sarvottam, K.; Yadav, R.; Netam, R.; Yadav, R. Ulnar Nerve Motor Conduction Velocity Correlates with Body Mass Index in Indian Young Healthy Subjects. *Natl. J. Physiol. Pharm. Pharmacol.* **2017**, *1*, 1–4. [CrossRef]
25. Saleh, M.S.M. Influence of Obesity on Proprioception of Knee and Ankle Joints in Obese Prepubertal Children. *Bull. Fac. Phys. Ther.* **2018**, *23*, 9–14. [CrossRef]
26. Wang, L.; Li, J.X.; Xu, D.Q.; Hong, Y.L. Proprioception of Ankle and Knee Joints in Obese Boys and Nonobese Boys. *Med. Sci. Monit.* **2008**, *14*, CR129–CR135.
27. Goble, D.J. Proprioceptive Acuity Assessment Via Joint Position Matching: From Basic Science to General Practice. *Phys. Ther.* **2010**, *90*, 1176–1184. [CrossRef]
28. Han, J.; Waddington, G.; Adams, R.; Anson, J.; Liu, Y. Assessing Proprioception: A Critical Review of Methods. *J. Sport Health Sci.* **2016**, *5*, 80–90. [CrossRef]
29. Shah, N.R.; Braverman, E.R. Measuring Adiposity in Patients: The Utility of Body Mass Index (BMI), Percent Body Fat, and Leptin. *PLoS ONE* **2012**, *7*, e33308. [CrossRef]
30. Wu, Y.; Li, D.; Vermund, S.H. Advantages and Limitations of the Body Mass Index (BMI) to Assess Adult Obesity. *IJERPH* **2024**, *21*, 757. [CrossRef]
31. Trang, L.T.; Trung, N.N.; Chu, D.-T.; Hanh, N.T.H. Percentage Body Fat Is As a Good Indicator for Determining Adolescents Who Are Overweight or Obese: A Cross-Sectional Study in Vietnam. *Osong Public Health Res Perspect* **2019**, *10*, 108–114. [CrossRef]
32. Loenneke, J.P.; Barnes, J.T.; Wilson, J.M.; Lowery, R.P.; Isaacs, M.N.; Pujol, T.J. Reliability of Field Methods for Estimating Body Fat. *Clin. Physiol. Funct Imaging* **2013**, *33*, 405–408. [CrossRef]
33. McCarthy, H.D.; Cole, T.J.; Fry, T.; Jebb, S.A.; Prentice, A.M. Body Fat Reference Curves for Children. *Int. J. Obes.* **2006**, *30*, 598–602. [CrossRef]

34. Silva-Moya, G.; Méndez-Rebolledo, G.; Valdes-Badilla, P.; Gómez-Álvarez, N.; Guzmán-Muñoz, E. Effects of Neuromuscular Training on Psychomotor Development and Active Joint Position Sense in School Children. *J. Mot. Behav.* **2022**, *54*, 57–66. [CrossRef] [PubMed]
35. Bruyneel, A.-V. Evaluación de la propiocepción: Pruebas de estatestesia y cinestesia en la práctica clínica. *EMC-Kinesiterapia-Med. Física* **2023**, *44*, 1–15. [CrossRef]
36. Eryıldız, E.; Delibaş, E.; Kesgin, İ.M.; Öztürk, İ.B.; Ersoz Huseyinsinoglu, B. Investigation of Physical Parameters Influencing Body Awareness in Adults with Obesity: An Observational, Controlled Study. *J. Bodyw. Mov. Ther.* **2024**, *40*, 1342–1348. [CrossRef] [PubMed]
37. Buschbacher, R.M. Body Mass Index Effect on Common Nerve Conduction Study Measurements. *Muscle Nerve* **1998**, *21*, 1398–1404. [CrossRef]
38. Pal, P.; Pal, G.; Balakumar, B.; Dutta, T.; Naik, B. Assessment of Motor Nerve Conduction in Healthy Obese Indian Population. *Int. J. Clin. Exp. Physiol.* **2014**, *1*, 277. [CrossRef]
39. Mignardot, J.-B.; Olivier, I.; Promayon, E.; Nougier, V. Obesity Impact on the Attentional Cost for Controlling Posture. *PLoS ONE* **2010**, *5*, e14387. [CrossRef]
40. Das, P. Gender Specific Effect of Obesity on Knee Proprioception in Weight Bearing and Non-Weight Bearing Position. *JK Practitioner* **2020**, *25*, 1–6.
41. Eiling, E.; Bryant, A.L.; Petersen, W.; Murphy, A.; Hohmann, E. Effects of Menstrual-Cycle Hormone Fluctuations on Musculotendinous Stiffness and Knee Joint Laxity. *Knee Surg. Sports Traumatol. Arthrosc.* **2007**, *15*, 126–132. [CrossRef] [PubMed]
42. Sharma, A.; Kaur, S.; Kumar, P.; Sharma, N. Unveiling the Menstrual Mind: Exploring Proprioceptive Proficiency, Reaction Responsiveness, and Motor Mastery in Premenstrual Syndrome versus Non-PMS Cohorts: A Two-Arm Parallel Non-Blinded Cross-Sectional Study. *Clin. Epidemiol. Glob. Health* **2024**, *29*, 101782. [CrossRef]
43. Mosavi Ghomi, M.; Shariati, M.; Mokhtari, M.; Ramezani Nowrozani, F. The Effects of Estrogen and Progesterone Hormones on Active Hip Joint Position Sense in Healthy Women in Different Phases of a Menstrual Cycle. *JMR* **2021**. [CrossRef]
44. Gilfarb, R.A.; Leuner, B. GABA System Modifications During Periods of Hormonal Flux Across the Female Lifespan. *Front. Behav. Neurosci.* **2022**, *16*, 802530. [CrossRef]
45. Glenister, R.; Sharma, S. Anatomy, Bony Pelvis and Lower Limb, Hip. 2023. Available online: https://www.ncbi.nlm.nih.gov/books/NBK526019/ (accessed on 3 February 2025).
46. Notarnicola, A.; Farì, G.; Maccagnano, G.; Riondino, A.; Covelli, I.; Bianchi, F.P.; Tafuri, S.; Piazzolla, A.; Moretti, B. Teenagers' perceptions of their scoliotic curves. An observational study of comparison between sports people and non-sports people. *Muscles Ligaments Tendons J.* **2019**, *9*, 225–235. [CrossRef]

Journal of
Functional Morphology and Kinesiology

MDPI

Review

Differences in Anthropometric and Body Composition Factors of Blind 5-a-Side Soccer Players in Response to Playing Position: A Systematic Review

Boryi A. Becerra-Patiño [1,2], Armando Monterrosa-Quintero [3], Jorge Olivares-Arancibia [4], José Francisco López-Gil [5,6,*] and José Pino-Ortega [7]

1 Programa de Doctorado en Ciencias de la Actividad Física y del Deporte, University of Murcia, San Javier, 30720 Murcia, Spain; boryialexander.becerrap@um.es
2 Facultad de Educación Física, Universidad Pedagógica Nacional, Bogotá 110221, Colombia
3 Altius Performance Laboratory, Physical Education and Sports Program, Universidad Surcolombiana, Neiva 410001, Colombia; adomonterrosa@gmail.com
4 AFySE Group, Research in Physical Activity and School Health, School of Physical Education, Faculty of Education, Universidad de las Américas, Santiago 7500975, Chile; jorge.olivares.ar@gmail.com
5 School of Medicine, Universidad Espíritu Santo, Samborondón 092301, Ecuador
6 Vicerrectoría de Investigación y Posgrado, Universidad de los Lagos, Osorno 5290000, Chile
7 Faculty of Sport Science, University of Murcia, 30100 Murcia, Spain; josepinoortega@um.es
* Correspondence: josefranciscolopezgil@gmail.com

Abstract: Background: Blind 5-a-side soccer is an intermittent sport that requires the integration of physiological and physical processes, where body composition (BC) is an influential and differentiating factor of the sporting level, according to the conclusions of some studies. However, to date, no systematic review has been reported comparing BC in players with visual impairment. **Objectives:** The aims of this study were to systematically synthesize the existing evidence on differences in anthropometric characteristics and body composition among blind 5-a-side football players according to playing position and to derive practical recommendations for researchers and coaches. **Methods:** The following databases were consulted: PubMed (Medline), Scopus, Web of Science, and Science. This systematic review uses the guidelines of the PRISMA declaration and the guidelines for conducting systematic reviews in sports science. PICO strategy was used for the selection and inclusion of studies in the present work, with a series of inclusion and exclusion criteria. The quality was methodologically assessed using the PEDro scale. **Results**: The 10 studies comprising this systematic review had a total sample size of 168 athletes. The main findings of this research were (1) the somatotype of blind 5-a-side soccer players tends toward meso-endomorphic; (2) there are differences in the variables of muscle mass, fat mass, and body weight in response to playing position and sporting level; (3) the players present a somatotypical profile with a predominance of the mesomorph component. **Conclusions**: The results of this review reveal a tendency to define BW as influencing the athletic performance of blind 5-a-side soccer players. However, it is not conclusive whether these improvements occur in response to each playing position. More studies are needed to analyze the effect of BW on athletic performance, especially when correlating BW with other physical, nutritional, technical, and tactical variables in training and competition.

Keywords: blind soccer; body composition; anthropometry; somatotype; body weight

1. Introduction

The study of body composition (BC) in sports aims to evaluate the body's reserves through different variables [1], including fat-free mass (FFM), fat mass (FM), and muscle mass (MM), to assess an individual's nutritional status and promote nutritional processes [2,3]. BC allows the determination of values that are important for athletic performance and athlete health [4–6], such as the amount of skeletal muscle, body density, MM, metabolic balance, muscle-fat ratio, skeletal index, and somatotype, among others [7,8].

Athletic performance is a multifactorial, complex, and dynamic process [9–11] that requires the integration of various physical factors, including strength to perform specific actions of acceleration, deceleration, changes in direction, and jumping [12–14]; speed to cover distances in short times [15,16]; endurance to maintain the quality of actions performed at high intensity levels [17,18]; and flexibility to improve the quality of actions performed [19,20]. Likewise, relevant factors at the physiological level highlight the processes of adaptation, recovery, energy systems, and BC [10,21]. Therefore, investigating the influence of these factors on sports is crucial for promoting increasingly specific training processes [22,23].

In this sense, BC has been established as a relevant and determining factor in athletic performance in team sports [24,25], especially in women's soccer [26,27], men's soccer [28,29], soccer players with cerebral palsy [30,31], and, consequently, blind soccer players [32–41]. Each sport has specific characteristics, requiring athletes to adapt to competitive demands, many of which are imposed by a series of actions that they must perform in response to their playing position [35,36,42]. Thus, blind 5-a-side is a physically, physiologically, technically, and tactically demanding sport [32,42–44]. Blind 5-a-side is an intermittent sport that combines the aerobic-anaerobic system, seeking to demonstrate optimal cardiovascular endurance, agility to quickly change direction and orientation [45], speed for dribbling, strength and power for shooting, acceleration, deceleration, and coordination for passing and positioning [46–48]. Thus, blind 5-a-side being a sport that promotes intense efforts with short recovery phases requires athletes to have high levels of MM, low levels of body fat percentage, and a mesomorphic somatotype profile [33,35,37,38,40]. Some studies have analyzed the BC of blind 5-a-side players, determining that each playing position has specific requirements and that these requirements are related to physical abilities [37,40]. However, other studies have reported that, although there is homogeneity in anthropometric and BC factors, the different playing positions do not differ [35,41]. Another determining variable when analyzing whether different playing positions express variations in BC is the inclusion of goalkeepers in the studies. Thus, the study by Gorla et al. [34], which included goalkeepers in the evaluation, determined significant differences compared with other positions.

To date, no systematic review has compiled the available evidence on the differences in anthropometric factors, BC characteristics, and somatotypic profiles of blind 5-a-side players, and little research has been conducted on this sport. Therefore, the aims of this study were to systematically synthesize the existing evidence on differences in anthropometric characteristics and body composition among blind 5-a-side football players according to playing position and to derive practical recommendations for researchers and coaches.

2. Materials and Methods

2.1. Design

This systematic review followed the guidelines of the Preferred Reporting Items for Systematic Reviews and Meta-Analyses (PRISMA) statement [49,50] and the sports science guidelines for conducting systematic reviews [51]. The review protocol was registered in International Platform of Registered Systematic Review and Meta-analysis Protocols (INPLASY) website on 18 June 2025 (ID 202560075).

2.2. Sources of Information

The search strategies considered the following characteristics. Date: All studies published up to 15 March 2025 were retrieved. The following databases were consulted: PubMed (Medline), Scopus, Web of Science, Science Direct, and SPORTDiscus. Google Scholar and ResearchGate were also searched. These databases were consulted for use in various reviews and were used to search databases and other sources.

2.3. Inclusion and Exclusion Criteria

Two authors searched independently (B.A.B.-P. and J.O.-A.). The purpose was to identify papers that met the criteria (Table 1). After the selected studies were identified, the comma-separated value (CSV) file was downloaded, and relevant criteria for study selection were defined (title, keywords, abstract, year, journal, citations received). Documents were screened to remove duplicates. Furthermore, if any documents were found and not captured by the search equation, they were added through external sources. For the selection and inclusion of studies in this study, a series of inclusion and exclusion criteria were established on the basis of the participants, interventions, comparison, and outcomes (PICO) strategy (Table 1). Any document that included a comparison between blind and sighted players within the research area was excluded. The inclusion criteria were as follows: (i) studies published without language restrictions and (ii) original studies. The exclusion criteria were as follows: (i) systematic reviews, meta-analyses, bibliometric analyses, narrative or literary reviews; (ii) abstracts, meetings, books, reviews, letters, and editorials; (iii) articles written without academic peer review; and (iv) studies without full access to the original text.

Table 1. Inclusion and exclusion criteria.

Population	Intervention	Comparison	Outcomes
Blind 5-a-side players aiming to train or improve their performance	Anthropometry Bioelectrical impedance (BIA) Dual-energy X-ray absorptiometry (DXA)	Measurement methods Equations Performance levels Blind players	Anthropometric characteristics (skinfolds, circumferences, diameters, lengths), somatotype, body composition, fat mass, fat-free mass, muscle mass

2.4. Search Strategy and Data Extraction

The 10 studies included in this systematic review included a total sample of 168 athletes. To design the search strategy, the P (population), I (intervention), C (comparison), and O (outcomes) strategies were applied, as suggested by the guidelines used for this systematic review [52]. The Boolean operators "AND" and "OR" were used to group the terms. A similar procedure was followed for each database. Before the final search phrase for each database was constructed, possible combinations were tested with the following list of words: ("Athletes of 5-a-side Football" [All fields]) OR ("blind soccer" [All fields]) OR ("FA5 for blind persons" [All fields]) OR ("w5-a-side football team Paralympic" [All fields]) OR ("5-a-side football team" [All fields]) AND ("body composition" OR somatotype OR anthropometry [All fields]). From these terms, the following search equation was constructed: ("Athletes of 5-a-side Football" OR "blind soccer" OR "FA5 for blind persons" OR "w5-a-side football team Paralympic" OR "5-a-side football team") AND ("body composition" OR somatotype OR anthropometry). This search string was adapted for the databases and the other methods. The controlled vocabulary search was performed with the keyword search to improve retrieval. Searches were conducted to identify studies without other restrictions regarding publication date, language, or study design. Citation searches were also performed for key included studies, with the goal of tracking other documents. When it was not possible to obtain the full texts of articles from institutional or open access subscriptions, attempts were made to contact the corresponding authors directly through the ResearchGate platform. Furthermore, if a document was found that did not appear in the search strategy, it was added through external sources.

All the retrieved articles were analyzed for duplicate entries. Two authors (B.A.B.-P. and J.O.-A.) independently reviewed the different searches to determine the terms that yielded the greatest number of documents related to the topic. Any disagreement (5% of the total documents) regarding the final inclusion/exclusion status was resolved through academic discussion, both in the selection and inclusion phases. During the discussion, the two independent authors simultaneously analyzed the articles following the criteria established in the order shown in Table 2. This process was systematized in Excel. The academic debates for the inclusion of the studies took into account the duplicate search by two authors on two different days to review the documents. In particular, the methodology (study design, variables, instruments, determination of fat percentage, and somatotype) was reviewed, as well as the results and main conclusions.

Table 2. Methodological quality of the studies evaluated with the PEDro scale.

	Items												
Studies (Author(s)–Year)	**1**	**2**	**3**	**4**	**5**	**6**	**7**	**8**	**9**	**10**	**11**	**Total PEDro**	
Hernández-Beltrán et al. [41]	1	0	0	1	0	0	0	1	1	1	1	6	Good
Lameira Oliveira et al. [40]	1	0	0	1	0	0	0	1	1	0	1	5	Moderate
Esatbeyoglu and Kin-İsler [39]	1	0	0	1	0	0	0	1	1	1	1	6	Good
Sancio et al. [38]	1	0	0	1	0	0	0	1	1	1	1	6	Good
Lameira Oliveira et al. [37]	1	0	0	1	0	0	0	1	1	0	1	5	Moderate
Lameira De Oliveira et al. [36]	1	0	0	1	0	0	0	1	1	0	1	5	Moderate
Lameira De Oliveira et al. [35]	1	0	0	1	0	0	0	1	1	1	1	6	Good
Gorla et al. [34]	1	0	0	1	0	0	0	1	1	1	1	6	Good
Durán-Agüero et al. [33]	1	0	0	1	0	0	0	1	1	1	1	6	Good
Castelli Correia de Campos et al. [32]	1	1	0	1	1	1	0	1	1	0	1	8	Good

3. Results

3.1. Identification and Selection of Studies

A total of 75 documents were identified. After an initial review of the final database, documents were eliminated because of duplication (n = 10), leaving four for the databases and six for the other methods. Documents that were not related to the topic after the title/abstract/keywords had been reviewed were excluded from the databases (n = 62) or other methods (n = 137). A total of 199 studies were excluded. Thirteen screened documents were analyzed in depth through a systematic reading (Figure 1). After this analysis, 10 studies met the eligibility criteria. Table 3 was compiled to contextualize the sample for each of the included studies.

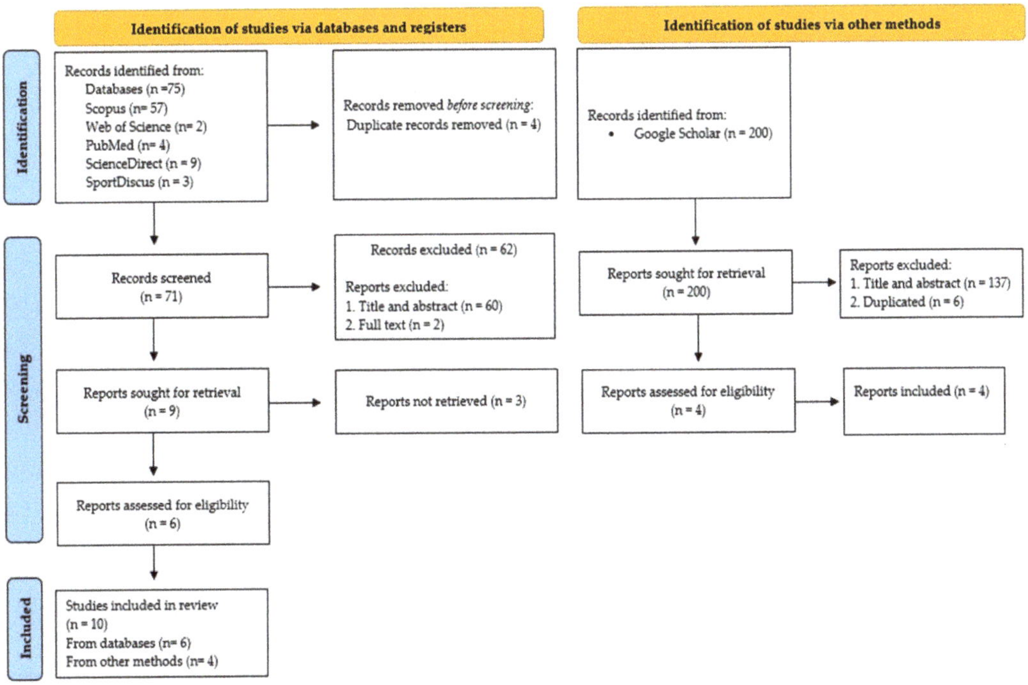

Figure 1. Flow diagram of the systematic review.

Table 3. Classification of the general variables of the selected studies.

Author(s)–Year	N° Part	S	Age (Yrs) and Func Class	Weight (kg)	Height (cm)	IMC (kg/m^2)	BF (%)	MM (%) * and LM ** (kg)	Clas
Hernández-Beltrán et al. [41]	12	♂	28.7 ± 8.8	73.8 ± 10.7	176.8 ± 9.0	23.57	12.55		
Lameira Oliveira et al. [40]	63	♂	28.0 ± 5.8 (B1)	74.8	170.0	24.9	19.3	80.7 **	Mes-End
Esatbeyoglu and Kin-İsler [39]	12	♂	23.2 ± 3.7 (B1)	79.8 ± 10.9	181 ± 0.08	24.3 ± 2.1	10.53 ± 3.6		
Sancio et al. [38]	8	♂	26.8 ± 6.5 (B1)	81.8 ± 15.7	170.3 ± 5.02		28.12 ± 6.6	43.63 ± 4.5 *	Mes-End
Lameira Oliveira et al. [37]	5	♂	32.6 ± 8.0 (B1)	70.9 ± 10.5	169 ± 7.7	25.1 ± 5.4	20.4 ± 5.1	39.5 ± 3.5 **	
Lameira De Oliveira et al. [36]	13	♂	27.0 ± 6.5 (B1)	71.7 ± 7.4	172.0 ± 6.1	24.1 ± 1.7	15.9 ± 2.9	43.6 ± 2.5 *	Mes-End
Lameira De Oliveira et al. [35]	15	♂	24 ± 5.6 (B1)	71.7 ± 7.4	172 ± 6.1	24.1 ± 1.7	15.9 ± 2.9		Mes-End
Gorla et al. [34]	23	♂	22.5 ± 31 (B1)	64.9–77.9	169–175	22.3–26	10.4–15.9		End-Mes
Durán-Agüero et al. [33]	11	♂	26.4 ± 9.8	71.4 ± 18.9	163.6 ± 16.0	25.1	25.8	45.6 *	Mes
Castelli et al. [32]	6	♂	27.3 ± 5.5 (B1)		1.72 ± 0.09	25.6 ± 1.3	15.9 ± 4.54		End

Note—Ref: Reference; N° Part: Number of participants; Man: S: sex: ♂; Yrs: years; kg: kilograms; cm: centimeters; m: meters; %: percentage; Clas: Classification; Mes: Mesomorph; End: Endomorf; B1: Paralympic classification based on medical criteria for sport for blind people; func class: functional classification; BF: body fat; MM: muscle mass; LM: lean mass. * Evaluation of lean mass in kg; ** Evaluation of muscle mass in %.

3.2. Methodological Quality

The methodological quality of the articles included in this review was assessed via the PEDro scale [53]. This scale is based on criteria that allow the identification of whether the studies have sufficient internal validity and statistical information to interpret the results (external validity (item 1), internal validity (items 2–9), and statistical information (items 10–11). Each item was classified as yes or no (1 or 0, respectively), depending on whether the criterion was met in the study. The total score considers items 2 to 11; therefore, the maximum score was 8 [32]. Regarding the quality of the evidence, scores < 4 are considered poor quality, scores ranging from 4–5 moderate quality, scores ranging from 6–8 good, and scores ranging from 9–10 excellent [41]. In this review, 100 items (97.5%) were assessed by agreement between two reviewers, and the remaining items were assessed according to the mean of the studies (Table 2). The methodological quality ranged from "moderate to good" since some studies did not present randomization in the selection of the sample, nor did they have a control group. Furthermore, the methodological quality was heterogeneous across all studies. Therefore, the methodological quality was defined by the consensus of the investigators as "moderate", indicating differences in the methodological rigor of the included studies [53].

3.3. Analysis of the Participants

The 10 studies comprising the sample of this systematic review included 168 athletes, all of whom were men. Table 3 specifies the characteristics of the sample selected.

3.4. Analysis of the Studies

Table 4 specifies the characteristics of the sample selected for each study (study aim, variables, results, instruments, determination of % fat and somatotype, conclusions).

Table 4. Classification of the methodological procedures of the studies.

Author(s)/Year	Study's Aim	Variables	Instruments	Determination of % Fat and Somatotype	Results	Conclusions
Hernández-Beltrán et al. [41]	Analyze BC based on laterality and playing position of the players of the Spanish FpC	National Team PP, laterality, weight, FM, FFM, body water, BMD, AEC/AET, trunk weight, left and right arm weight, left and right leg weight	Tanita BC-601 BC monitor (Tokyo, Japan), SECA wall-mounted height rod (Hamburg, Germany).	NA	Laterality does not differ in playing position in blind 5-a-side players. BC was found to differ in response to playing positions, so determining it can be key when selecting players for a specific position.	BC influences players' performance and, in turn, is associated with improved health. Low levels of MM increase the likelihood of injury. Therefore, determining players' BC will allow for the development of specific training sessions aimed at increasing muscle strength.
Lameira Oliveira et al. [40]	To compare BC and somatotype of high-performance blind 5-a-side athletes from different playing positions	PP, skinfolds, body perimeters, bone diameters, height and BW, somatotype	Cescorf caliper (Porto Alegre, Brazil), precision 0.1 mm, Cardiomed pachymeter (Brasília, Brazil), precision 0.1 cm, Sanny Medical flexible metal tape (São Paulo, Brazil), precision 0.1 cm, Soehnle scale (Backnang, Germany), precision 0.1 kg, Soehnle stadiometer (Backnang, Germany), precision 0.1 cm.	Siri formula for body fat percentage [54], Heath–Carter method [55] for somatotype	Wing players presented lower values in body fat percentage (%F = 17.4%) compared to the Closer (23.1%) and Center (21.5%) positions ($p < 0.05$). There is a predominance of the muscular component and a meso-endomorphic somatotype profile overall and for each of the playing positions evaluated.	PP in blind 5-a-side football expresses various variations linked to the specific physical demands of the sport, where BC has been shown to vary in response to playing position. Information on the overall somatotype profile and by PP in blind 5-a-side football can support the development of specific training processes.

Table 4. *Cont.*

Author(s)/Year	Study's Aim	Variables	Instruments	Determination of % Fat and Somatotype	Results	Conclusions
Esatbeyoglu and Kin-İsler [39]	To determine sex differences in variables related to BP, BMI, BC, and postural balance in athletes with VI	Balance, BP level, BC, % fat, FM, FFM, BW	International Physical Activity Questionnaire short version, Modified Sensory Integration and Balance Clinical Test Tool, Tanita TBF401A scale (Tokio, Japan), accuracy 0.1 kg, Holtain wall stadiometer (Crosswell, UK).	NA	No statistically significant differences were reported in BC indicators, especially in FM and FFM when comparing sighted athletes with athletes with VI.	Male athletes with VI expressed a higher BMI than women. The BP level demonstrates that VI is not a barrier to maintaining optimal BP levels, while there is a difficulty in balance, even if their PF levels are acceptable. It is suggested to incorporate balance into training sessions in this population group.
Sancio et al. [38]	To analyze the anthropometric profile and its relationship with ball transfer speed in players of the Argentine National futsal Team for the Blind	Weight, height, length of lower limbs, % adipose fat, skinfolds, % MM, muscle adipose ratio, skeletal index, somatotype	Omron® scale, model HBF500INT (Kyoto, Japan), accuracy 0.1 kg; wall-mounted acrylic stadiometer, brand Calibres Argentinos (Rosario, Argentina); Harpenden skinfold caliper, accuracy 0.2 mm; Calibres Argentinos metallic anthropometric tape (Rosario, Argentina)	Heath–Carter method [55] for somatotype	The results express that there is a high correlation between transfer speed and skeletal index (r 0.85 $p < 0.01$), and a moderate correlation with the length of lower limbs (r 0.69) and with variables related to muscle tissue, especially with mesomorphism (r 0.59), kg MM (r 0.57), thigh muscle area (r 0.56) and calf (r 0.55).	Ball transfer speed is related to the anthropometric profile, primarily by the length of the lower limbs and their relationship to trunk length. This allows coaches to consider these variables within the process of selecting and developing players in different short- and long-term sporting processes.

Table 4. *Cont.*

Author(s)/Year	Study's Aim	Variables	Instruments	Determination of % Fat and Somatotype	Results	Conclusions
Lameira Oliveira et al. [37]	To describe the anthropometric characteristics and aerobic fitness of blind 5-a-side football players	Skinfolds, bone diameters, body circumferences, BW, height, BMI, % FM, % LM, % bone mass	Cescorf caliper (Porto Alegre, Brazil), accuracy 0.1 mm; Cardiomed pachymeter (Brasília, Brazil), accuracy 0.1 cm; Sanny Medical flexible metal tape (São Paulo, Brazil), accuracy 0.1 cm; Soehnle digital scale (Backnang, Germany), accuracy 0.1 kg; Soehnle stadiometer (Backnang, Germany), accuracy 0.1 cm.	Siri formula for fat percentage [54], Rocha modified von Doblen equation for bone mass [56]	Anthropometric characteristics were consistent with the specificities of the sport and sport level. Likewise, body fat percentage (%F = 20.4) and average VO2 max value. (36.3 ± 4.7 mL^{-1} kg^{-1} min) are lower than those reported by elite athletes in blind 5-a-side football.	Anthropometric characteristics and aerobic fitness are crucial in blind 5-a-side football, as high levels of MM and good aerobic fitness allow athletes to adapt to the physical, technical, and tactical demands of the game. This, in turn, helps improve training processes and athletic performance.
Lameira De Oliveira et al. [36]	To analyze the dermatoglyphic characteristics and BC of blind 5-a-side football players belonging to the Brazilian National Team	Fingerprints, skinfolds, bone diameters, body perimeters, BW, height, BMI, % BF, % LM, % bone mass	Cescorf caliper (Porto Alegre, Brazil), precision 0.1 mm, Cardiomed pachymeter (Brasília, Brazil), precision 0.1 cm, Soehnle digital scale (Backnang, Germany), precision 0.1 kg.	Siri formula for fat percentage [54], Jackson & Pollock equation for BD [57], Heath–Carter method [55] for somatotype	There is a proximity in the reported values in BC between goalkeepers and full-backs with a meso-endomorph profile and between defenders and pivots characterized by a balanced mesomorph profile.	The somatotypic profile of blind 5-a-side football players leans toward the independent muscle component of the PP that they occupy, and this, in turn, is related to the dermatoglyphic characteristics of speed and strength. These genetic and morphological relationships are key to identifying and supporting preparation processes in response to the specific demands of the sport.

Table 4. *Cont.*

Author(s)/Year	Study's Aim	Variables	Instruments	Determination of % Fat and Somatotype	Results	Conclusions
Lameira De Oliveira et al. [35]	To analyze the BC and somatotype of the Brazilian Paralympic futsal team athletes at Rio 2016 in response to the playing position	Somatotype, BC, % FM, BD	Cescorf caliper (Porto Alegre, Brazil), precision 0.1 mm, Cardiomed pachymeter (Brasília, Brazil), precision 0.1 cm, Soehnle digital scale (Backnang, Germany), precision 0.1 kg, Soehnle stadiometer (Backnang, Germany), precision 0.1 cm.	Siri formula for fat percentage [54], Jackson & Pollock equation for BD [57], Heath–Carter method [55] for somatotype	The study did not report statistically significant differences in any of the anthropometric variables or BC. Regarding the somatotypic profile, the group was classified as meso-endomorph. The defenders (2.6-4.4-2.4) and the pivots (2.2-5.6-2.3) had a balanced mesomorphic profile, while the goalkeepers (3.2-5.8-1.6) and wings (3.2-5.7-1.6) presented a meso-endomorphic profile.	The team was characterized by its homogeneity in terms of anthropometry and BC, where no differences were reported in response to the playing position. Blind 5-a-side football players present a predominance of the muscular component in the somatotypic profile at a general level and in each of the playing positions.
Gorla et al. [34]	To determine the somatotypic profiles and BC of the Brazilian national blind 5-a-side football team players	BMI, % FM, somatotype, skinfolds, bone diameters, BD	Bascula Plena Acqua® model, WCS wall-mounted stadiometer, Harpenden caliper (Crosswell, UK), precision 0.2 mm, Cardiomed pachymeter (Brasília, Brazil), precision 0.1 cm.	Siri formula for body fat percentage [54], Heath-Carter method [55] for somatotype	Goalkeepers express a statistically significant difference ($p \leq 0.05$) in the anthropometric variable of CM (82.3 kg) and in the BC variables: %GC (21.5%) and ∑9DC (169.5) compared to the other positions. Regarding the somatotypic profile, there were no statistically significant differences ($p \leq 0.05$). However, there is a trend toward an endo-mesomorphic profile.	There is a difference in the somatotype of the GK compared to other PPs, which leads to defining the specific training characteristics that each player must receive to respond to the specific demands of the game. The tendency toward an endo-mesomorphic somatotype does not favor high-intensity actions, so it is necessary to focus on reducing FM.

Table 4. *Cont.*

Author(s)/Year	Study's Aim	Variables	Instruments	Determination of % Fat and Somatotype	Results	Conclusions
Durán-Agüero et al. [33]	To determine the anthropometric profile of elite Chilean Paralympic athletes by means of BC and somatotype	BC, somatotype	Scale Scale-tronix somatotype (Batesville, USA), precision 0.1 kg, SECA stadiometer (Hamburg, Germany), precision 0.1 cm, Rosscraft anthropometer (Minneapolis, USA), precision 0.1 mm, Sanny measuring tape (São Paulo, Brazil), precision 0.1 mm, Harpenden caliper (Crosswell, UK), precision 0.2 mm.	Determination of BC using the Kerr model [58], the pentacompartmental method. and the Heath–Carter method [55] for the somatotype	The athletes express a predominance toward the meso-endomorph somatotype. Likewise, blind 5-a-side football players have a predominance toward the MM component (45.6%) and bone mass (12.1%), as well as low levels of residual mass (11.6%) compared to other Paralympic athletes.	Chilean elite futsal players present a meso-endomorphic somatotype profile with a predominance of the lower limbs and elevated levels of FM compared to other Paralympic athletes (swimming, wheelchair tennis, and powerlifting). It is necessary to develop training programs that improve BC and nutritional habits.
Castelli Correia de Campos et al. [32]	To analyze the effect of 16 weeks of training on physical fitness (PF) and body composition (BC) in blind 5-a-side football athletes on the Brazilian national team	Skinfolds, widths, body density (BD), body fat percentage, somatotype profile, FM, FFM.	Instruments used included the Plena scale Acqua® model, WCS wall stadiometer, Harpenden caliper (Crosswell, UK) with 0.2 mm precision, and Cardiomed pachymeter (Brasília, Brazil) with 0.1 cm precision.	The Siri formula [54] was used to estimate body fat percentage, the Heath–Carter method [55] and Hebbelinck [59] for somatotype, and the Jackson & Pollock equation [57] for BD	No significant differences ($p \leq 0.05$) were reported between the absolute values of body weight before (77.08 ± 7.73 kg) and after (76.16 ± 8.38 kg) the tests. The same trend was observed in the BC and somatotype of blind 5-a-side football players.	Sixteen weeks of training are effective for improving physical fitness, although they do not appear to be sufficient to produce changes in body composition indicators in blind 5-a-side football players.

Note. BC: Body composition; BD: body density; BMD: bone mineral density; BMI: body mass index; BW: body weight; cm: centimeters; FM: fat mass; FFM: fat-free mass; GK: goalkeeper; kg: kilograms; LM: Lean mass; mm: millimeters; MM: muscle mass; NA: Not applicable; PA: physical activity; PF: physical fitness; PP: playing position; VI: visual impairment.

4. Discussion

To our knowledge, this is the first systematic review to analyze BC in blind 5-a-side footballers both in general and in response to playing position. The main findings of this study were as follows: (1) the somatotype of blind 5-a-side football players tends toward meso-endomorphic [35,36,38,40]; (2) there are differences in MM, FM, and BW variables in response to playing position and sporting level; (3) the players present a somatotypic profile with a predominance of the muscular component; (4) different formulas are used in the studies, although the most common are the Siri formula to determine body fat percentage on the basis of body density from other equations, the Jackson & Pollock [57] equation for BD, and the Heath–Carter method for somatotyping [55]; and (5) no significant differences were observed in the absolute values of body mass, BC, and somatotype after 16 weeks of training.

4.1. Body Composition and Anthropometric Factors in Blind 5-a-Side Football Players

In blind 5-a-side football, different playing positions require specific demands [60–62]. However, most studies on blind 5-a-side football have analyzed BC in cross-sectional studies [33–41], and only one longitudinal study has been reported, concluding that 16 weeks of training did not produce significant changes in the players' BC or somatotype [32].

Among the anthropometric results, it is noteworthy that blind 5-a-side football players have a body fat percentage ranging from 10.4% to 15.9%, and the sum of the nine skinfold measurements reveals a percentage ranging from 89.7% to 121.8% in blind Brazilian national team football players [34]. On the other hand, the study conducted by Durán-Agüero et al. [33] revealed that Chilean elite Paralympic 5-a-side football players have a body fat percentage of 25.8%, an MM of 45.6%, and a bone mass of 12.1%. Another study reported that the body fat percentage was 16.23%, FM was 11.08 kg, and FFM was 57.79 kg [39]. Another study reported that the body fat mass of blind 5-a-side football players is 11.5 ± 2.7 kg, the bone mass is 11.7 ± 1.1 kg, the bone mass percentage is 16.3 ± 1.2%, and the body fat percentage is 15.9 ± 2.9 [35]. Other studies reported that blind 5-a-side footballers had body fat percentage values of 15.9 ± 2.9 mm, and the sum of the nine skinfolds was 76.9 ± 17.1 mm [36]. A study that analyzed the body components of blind 5-a-side players at the national level reported higher body fat percentage values (20.4 ± 5.1 mm), a sum of the nine skinfolds of 119.4 ± 5 mm, and a bone mass percentage similar to that reported in another study (16.1 ± 0.9 mm). On the other hand, the MM percentage was low (39.5 ± 3.5 mm) [37].

A study conducted by Gorla et al. [34] revealed that goalkeepers had the highest body weight (82.3 kg), followed by pivots (71.4 kg), defenders (70.8 kg), and wings (68.5 kg). The BF of goalkeepers was higher (21.5%) than that of wings, who presented the lowest BF (10.6%). In another context, with blind 5-a-side football players of the Spanish national team, the FM was 12.55 ± 6.21 kg and the FFM was 61.34 ± 6.36 kg [41].

4.2. Somatotype Values in Response to Playing Position

With respect to somatotype, there is a clear tendency for blind 5-a-side footballers to exhibit a meso-endomorphic profile. However, these approaches also reflect that, methodologically, some studies analyze somatotype generally, whereas others specify it by playing position. In the study conducted by Durán-Agüero et al. [33], there was a tendency toward a mesomorphic somatotype profile, followed by endomorphism in elite Chilean players.

With respect to playing positions, the study by Gorla et al. [34] revealed that goalkeepers ($n = 4$) tended to endomorph–mesomorph, wing ($n = 7$) toward endo-mesomorph, defender ($n = 6$) toward balanced mesomorph, pivot ($n = 6$) toward endo-mesomorph, and at a general level ($n = 23$) toward endo-mesomorph. In this same study, no significant differences were found ($p \leq 0.05$) between the components in terms of ectomorphy, mesomorphy, and endomorphy related to playing position. When stratifying them by playing position, goalkeepers present an endo-mesomorphic profile, whereas defenders present a balanced mesomorphic profile [34]. A study analyzing the dermatoglyphic profile and BC in blind Brazilian national football team players revealed that goalkeepers tend toward meso-endomorphic characteristics, fix and pivot players toward balanced mesomorphism, and, finally, wing players toward meso-endomorphic characteristics [36].

A study analyzing the somatotype frequency distribution in response to playing position in blind football players revealed that goalkeepers have a 100% meso-endomorphic distribution, with 66.6% balanced mesomorphic characteristics and 33.3% meso-endomorphic characteristics, wing players with 71.5% meso-endomorphic characteristics and 28.5% balanced mesomorphic characteristics, and pivot players with 100% balanced mesomorphic characteristics [35]. Moreover, at a general level, the distribution of the 15 players evaluated was 60% meso-endomorphic and 40% balanced mesomorphic [35]. Moreover, a recent study with players ($n = 63$) from various high-performance 5-a-side football teams revealed that 69.2% of pivot players ($n = 13$) presented with meso-endomorphs, 15.4% with balanced mesomorphs, and 15.4% ectomorph–mesomorph. The wing players ($n = 24$) presented with 58.3% mesomorphs, 20.6% with endomorphs and mesomorphs, 15.9% with balanced mesomorphs, 4.8% with ecto-mesomorphs, 3.2 with endo-mesomorphs, 1.6% with meso-ecotomorphs, and 1.6% with ectomorphs [40].

4.3. Influence of BC on the Athletic Performance of Blind 5-a-Side Football Players

The skeletal index was correlated with the ball-handling speed ($r = 0.85$, $p = 0.01$). It has a moderate correlation with lower limb length ($r = 0.69$), as well as with other anthropometric variables in which muscle tissue stands out: mesomorphism ($r = 0.59$), MM ($r = 0.57$), thigh muscle area ($r = 0.56$), and calf muscle area ($r = 0.55$) [38]. Similarly, there is a relationship between dermatoglyphic characteristics and BC in blind 5-a-side football players, as evidenced by a somatotype profile with a predominance of the muscular component, which is related to the genetic predisposition toward the development of speed and strength [36]. Among the roles played by blind 5-a-side football players, laterality has been reported to be related to BC, with ambidextrous players showing lower values than other players (left- or right-handed) [41].

In this regard, and following other systematic reviews on BC analysis in football, the use of different measurement protocols can lead to significant differences in the data reported for the groups ($p < 0.001$) [6]. This is similar to the case reported in the present study, where different anthropometric formulas, equipment, and indicators are considered in each study in a heterogeneous manner. Moreover, the performance of anthropometric measurements under protocols established by international organizations such as those of the International Society for the Advancement of Kineanthropometry (ISAK) is not evident in the manuscripts studied [63,64].

Furthermore, when the playing position was used as a reference, it was determined that there were significant differences in response to BW, the sum of the skinfolds, the MM (kg), and the FFM ($p = 0.001$). No differences were reported in the percentages of MM, FM, or bone mass or in the somatotype [29]. These results are in line with those reported in the present review with blind 5-a-side futsal players. Finally, what was reported in futsal players with cerebral palsy reveals that the anthropometric and BC profiles do not vary with the functional classification, and the expressed somatotypic profile is meso-endomorphic, a case similar to that reported in futsal players with visual impairment, where there is also a tendency toward endo-mesomorphic [65]. These findings reveal the need to continue researching the effects of BC on futsal players with disabilities.

4.3.1. Limitations and Strengths

This study has several limitations, which are outlined below. There is diversity in the number of participants evaluated: the most commonly used study designs have been cross-sectional, with only one longitudinal study reported. Another limitation is that not all studies include goalkeepers, making it difficult to generate a deep understanding of BC in response to playing positions. Similarly, there is diversity in the variables that determine anthropometric and BC factors. This prevented comparative measurements for each of the variables analyzed in the studies included in this review. The studies included in this systematic review were too heterogeneous and of moderate methodological quality, making it impossible to conduct a meta-analysis. Although this type of study, which was conducted to analyze blind 5-a-side football, does not allow for solid conclusions, the information contained in Table 4 reflects important information from each study that could be further explored by the scientific community.

4.3.2. Future Recommendations

Future research directions for studying BC in blind 5-a-side football should seek associations with other performance indicators, especially nutritional, physical, and external load variables in competitions. Likewise, longitudinal studies, randomized controlled trials, or principal component analyses based on standardized and recognized protocols are needed to understand how different variables are related to playing position. Furthermore, the findings of this study should be carefully analyzed to be incorporated into different training processes and the practical work of coaches and athletes. Finally, it would be important to conduct BC assessments in blind female 5-a-side football players, as no studies analyzing these factors in this population sample have been reported. This will allow us to further expand the horizons of this sport because it is still a practice that has produced low scientific productivity and requires an increasing amount of research [66].

4.3.3. Practical Applications

A particularly noteworthy result is the lack of significant somatotype differentiation across player positions, with a general dominance of the mesomorphic–endomorphic type. This finding holds important implications for training methodology, talent identification, and athlete development in this sport. Furthermore, variations in fat mass observed between different subgroups raise new questions, potentially linked to training regimens, selection policies, or regional disparities—warranting further investigation.

5. Conclusions

This systematic review provides information that may be useful to technical and medical staff who develop sports programs with blind 5-a-side football players, thus facilitating an adequate assessment of BC and anthropometric factors. The results of this review reveal a tendency toward defining BC as influencing the athletic performance of blind 5-a-side football players. However, it is not conclusive that these improvements occur in response to each playing position. Further studies are needed to analyze the effects of BC on athletic performance, especially when BC is related to other physical, nutritional, technical, and tactical variables during training and competition.

Author Contributions: Introduction: B.A.B.-P. and J.O.-A.; methods: B.A.B.-P. and J.O.-A.; analysis: B.A.B.-P. and J.O.-A.; discussion and conclusions: B.A.B.-P., J.O.-A., A.M.-Q., J.F.L.-G., and J.P.-O.; writing and preparation of the paper: B.A.B.-P., J.O.-A., A.M.-Q., J.F.L.-G., and J.P.-O.; revision and editing: B.A.B.-P., J.O.-A., A.M.-Q., J.F.L.-G., and J.P.-O. All authors have read and agreed to the published version of the manuscript.

Funding: This research received no external funding.

Conflicts of Interest: The authors declare no conflicts of interest.

References

1. Högström, G.M.; Pietilä, T.; Nordström, P.; Nordström, A. Body Composition and Performance: Influence of Sport and Gender among Adolescents. *J. Strength Cond. Res.* **2012**, *26*, 1799–1804. [CrossRef] [PubMed]
2. Thibault, R.; Genton, L.; Pichard, C. Body Composition: Why, When and for Who? *Clin. Nutr.* **2012**, *31*, 435–447. [CrossRef] [PubMed]
3. Giovanelli, L.; Biganzoli, G.; Spataro, A.; Malacarne, M.; Bernardelli, G.; Spada, R.; Pagani, M.; Biganzoli, E.; Lucini, D. Body Composition Assessment in a Large Cohort of Olympic Athletes with Different Training Loads: Possible Reference Values for Fat Mass and Fat-Free Mass domains. *Acta Diabetol.* **2024**, *61*, 361–372. [CrossRef]
4. Mathisen, T.F.; Ackland, T.; Burke, L.M.; Constantini, N.; Haudum, J.; Macnaughton, L.S.; Meyer, N.L.; Mountjoy, M.; Slater, G.; Sundgot-Borgen, J. Best Practice Recommendations for Body Composition Considerations in Sport to Reduce Health and Performance Risks: A Critical Review, Original Survey and Expert Opinion by a Subgroup of the IOC Consensus on Relative Energy Deficiency in Sport (REDs). *Br. J. Sports Med.* **2023**, *57*, 1148–1158. [CrossRef]
5. Toselli, S. Body Composition and Physical Health in Sports Practice: An Editorial. *Int. J. Environ. Res. Public Health* **2021**, *18*, 4534. [CrossRef]
6. Sebastiá-Rico, J.; Soriano, J.M.; González-Gálvez, N.; Martínez-Sanz, J.M. Body Composition of Male Professional Soccer Players Using Different Measurement Methods: A Systematic Review and Meta-Analysis. *Nutrients* **2023**, *15*, 1160. [CrossRef]
7. Kuriyan, R. Body composition techniques. *Indian J. Med. Res.* **2018**, *148*, 648. [CrossRef] [PubMed]
8. Campa, F.; Toselli, S.; Mazzilli, M.; Gobbo, L.A.; Coratella, G. Assessment of Body Composition in Athletes: A Narrative Review of Available Methods with Special Reference to Quantitative and Qualitative Bioimpedance Analysis. *Nutrients* **2021**, *13*, 1620. [CrossRef]
9. Becerra Patiño, B. *Soccer as a Synergetic Model: Complexity of the Player-Game (Fútbol Como Modelo Sinérgico: Complejidad del Juego-Jugador)*; Kinesis: Armenia, Colombia, 2021.
10. Massidda, M.; Calò, C.M.; Cięszczyk, P.; Kikuchi, N.; Ahmetov, I.I.; Williams, A.G. Genetics of Team Sports. Chapter Five. *Sports Exerc. Nutr. Genom.* **2019**, 105–128. [CrossRef]
11. Gomez-Ruano, M.A.; Ibáñez, S.J.; Leicht, A.S. Editorial: Performance Analysis in Sport. *Front. Psychol.* **2020**, *11*, 611634. [CrossRef]
12. Roso-Moliner, A.; Gonzalo-Skok, O.; Villavicencio Álvarez, V.E.; Calero-Morales, S.; Mainer-Pardos, E. Analysing the Influence of Speed and Jumping Performance Metrics on the Percentage Change of Direction Deficit in Adolescent Female Soccer Players. *Life* **2024**, *14*, 466. [CrossRef]
13. Kalata, M.; Williams, C.A.; Verbruggen, F.F.; De Ste Croix, M.; Lehnert, M.; Zahalka, F.; Maly, T. Strength and Jumping Performance in Youth Athletes: Do Sport Specialization and Age Categories Affect Strength Asymmetry? *Res. Sports Med.* **2025**, *33*, 334–351. [CrossRef] [PubMed]
14. Montenegro Bonilla, A.D.; Rodríguez Pachón, S.D.; Hernández-Beltrán, V.; Gamonales, J.M.; Rico-González, M.; Pino-Ortega, J.; Olivares-Arancibia, J.; Yánez-Sepúlveda, R.; López-Gil, J.F.; Becerra Patiño, B.A. Comparative Analysis of the Physical, Tactical, Emotional, and Mood Characteristics of Under-13 Soccer Players by Performance Level. *J. Funct. Morphol. Kinesiol.* **2024**, *9*, 237. [CrossRef]
15. Gualtieri, A.; Rampinini, E.; Dello Iacono, A.; Beato, M. High-Speed Running and Sprinting in Professional Adult Soccer: Current thresholds Definition, Match Demands and Training Strategies. A Systematic Review. *Front. Sports Act. Living* **2023**, *5*, 1116293. [CrossRef] [PubMed]
16. Cotteret, C.; González-de-la-Flor, Á.; Prieto Bermejo, J.; Almazán Polo, J.; Jiménez Saiz, S.L. A Narrative Review of the Velocity and Acceleration Profile in Football: The Influence of Playing Position. *Sports* **2025**, *13*, 18. [CrossRef]
17. Hofmann, P.; Tschakert, G. Intensity- and Duration-Based Options to Regulate Endurance Training. *Front. Physiol.* **2017**, *8*, 337. [CrossRef]
18. Birnbaumer, P.; Weiner, L.; Handl, T.; Tschakert, G.; Hofmann, P. Effects of Different Durations at Fixed Intensity Exercise on Internal Load and Recovery—A Feasibility Pilot Study on Duration as an Independent Variable for Exercise Prescription. *J. Funct. Morphol. Kinesiol.* **2022**, *7*, 54. [CrossRef] [PubMed]
19. Díaz-Cortés, J.C.; Hernández-Beltrán, V.; Solano-Ruiz, L.F.; Cepeda-Hernández, J.A.; Méndez-Castro, G.E.; Becerra-Patiño, B.A.; Gamonales, J.M. Decrease in Muscle Shortening and Effect on Strength and Speed in Adolescent Soccer Players (u-13). *Sport. Sci. J. Sch. Sport. Phys. Educ. Psychomot.* **2024**, *10*, 735–761. [CrossRef]
20. Alimoradi, M.; Sahebozamani, M.; Hosseini, E.; Konrad, A.; Noorian, S. The Effect on Flexibility and a Variety of Performance Tests of the Addition of 4 Weeks of Soleus Stretching to a Regular Dynamic Stretching Routine in Amateur Female Soccer Players. *Sports* **2023**, *11*, 138. [CrossRef]

21. Rueda-Cordoba, M.; Martin-Olmedo, J.J.; Espinar, S.; Ruiz, J.R.; Jurado-Fasoli, L. Multidimensional Differences Between Athletes of Endurance, Strength, and Intermittent Sports: Body Composition, Diet, Resting Metabolic Rate, Physical Activity, Sleep Quality, and Subjective Well-Being. *Nutrients* **2025**, *17*, 1172. [CrossRef]
22. Sannicandro, I.; Agostino, S.; Abate Daga, M.; Veglio, F.; Abate Daga, F. Developing the Physical Performance in Youth Soccer: Short-Term Effect of Dynamic–Ecological versus Traditional Training Approach for Sub-Elite U13 Players—An Ecological Exploratory Cluster Randomised Trial. *J. Funct. Morphol. Kinesiol.* **2024**, *9*, 83. [CrossRef]
23. Burke, L.M.; Hawley, J.A. Swifter, higher, stronger: What's on the menu? *Science* **2018**, *362*, 781–787. [CrossRef] [PubMed]
24. Castillo, M.; Martínez-Sanz, J.M.; Penichet-Tomás, A.; Sellés, S.; González-Rodriguez, E.; Hurtado-Sánchez, J.A.; Sospedra, I. Relationship between Body Composition and Performance Profile Characteristics in Female Futsal Players. *Appl. Sci.* **2022**, *12*, 11492. [CrossRef]
25. Hernandez-Martinez, J.; Perez-Carcamo, J.; Coñapi-Union, B.; Canales-Canales, S.; Negron-Molina, M.; Avila-Valencia, S.; Cid-Calfucura, I.; Herrera-Valenzuela, T.; Cisterna, D.; Branco, B.H.M.; et al. Relationship between Body Composition and Physical Performance by Sex in Professional Basketball Players. *Appl. Sci.* **2024**, *14*, 9165. [CrossRef]
26. Randell, R.K.; Clifford, T.; Drust, B.; Moss, S.L.; Unnithan, V.B.; De Ste Croix, M.B.; Datson, N.; Martin, D.; Mayho, H.; Carter, J.M.; et al. Physiological Characteristics of Female Soccer Players and Health and Performance Considerations: A Narrative Review. *Sports Med.* **2021**, *51*, 1377–1399. [CrossRef] [PubMed]
27. Becerra Patiño, B.A.; Sarria Lozano, J.C.; Prada Clavijo, J.F. Morphofunctional Characteristics by Position in U-15 Female Soccer Players from Bogota. *Retos* **2022**, *45*, 381–389. [CrossRef]
28. Zambrano-Villacres, R.; Frias-Toral, E.; Maldonado-Ponce, E.; Poveda-Loor, C.; Leal, P.; Velarde-Sotres, A.; Leonardi, A.; Trovato, B.; Roggio, F.; Castorina, A.; et al. Exploring Body Composition and Somatotype Profiles Among Youth Professional Soccer Players. *Mediterr. J. Nutr. Metab.* **2024**, *17*, 241–254. [CrossRef]
29. Sebastiá-Rico, J.; Martínez-Sanz, J.M.; González-Gálvez, N.; Soriano, J.M. Differences in Body Composition between Playing Positions in Men's Professional Soccer: A Systematic Review with Meta-Analysis. *Appl. Sci.* **2023**, *13*, 4782. [CrossRef]
30. Doménech, C.; Roche, E.; Reina, R.; Sarabia, J.M. Somatotype Analysis of International Football Players with Cerebral Palsy: A Comparison with Non-Disabled Football Players. *J. Funct. Morphol. Kinesiol.* **2023**, *8*, 166. [CrossRef]
31. Sarabia, J.M.; Doménech, C.; Roche, E.; Vicente-Salar, N.; Reina, R. Anthropometrical Features of Para-Footballers According to Their Cerebral Palsy Profiles and Compared to Controls. *Int. J. Environ. Res. Public Health* **2020**, *17*, 9071. [CrossRef]
32. Castelli Correia de Campos, L.F.; de Athayde Costa e Silva, A.; Teixeira Fabrício dos Santos, L.G.; Trevisan Costa, L.; Montagner, P.C.; Borin, J.P.; Ferreira de Araújo, P.; Irineu Gorla, J. Effects of Training in Physical Fitness and Body Composition of the Brazilian 5-a-Side Football Team. *Rev. Andal. Med. Deporte* **2013**, *6*, 91–95. [CrossRef]
33. Durán-Agüero, S.; Valdés-Badilla, P.; Varas-Standen, C.; Arroyo-Jofre, P.; Herrera-Valenzuela, T. Perfil antropométrico de deportistas paralímpicos de élite chilenos. *Rev. Española Nutr. Humana Dietética* **2016**, *20*, 307–315. [CrossRef]
34. Gorla, J.I.; de Athayde Costa e Silva, A.; Castelli Correia de Campos, L.F.; Ferreira dos Santos, C.; Gavião de Almeida, J.J.; Duarte, E.; Queiroga, M.R. Composição Corporal e Perfil Somatotípico de Atletas da Seleção Brasileira de Futebol de 5. *Rev. Bras. Ciências Esporte* **2017**, *39*, 79–84. [CrossRef]
35. de-Oliveira, G.L.; de Pinho-Gonçalves, P.S.; de-Oliveira, T.A.P.; Valentim-Silva, J.R.; Roquetti-Fernandes, P.; Fernandes-Filho, J. Body Composition and Somatotype of Athletes of Brazilian w5-a-Side Football Team: Paralympic Team Rio 2016. *Rev. Fac. Med. Univ. Nac. Colomb.* **2018**, *66*, 25–29. [CrossRef]
36. De Oliveira, G.L.; De Oliveira, T.A.P.; Valentim-Silva, J.R.; Fernandes, J. Dermatoglyphic Profile and Body Composition of Athletes from the Brazilian Five-a-Side National Football Team. *Int. J. Sports Sci.* **2018**, *8*, 78–82. [CrossRef]
37. Lameira Oliveira, G.; Perini Oliveira, T.; Maioli Penello, F.; Fernandes Filho, J. Características antropométricas y aptitud aeróbica de los atletas ciegos de fútbol 5. *Rev. Peru. Cienc. Act. Fis. Deporte* **2020**, *7*, 10. [CrossRef]
38. Sancio, D.; Arcodia, J.L.; Roselló, M. Perfil antropométrico y velocidad con balón en jugadores argentinos profesionales de futbol sala para ciegos. *Rev. Peru. Cienc. Act. Fis. Deporte* **2021**, *8*, 12. [CrossRef]
39. Esatbeyoglu, F.; Kin-İsler, A. Gender differences in postural balance, physical activity level, BMI, and body composition in athletes with visual impairment. *Br. J. Vis. Impair.* **2022**, *40*, 530–544. [CrossRef]
40. Lameira Oliveira, G.; Adão Perini Oliveira, T.; Braz, M.; Irineu, J.I.; Fernandes Filho, J. Body composition and somatotype of Blind Athletes of 5-a-side Football of High-Performance. *Retos* **2023**, *48*, 873–877. [CrossRef]
41. Hernández-Beltrán, V.; Gámez-Calvo, L.; Castelli Correia de Campos, L.F.; Bertu, F.; Gamonales, J.M. Analysis of the Body Composition of the Players of the Spanish Blind Football Team. *Rev. Andal. Med. Deporte* **2023**, *16*, 100–106. [CrossRef]
42. Gamonales, J.M.; Muñoz-Jiménez, J.; León-Guzmán, K.; Ibáñez, S.J. 5-A-Side Football for Individuals with Visual Impairments: A review of the Literature. *Eur. J. Adapt. Phys. Act.* **2018**, *11*, 4. [CrossRef]
43. Gamonales, J.M.; Muñoz Jiménez, J.; Mancha-Triguero, D.; Ibáñez, S.J. The influence of the competition phase and the result of the match on the competitive demands in football 5-a-side for the visually impaired. *Int. J. Perform. Anal. Sport.* **2020**, *21*, 1–11. [CrossRef]

44. Gamonales, J.M.; León, K.; Rojas-Valverde, D.; Sánchez-Ureña, B.; Muñoz-Jiménez, J. Data Mining to Select Relevant Variables Influencing External and Internal Workload of Elite Blind 5-a-Side Soccer. *Int. J. Environ. Res. Public Health* **2021**, *18*, 3155. [CrossRef] [PubMed]
45. Nygaard Falch, H.; Guldteig Rædergård, H.; van den Tillaar, R. Effect of Different Physical Training Forms on Change of Direction Ability: A Systematic Review and Meta-analysis. *Sports Med. Open* **2019**, *5*, 53. [CrossRef]
46. Hostrup, M.; Bangsbo, J. Performance Adaptations to Intensified Training in Top-Level Football. *Sports Med.* **2023**, *53*, 577–594. [CrossRef]
47. Tafah, M.; Majelan, A.S. Identification and Priorization of Important Attributes for Talent Identification in Blind Football. *Stud. Sport. Med.* **2024**, *15*, 65–90. [CrossRef]
48. Finocchietti, S.; Gori, M.; Souza Oliveira, A. Kinematic Profile of Visually Impaired Football Players During Specific Sports Actions. *Sci. Rep.* **2019**, *9*, 10660. [CrossRef]
49. Liberati, A.; Altman, D.G.; Tetzlaff, J.; Mulrow, C.; Gøtzsche, P.C.; Ioannidis, J.P.; Clarke, M.; Devereaux, P.J.; Kleijnen, J.; Moher, D. The PRISMA Statement for Reporting Systematic Reviews and Neta-analyses of Studies that Evaluate Health Care Interventions: Explanation and Elaboration. *PLoS Med.* **2009**, *2*, e1000100. [CrossRef]
50. Page, M.J.; McKenzie, J.E.; Bossuyt, P.M.; Boutron, I.; Hoffmann, T.C.; Mulrow, C.D.; Shamseer, L.; Tetzlaff, J.M.; Akl, E.A.; Brennan, S.E.; et al. The PRISMA 2020 Statement: An Updated Guideline for Reporting Systematic Reviews. *BMJ* **2021**, *372*, n71. [CrossRef]
51. Rico-González, M.; Pino-Ortega, J.; Clemente, F.M.; Los Arcos, A. Guidelines for Performing Systematic Reviews in Sports Science. *Biol. Sport* **2022**, *39*, 463–471. [CrossRef]
52. Frandsen, T.F.; Bruun Nielsen, M.F.; Lindhardt, C.L.; Eriksen, M.B. Using the full PICO Model as a Search Tool for Systematic Reviews Resulted in Lower Recall for Some PICO Elements. *J. Clin. Epidemiol.* **2020**, *127*, 69–75. [CrossRef]
53. Cashin, A.G.; McAuley, J.H. Clinimetrics: Physiotherapy Evidence Database (PEDro) Scale. *J. Physiother.* **2020**, *66*, 59. [CrossRef] [PubMed]
54. Siri, W.E. Body Composition from Fluid Space and Density. In *Techniques for Measuring Body Composition*; Brozek, J., Hanschel, A., Eds.; National Academy of Science: Washington, DC, USA, 1961; pp. 223–224.
55. Heath, B.H.; Carter, J.E. A modified Somatotype Method. *Am. J. Phys. Anthropol.* **1967**, *27*, 57–74. [CrossRef]
56. Rocha, M.L. Peso Ósseo Do Brasileiro de Ambos os Sexos de 17 a 25 Anos. *Arq. Anat. Antropol.* **1975**, *1*, 444–451.
57. Jackson, A.S.; Pollock, M.L. Generalized Equations for Predicting Body Density of Men. *Br. J. Nutr.* **1978**, *40*, 497–504. [CrossRef] [PubMed]
58. Kerr, D.A. *An Anthropometric Method for the Fractionation of Skin, Adipose, Bone, Muscle and Residual Tissue Masses in Males and Females Age 6 to 77 Years*; Simon Fraser University: Burnaby, BC, Canada, 1988.
59. Hebbelinck, M.; Carter, L.; De Garay, A. Body Build and Somatotype of Olympic Swimmers, Divers, and Water Polo Players. In *Swimming II*; Lewillie, L., Clarys, J.P., Eds.; University Park Press: Baltimore, MD, USA, 1975; pp. 285–305.
60. Castelli Correia de Campos, L.F.; Borin, J.P.; Santos, L.G.T.F.; Souza, T.M.F.; Paranhos, V.M.S.; Tanhoffer, R.A. Avaliação Isocinética em Atletas da Seleção Brasileira de Futebol de 5. *Rev. Bras. Med. Esporte* **2015**, *21*, 220–223. [CrossRef]
61. Souza, R.P.; Alves, J.M.V.M.; Gorla, J.I.; Novaes, G.; Cabral, S.I.C.; Neves, E.B.; Nogueira, C.D. Characterization of the Intensity of Effort of Blind Athletes from the Brazilian Football 5-A-Side National Team. *J. Health Biol. Sci.* **2016**, *4*, 218–226. [CrossRef]
62. Papadopoulos, C.; Michailidis, Y.; Metaxas, T.I.; Mandroukas, A.; Fotiadou, E.G.; Giagazoglou, P.; Christoulas, K.; Tsimaras, V. Physiological Profile and Correlations between VO_{2max} and Match Distance Running Performance of Soccer Players with Visual Impairment. *Appl. Sci.* **2023**, *13*, 10762. [CrossRef]
63. Petri, C.; Campa, F.; Holway, F.; Pengue, L.; Arrones, L.S. ISAK-Based Anthropometric Standards for Elite Male and Female Soccer Players. *Sports* **2024**, *12*, 69. [CrossRef]
64. Stewart, A.; Marfell-Jones, M.; Olds, T.; De Ridder, H. *International Standards for Anthropometric Assessment*; International Society for the Advancement of Kinanthropometry: Lower Hutt, New Zealand, 2011; Volume 115.
65. Gorla, J.I.; Nogueira, C.D.; Gonçalves, H.R.; De Faria, F.R.; Buratti, J.R.; Nunes, N.; Pereira do Rêgo, J.T.; Borges, M.; Vieira, I.B.; Labrador Roca, V. Body Composition and Somatotype Profile of Football-Seven Brazilian p. *Retos* **2019**, *35*, 326–328. [CrossRef]
66. Becerra Patiño, B.A.; Pino-Ortega, J.; Olivares-Arancibia, J. Exploratory Review of the Scientific Production of 5-Blind Soccer with the Bibliometrix Tool. *Retos* **2025**, *68*, 1025–1047. [CrossRef]

MDPI AG
Grosspeteranlage 5
4052 Basel
Switzerland
Tel.: +41 61 683 77 34

Journal of Functional Morphology and Kinesiology Editorial Office
E-mail: jfmk@mdpi.com
www.mdpi.com/journal/jfmk

www.ingramcontent.com/pod-product-compliance
Lightning Source LLC
Chambersburg PA
CBHW061838210926
51850CB00006B/920

* 9 7 8 3 7 2 5 8 7 6 2 1 1 *